RECEPTOR CLASSIFICATION

THE INTEGRATION OF OPERATIONAL, STRUCTURAL, AND TRANSDUCTIONAL INFORMATION

ANNALS OF THE NEW YORK ACADEMY OF SCIENCES

Volume 812

RECEPTOR CLASSIFICATION

THE INTEGRATION OF OPERATIONAL, STRUCTURAL, AND TRANSDUCTIONAL INFORMATION

Edited by D. G. Trist, P. P. A. Humphrey, P. Leff, and N. P. Shankley

The New York Academy of Sciences
New York, New York
1997

∞ *The paper used in this publication meets the minimum requirements of American National Standard for Information Sciences—Permanence of Paper for Printed Library Materials, ANSI Z39.48-1984.*

Cover art: The softcover illustration shows the now-legendary balcony where Juliet called to her lover, Romeo—Casa Di Giulietta in Via Cappello, Verona, Italy

Library of Congress Cataloging-in-Publication Data

Receptor classification: the integration of operational, structural,
and transductional information/edited by D. G. Trist . . . [et al.].
p. cm. —(Annals of the New York Academy of Sciences, ISSN
0077-8923; v. 812).
Result of conference entitled Receptor classification held in
Verona, Italy on Sept. 21–22, 1995.
Includes bibliographical references and index.
ISBN 1-89766-987-8 (cloth). — ISBN
1-89766-988-6 (pbk.)
1. Neurotransmitter receptors—Congresses. I. Trist, D. G.
II. Series.
[DNLM: 1. Receptors, Cell Surface—classification—congresses.
2. Receptors, Drug—classification—congresses. W1 AN626YL v. 812
1997/QU 15 R295 1997]
Q11.Nb vol. 812
[QP364.7]
500 s—dc21
[573.8′19]
DNLM/DLC
for Library of Congress

97-6552
CIP

BiC/PCP
Printed in the United States of America
ISBN 1-89766-987-8 (cloth)
ISBN 1-89766-988-6 (paper)
ISSN 0077-8923

ANNALS OF THE NEW YORK ACADEMY OF SCIENCES

Volume 812
May 30, 1997

RECEPTOR CLASSIFICATION

THE INTEGRATION OF OPERATIONAL, STRUCTURAL, AND TRANSDUCTIONAL INFORMATION[a]

Editors
D. G. TRIST, P. P. A. HUMPHREY, P. LEFF, AND N. P. SHANKLEY

Conference Organizing Committee
P. P. A. HUMPHREY, P. LEFF, G. R. MARTIN, N. P. SHANKLEY, J. SHAW, M. SPEDDING, AND D. G. TRIST

CONTENTS

[a] This volume is the result of a conference entitled **Receptor Classification: The Integration of Operational, Structural, and Transductional Information,** supported by the International Union of Pharmacology (IUPHAR) Committee on Receptor Nomenclature and Drug Classification and by the Federation of European Pharmacological Societies (EPHAR) and held in Verona, Italy on September 21–22, 1995.

Part III. Influence of Recent Theoretical Advances

Part IV. Oral Communications

Part V. Poster Presentations

Financial support was received from:

- Amersham International plc
- Fisons Pharmaceuticals
- Glaxo S.p.A.
- Institut Recherches Servier
- James Black Foundation
- Merck Sharp & Dohme
- Pfizer Central Research
- Sandoz Ltd.
- SmithKline Beecham Ltd.
- Syntex Research IOP
- Wellcome Foundation plc
- Wyeth Research
- Zeneca Pharmaceuticals

Preface and Introduction

P. P. A. HUMPHREY, P. LEFF, N. SHANKLEY,
AND D. G. TRIST

The allocation of receptors for hormones and transmitters into groups or types has a long tradition. The need to know how different drugs work, both in physiology and in pathology, has driven the process of classifying receptors through which they act. This is not only of academic interest but also has practical consequences. Differences in receptors are potentially exploitable to develop specific mimics of their natural ligands (agonists) or to find specific receptor blockers (antagonists), which antagonize the action of agonists. Many of our important existing medicines have arisen from the knowledge that agonists can act through multiple receptors. The great advantage of using synthetic chemicals, selective for one receptor against the other, is that potent medicines with little or no side-effects can be developed.

Early proponents of receptor characterization like Clark, Gaddum and Schild, lacked most of the important biomolecular information that has been generated over the last 5 to 10 years. They had no knowledge of the structure of receptors nor of the biochemical events involved in linking receptor occupation with the effect mediated. Thus, they looked for surrogates of the missing information on structure and transduction. The operational methods, which they developed, relied on the use of agonists and antagonists with specific properties to discriminate one receptor from another. Since the potency of an agonist is described by both its affinity and its efficacy in the particular systems being compared, antagonists, which were considered to have zero efficacy have been the preferred tools for functional characterization of receptors. However, this approach does have a major potential weakness, particularly in the light of current knowledge about inverse agonists and multiple binding sites on a single receptor macromolecule. Thus receptors were classified on the basis of antagonist affinity and at the same time the antagonists were defined themselves by the receptors with which they interact. Obviously and inevitably, there is much circularity in the argument. Nevertheless, now that structural information has become available for many of the common receptors already classified pharmacologically, it is apparent that in the majority of the cases the operational approach was successful.

For many years pharmacologists interested in classification (be it for pragmatic reasons or Darwinian inclinations) studied a *black box* encompassing receptor occupation and mediated effect and the need to illuminate the contents of this box did not seem an *a priori* necessity. Over the last 20–30 years biochemists have been providing information to make the black box less black, if as yet not completely white. Knowledge of receptor protein structure and their corresponding genes, and how these molecules are involved in neurotransmitter signaling, has given much insight into the way receptor diversity has evolved. Importantly, it has given the pharmacologist two new approaches to use, namely, structural and transductional. However, as in all things, new knowledge has also brought with it new questions. For example, can we classify receptors on structure alone or do we need also to have functional information? Can we use transductional information when it is difficult to determine which might be the physiological transducer pathway in whole tissues? The realization that receptors with hepta-helical transmembrane spanning loops can be promiscuous and combine with different G-proteins suggests that it is potentially difficult to identify the relevant transducer using recombinant expression systems. These and many other questions suggest that today we are moving into new territory where the old operational approach is deemed not to be sufficient, but as yet we have not reached an understanding about precisely what alternative we might use. For this reason an international meeting on receptor classification was held to build on the initiatives begun by the Technical Subcommittee of the IUPHAR Nomenclature and Receptor Classification Committee (NC-IUPHAR). The result of two days of presentations and discussions are published in this volume of the *Annals* of the New York Academy of Sciences. The debate was enthusiastic and even if no definitive, or even consensual, conclusion was reached (as expected), the issues have been clearly identified for future focus. It is hoped that this initiative will be repeated at regular intervals as an aid towards arriving at a universally accepted scheme in which receptors can be characterized and classified.

The Characterization and Classification of Neurotransmitter Receptors

PATRICK P. A. HUMPHREY

Glaxo Institute of Applied Pharmacology
Department of Pharmacology
University of Cambridge
Tennis Court Road
Cambridge CB2 1QJ, United Kingdom

INTRODUCTION

Pharmacologists have been interested in the characterization of receptors ever since Ehrlich and Langley put forward the idea, at the beginning of this century, that inter-cellular chemical signaling was mediated by receptive sites on cell surfaces.[1] However, it was to take another seventy-five years or so before the first primary structure of one such site of major pharmacological interest, the nicotinic receptor for acetylcholine, became known.[2] In this period leading up to the sea change effect of molecular biology on receptor pharmacology, pharmacologists were able to determine much about the variety of receptor types, their recognitory characteristics and their degree of selectivity for different chemicals, both endogenous and synthetic, but they were only able to guess at their substance and structure. Amusingly, De Jongh likened this process to a man writing to and exchanging letters with a beautiful lady whom he does not know; with time he could learn much about her, despite not being able to see her.[3] Today, in reference to De Jongh's metaphor, the "picture" more frequently comes first; receptor amino acid sequences (deduced from the nucleotide sequence of cloned cDNA or genes) now abound for receptors, not only those which are familiar but also those for which there are no good ligands or indeed the endogenous ligand is unknown, so-called "orphan receptors."[4,5] With this explosion of new information about receptor structure and even their distribution (*e.g.*, indirectly from *in situ* hybridization studies) it becomes essential to reconsider the way pharmacologists approach the characterization and classification of receptors for neurotransmitters, hormones, and autacoids. For the purposes of this article, only two receptor super-families will be considered, the heptahelical G-protein-linked receptors and the ligand-gated ion channel receptors, which in mediating the effects of multiple neurotransmitters have been the most studied by pharmacologists to date.

It is important to consider first the nature and hence definition of a receptor. With the information about receptor structure now available, Stephenson's long established definition of "that spatial arrangement of atoms to which a substance endogenous to the organism attaches itself as an essential step in modifying cellular function" no longer seems entirely appropriate.[6] Certainly it can be argued that it is still essential to classify receptors according to their natural ligands, because of the critical importance of such information towards understanding the functional role of a given receptor or receptors in the whole body. However, a receptor must now be regarded as a macro-molecule, which may or may not be a single molecular entity, with multiple sites of interaction.[7–9] The individual proteins can be defined in terms of their corresponding genes but there may be local molecular differences according to the host cell, in terms of variable levels of glycosylation and different translational splice variants.[10–14] A modern definition of a receptor might therefore be "the entire protein molecule or cluster of molecules which can selectively recognize and be activated by an endogenous ligand (agonist) to mediate a cellular event." Synthetic drugs may recognize different (amino acid) sites on the same receptor and even if only a very localized region on the macromolecule is required for agonist binding, attendant sub-units (specific G-proteins or ion channel components) will be essential for transducing the effect produced by agonist binding.[15–20]

CRITERIA FOR RECEPTOR CHARACTERIZATION

Receptors for hormones and neurotransmitters have been well characterized over many decades by studying the actions of synthetic ligands or drugs, with selective actions at one or more receptors. Such ligands display either agonist or antagonist actions which can be quantified to define the characteristics of receptor function, often providing powerful discriminatory data between receptors of interest. By this approach good evidence for different types of receptor for a single hormone was provided, long before the molecular structures of receptors were known. As early as 1948, Raymond Ahlquist suggested that noradrenaline acted at two types of receptor, which he called α and β, on the basis of two distinct rank orders of agonist potencies for noradrenaline, adrenaline and some synthetic analogues in various tissue preparations.[21] This proposal found wider acceptance with the discovery of selective α- and β-adrenoceptor blocking drugs (antagonists), which further distinguished between the two types of receptor for noradrenaline.[22–25] Agonists were also shown to differentiate between putative β_1- and β_2- adrenoceptor sub-types and hence there was no surprise when their molecular structures were determined and found to be distinct but related, sharing about 50% homology.[26,27] In many other areas of pharmacology too, receptor sub-types have been identified with the aid of selective drug tools, well in advance of identifying the nature and amino acid composition of the receptors. Examples are numerous and include receptors for histamine, 5-hydroxytryptamine (5-HT), prostaglandins and adenosine.[28–32] Clearly, functional characteristics are predictive of structural differences and therefore scientifically valuable, but also importantly provide invaluable information towards drug discovery.

However, with the ready availability of structural information today, the approach to receptor characterization and classification needs to be revised. This exercise has been systematically carried out over more than a decade with regard to receptors for 5-HT and some interesting considerations have come to light.[30,33–37] Many of the known receptor types for 5-HT had been identified functionally prior to their gene cloning (*e.g.*, $5\text{-HT}_{1B/D}$, 5-HT_2, 5-HT_3, 5-HT_4), while others had not. In order to simplify a potentially confused literature regarding the true identities of the growing number of receptors for 5-HT, an attempt was made to establish a framework of classification which would rationalize existing data and yet provide a basis for incorporation of newly identified recombinant 5-HT receptor types for which there might be little or no functional data.[35] This led to the concept that three not two critical sets of data are necessary for receptor characterization, not just *functional* (recognitory) and *structural* information but also *transductional* information provide important identifying characteristics (see FIG. 1). It has been proposed that all neurotransmitter receptor families could be classified using a similar integrated approach to that used for 5-HT receptors.[38]

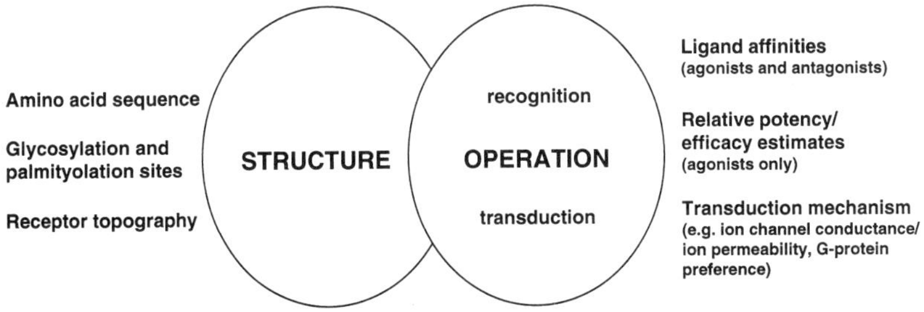

FIGURE 1. An integrated approach towards receptor characterization. It has been proposed that three distinct sets of "tripartite" data (structural, operational and transductional) are necessary for receptor characterization in an integrated scheme (Humphrey *et al.*, 1993[35]). Recent deliberations have led to a modified proposal that transductional data should be considered with recognitory data as a component of receptor operation or function (see text).

RECEPTOR OPERATION

Unquestionably data on receptor function is important, and although function is often indicative of receptor structure the corollary is not always so: a single amino acid difference can make a major difference in the functional characteristics of a receptor.[15,17,39,40] However, it must be recognized that qualitative functional data *per se* are not what is critical to *receptor characterization*, rather it is the quantitative drug recognition characteristics (*e.g.*, equilibrium dissociation constants for antagonists) which provide essential information about the *chemical* identity of the receptor. Equivalent data for both agonists and antagonists can be obtained from radioligand binding studies, but these cannot be referred to as "functional studies" as only binding interactions are involved. The term "operational" was therefore introduced to collectively describe data from both functional as well as radioligand binding studies, which together provide estimates of drug affinity, which reflect the recognitory characteristics of a given receptor.[35,38] Parenthetically the reader is reminded that attempts to measure affinity constants (more usually their reciprocal as equilibrium dissociation constants) are fraught with difficulties at both the technical and the theoretical level, requiring rigor and understanding of the potential pitfalls in the determination of meaningful estimates of any such parameter of drug-receptor interaction. Detailed considerations of such problems will not be considered here as they have been well reviewed in the literature and continue to occupy the minds of pharmacologists interested in drug-receptor theory.[41–43]

RECEPTOR STRUCTURE

The value of *structural* information is obvious. However, it should be acknowledged that the receptor protein composition is invariably deduced from the cloned cDNA of the gene and that local tissue variations may pertain such as differential glycosylation, translational splice variations or differential polymerization patterns for heteromeric receptors.[10,11,44,45] The latter particularly relates to multi-unit ion channel receptors, but might also conceptually include heptahelical receptor-G protein couplings.

Nevertheless, at the level of the gene a definitive "fingerprint" can be provided which unequivocally defines the receptor identity. Such information may provide useful clues to receptor function (*e.g.*, whether it is likely to be G-protein linked or a ligand-gated ion channel) but little about ligand recognition, unless other closely related receptors are already well known. Contemporary pharmacologists now find it necessary to study the pharmacology of recombinant receptors in cells (transfected with the appropriate DNA sequences) to determine their functional characteristics for comparison with similar or the same receptors expressed endogenously in whole tissues.

Structural information is not only valuable for receptor characterization purposes but also in their classification (see FIG. 2). Structural homologies can be quantified to provide information about inter-species differences as well as differences between receptor families (defined by the endogenous transmitter substance) and within families between types (or groups) and subtypes. Unfortunately, there seem to be no definitive quantitative demarcations between these divisions. However, species variants of receptors (the products of orthologous genes) commonly display 85–95% homology, while inter-family homologies are usually in the 15–35% region.[44,46–48] Unfortunately, no clear structural demarcations emerge to differentiate

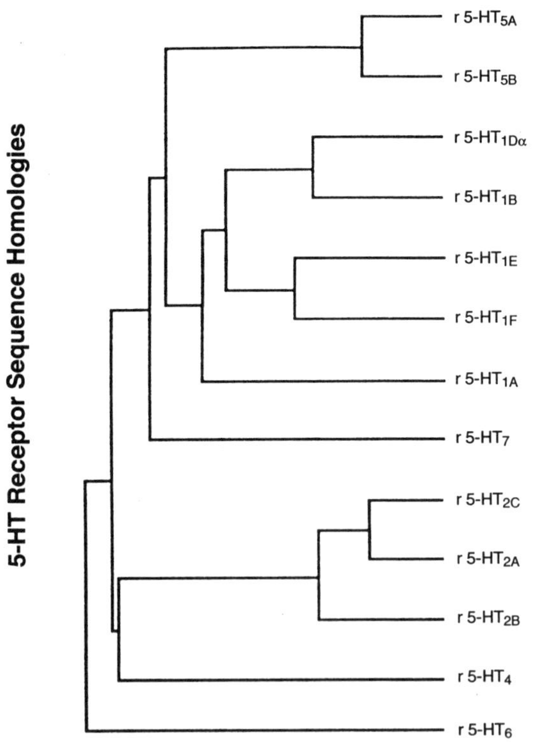

FIGURE 2. The relative sequence homologies of rat 5-hydroxytryptamine receptor proteins. Dendogram of rat 5-HT receptor family of G-protein coupled receptors. The protein sequences were compared and clustered according to sequence homology using the Pileup program by Gerald and colleagues (Gerald *et al.*, 1995[13]). Note the close structural homology within but not between the groups (5-HT_1 and 5-HT_2) which can be similarly classified on the basis of receptor operation. The 5-HT_4, 5-ht_6 and 5-HT_7 receptors are distinct structurally and operationally but share a common transduction pathway, *i.e.*, activation of adenylate cyclase. The 5-ht_{5a} and 5-ht_{5b} receptors are also distinct structurally and operationally (ligand affinities) but the transduction pathway is unknown (see Martin and Humphrey, 1994[36]).

between receptor types and subtypes. In the case of 5-HT receptors, there is between 53 and 80% identity between subtypes within a given receptor group (see TABLE 1). However, this will not always be so, as for example with the proposed classification of somatostatin receptors into $SRIF_1$ and $SRIF_2$ groups[49] where the receptor identity within a group (46–61%) is not much greater than the identities of subtypes between the two groups (44–50%).[50] However, the structure of the hydrophobic transmembrane regions of G-protein coupled receptors are highly conserved,[44,46] such that, for example, there is 72–96% homology within either the $SRIF_1$ or $SRIF_2$ receptor group and somewhat less (59–64%) between the groups in the second and third transmembrane regions of the five different recombinant somatostatin receptor types.[49] In an integrated scheme other criteria are also considered in deciding on a pragmatic and practical scheme of sub-classification into types and subtypes.

RECEPTOR TRANSDUCTION

It is polemic as to whether "transductional" data can meaningfully be used for receptor characterization and hence receptor classification purposes. However, such data have been useful in the sub-classification of 5-HT receptors. It remains to be seen whether this is coincidental or whether the concept is generally applicable. At the very least it can be argued that the functional determination of whether a receptor is G-protein linked (GTP-sensitive) or behaves as a ligand-gated ion channel receptor (current peak <0.1s) immediately provides information about receptor structure. Such considerations allowed, for example, the differentiation between 5-HT_2 (G-protein coupled) and 5-HT_3 (non-specific cation channel) receptors before their genes were cloned.[34]

The definition of transductional data for receptor characterization purposes is that of information relating to integral receptor activation processes and not just to down-stream, second messenger events, although they may give an indication of earlier events in the transduction cascade. For G-protein linked receptors it is important to know which specific G-proteins are activated and to understand their functional stoichiometry.[41,42,51–53] It remains to be seen to what degree the intracellular receptor structure and topography dictates G-protein preference. There is certainly good evidence using mRNA antisense techniques to indicate functional preferences for all three G-protein sub-units for different endogenous neurotransmitter receptors in the same cells (see FIG. 3).[54] The general applicability of these findings will dictate the value of transductional information for G-protein receptor classification. For ion channel receptors much of the transductional data will be determined from electrophysiological studies. This will provide information on the consequences of receptor activation regarding ion fluxes and associated charge carried, resulting from conformational changes and molecular re-arrangements of the component channel subunits. Analogously, it might be argued that for activation of heptahelical receptors, the essential G-protein involvement allows one to consider the molecular combination as an ephemeral tetrameric receptor (*i.e.*, receptor protein plus three complexed G-protein molecules, α, β and γ).

Regardless, pharmacologists will generate information on transductional data if only to understand how receptors mediate the cellular effects of the agonists which activate them. Much of this work will be carried out on recombinant receptors transfected into stable cell lines.[52,55–57] This could theoretically provide data of limited relevance to endogenous receptors in whole tissues because of differences in receptor densities, transfection into unnatural cell environments, and other consequences

TABLE 1. Percentage Amino Acid Identity in Transmembrane Domains of Rat 5-Hydroxytryptamine Receptors

	5-HT$_{1A}$	**5-HT$_{1B}$**	**5-HT$_{1D}$**	**5-ht$_{1e}$**	**5-ht$_{1f}$**	**5-HT$_{2A}$**	**5-HT$_{2B}$**	**5-HT$_{2C}$**	**5-HT$_{4}$**	**5-ht$_{5a}$**	**5-ht$_{5b}$**	**5-ht$_{6}$**	**5-HT$_{7}$**
5-HT$_{1A}$	**100**												
5-HT$_{1B}$	**54**	**100**											
5-HT$_{1D}$	**58**	**73**	**100**										
5-ht$_{1e}$	**53**	**63**	**64**	**100**									
5-ht$_{1f}$	**54**	**59**	**63**	**68**	**100**								
5-HT$_{2A}$	39	40	38	42	42	**100**							
5-HT$_{2B}$	43	40	40	45	38	**68**	**100**						
5-HT$_{2C}$	39	40	43	44	42	**80**	**71**	**100**					
5-HT$_{4}$	46	43	41	42	40	43	40	43	100				
5-ht$_{5a}$	49	48	48	48	49	38	37	40	36	**100**			
5-ht$_{5b}$	48	49	49	49	45	37	39	41	38	**84**	**100**		
5-ht$_{6}$	40	38	40	37	36	43	41	42	35	39	37	100	
5-HT$_{7}$	50	52	49	53	49	40	37	39	46	44	44	39	100

The data are modified from Gerald *et al.*, 1995.[13] Note the bold numbering to highlight the high homology within the larger receptor groups with subtypes (5-HT$_1$, 53–73%; 5-HT$_2$, 68–80%) and the range of 36–53% homology between the different groups or receptor types. Note too the use of lower case letters to denote recombinant receptor types whose full operational profile has yet to be determined fully in whole tissues (see Vanhoutte *et al.*, 1996[59]).

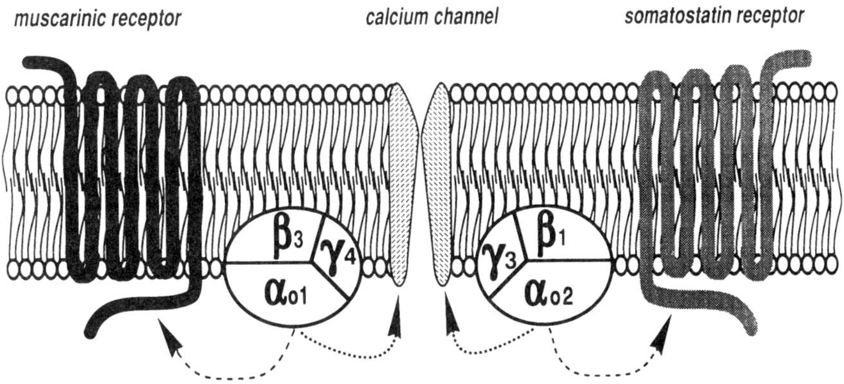

FIGURE 3. Selective transduction mechanisms for muscarinic and somatostatin receptors. Schematic representation of the G-proteins that couple and mediate receptor transduction for endogenous mucarinic and somatostatin receptors in pituitary GH_3 cells. Electrophysiological studies using anti-sense indicate a unique trimeric G-protein requirement for the function of each inhibitory neurotransmitter receptor (Kleuss *et al.*, 1993[54]).

of heterologous transfection (see below).[41,42] However, it is evident that it will provide important information about how receptors function and are activated and could ultimately lead to an understanding of efficacy at the molecular level.[58] It should be acknowledged that transductional data are much more difficult to obtain for endogenous receptors in tissues, which may limit the value for classification purposes. Thus, as defined above, transductional data do not specifically include the cataloguing of second messenger involvement but rather the definition of trimeric G-protein involvement in the cellular response, which must restrict the availability of such data from whole tissue studies. Nevertheless, the circumspect use of data on second messenger activation (*i.e.*, activation of adenylate cyclase or protein kinase C) may give a good indication of the G-protein involvement, albeit not a definitive answer. In the case of ligand-gated ion channel receptors, transductional data may be more easily obtained directly from electrophysiological studies.

AN INTEGRATED APPROACH TO RECEPTOR CLASSIFICATION

The use of an integrated approach to receptor classification based on recognitory, structural and transductional criteria has proved useful in the classification of 5-HT receptors allowing receptors to be grouped appropriately together in advance of their structure being known. Thus the 5-HT_4 receptor was accepted as a new type of receptor on the basis of operational and transductional data in advance of its gene being cloned and the structural distinction of the receptor protein confirmed.[13,30] Similarly, despite significant operational differences, the 5-ht_{1e} and 5-ht_{1f} receptors were included in the 5-HT_1 group on the basis of similar structural and transductional characteristics to the rest of the subtypes.[35]

More recently it has been agreed that the 5-$HT_{1D\beta}$ receptor should be called the human 5-HT_{1B} receptor,[37] despite the fact that the 5-$HT_{1D\beta}$ receptor has different operational characteristics (*i.e.*, not blocked by propranolol and cyanopindolol) to the rat 5-HT_{1B} receptor (which is blocked potently by β-adrenoceptor blockers). This is based on the understanding we have now that species variants of receptors frequently have different operational characteristics and that the structural homology between the rat 5-HT_{1B} and the human 5-$HT_{1D\beta}$ receptors is very high at 97%.[37,48] As a result of such considerations, the IUPHAR Receptor Nomenclature and Drug Classification Committee has sanctioned the idea of the primacy of the human genes and their products in any receptor classification scheme.[59] It is hoped that the principles which have been formulated and applied to the classification of 5-HT can similarly be applied to other receptor families.[38]

The "tripartite" approach to receptor classification has been used also to classify the adrenoceptor and prostaglandin receptor families.[31,60] Perhaps not surprisingly there is good agreement between the structural and operational data but in addition, as with the 5-HT receptor family, the transductional characteristics also appear to be broadly diagnostic for the intra-family receptor groups (see TABLE 2). Thus in the case of adrenoceptors, the division into α_1, α_2 and β-adrenoceptor groups is supported by the findings that in whole tissues α_1-adrenoceptors mediate increases

TABLE 2. Receptor Types and Their G-Protein Coupling

Receptor	Subtype	G_s	G_q/G_{11}	G_i/G_o
Adrenoceptor α_1	α_{1A}		✓	
	α_{1B}		✓	
	α_{1D}		✓	
Adrenoceptor α_2	α_{2A}			✓
	α_{2B}			✓
	α_{2C}			✓
Adrenoceptor β	β_1	✓		
	β_2	✓		
	β_3	✓		
Dopamine	D_1	✓		
	D_2			✓
	D_3			✓
	D_4			✓
	D_5	✓		
5-Hydroxytryptamine	5-HT_{1A}			✓
	5-HT_{1B}			✓
	5-HT_{1D}			✓
	5-HT_{2A}		✓	
	5-HT_{2C}		✓	
Prostanoid	DP	✓		
	FP		✓	
	IP	✓		
	TP		✓	
	EP_1		✓	
	EP_2	✓		
	EP_3	✓	✓	✓
	EP_4	✓		

Heptahelical receptor subtypes and their apparent G-protein coupling preference in whole tissues (data from Trends in Pharmacological Sciences Receptor Nomenclature Supplement, 1996; Coleman *et al.*, 1994[31]; Hoyer *et al.*, 1994[30]). Note that receptor groups (types) can be broadly differentiated on the basis of such transductional information (but see text).

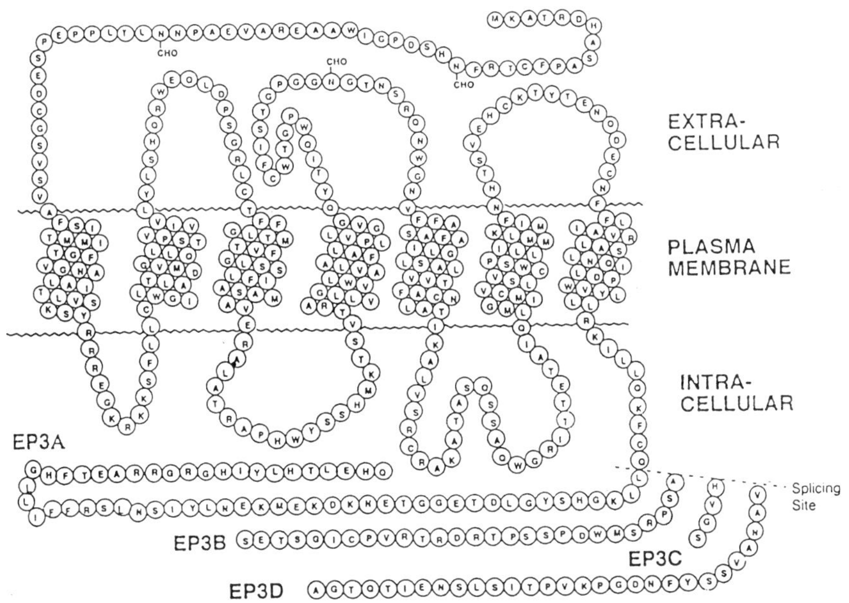

FIGURE 4. Prostaglandin EP_3 receptor splice variants. A schematic two dimensional representation of the amino sequence of the bovine prostaglandin EP_3 receptor to show the four alternative splice variants, $EP_{3(a)}$, $EP_{3(b)}$, $EP_{3(c)}$ and $EP_{3(d)}$, which only differ in the C-terminal intra-cellular tail (from Coleman *et al.*, 1994[31] with permission). Each receptor variant appears to activate different transduction mechanisms such that they can be differentiated functionally.

in intracellular calcium while α_2-adrenoceptors inhibit adenylate cyclase and β-adrenoceptors activate adenylate cyclase.[60] However, although α_2-adrenoceptor activation is consistently associated with inhibition of adenylyl cyclase it does not always appear to represent the signal transduction mechanism responsible for the cellular response mediated. Despite the distinct transductional characteristics for each of the three groups there are no obvious differences within each group of adrenoceptor.[60] This might either suggest that transductional data do not provide a sensitive enough indicator to differentiate between closely related receptor subtypes or alternatively the problems of collecting the appropriate data in whole tissues, alluded to above, may mask any transductional differences.

In the case of prostanoid receptors, each receptor group has different transductional characteristics (see TABLE 2). However, unusually in the EP_3 receptor group multiple second messenger systems appear to be activated, but each appears to be differentially mediated by distinct splice variants of which there are four (see FIG. 4). It would appear that the variations, which all occur in the C-terminal tail, lead to differential G-protein coupling.[31] Such splice variants for both G-protein linked receptors (*e.g.*, 5-HT_4, somatostatin sst_2 as well as EP_3) and ion channel receptors (*e.g.*, 5-HT_3) cannot always be distinguished on the basis of agonist or antagonist potencies but can be distinguished on the basis of transductional information.[12,13,31,61] This alone must justify the pharmacological importance of studying transductional mechanisms.

FUTURE DEVELOPMENTS

There is an emerging consensus amongst molecular biologists and pharmacologists interested in receptor classification that receptors can be characterized and classified using an integrated approach based both on receptor structure and function. As *structural* information is definitive it will always be pre-eminent in terms of receptor identification and archival reference. If only to avoid ambiguity, the primary archival data will be the structure of the receptor gene or genes (for multimeric receptors) and the inferred amino acid sequences of the corresponding protein(s). The recognitory and functional characteristics of the receptor (*operational* data) will also be essential and diagnostic for a given receptor. Whether or not *transductional* data should be considered as a third category of essential data remains to be decided, but there are good reasons to consider integrating transductional with operational data, which collectively relate to receptor activation and function (see FIG. 1).

Thus there is increasing evidence that transduction mechanisms are much more complex than was first thought. First, although the intra-cellular domain of each distinct receptor type is unique, it remains to be demonstrated whether or not there is sufficient repertoire of trimeric G-protein permutations to allow distinct transduction pathways for each. It would seem better to refer to G-protein preference(s) rather than selectivity, for as with ligand-receptor interactions selectivity is not absolute and stoichiometric considerations are paramount.[41,42,62,63] Secondly, the fundamentals of agonist efficacy remain to be understood and the concept of a two-state conformational model for receptor activation now seem in doubt. Indeed, there is now evidence to suggest that there may be multiple conformational states, differentially dictated by different agonists, as well as antagonists and inverse agonists. These different states might theoretically command different G-protein preferences and hence initiate different transduction pathways.[42,57] Given the controversy already alluded to concerning the use of transductional data to characterize receptors, it would seem prudent to include such data under operational characteristics as further descriptors of receptor operation. This in no way devalues the importance of pharmacologists studying transduction mechanisms. Indeed they will be different for different types of receptor and their detailed study with proper application of the principles of sound quantitative pharmacology will lead us closer to the "Holy Grail" of pharmacology, namely, a true understanding of the molecular interactions, which explain the phenomenon of agonist receptor activation, that we abstractly refer to as "efficacy."

REFERENCES

1. HIMMELWEIT, F. 1960. The collected papers of Paul Ehrlich, Vol. III, Chemotherapy. Edited and compiled by F. Himmelweit.
2. CLAUDIO, T. 1986. Recombinant DNA technology in the study of ion channels. Trends Pharmacol. Sci. **7:** 308–312.
3. DE JONGH, D. K. 1964. A molecular approach to general pharmacology. *In* Molecular Pharmacology—The Mode of Action of Biologically Active Compounds, Vol. I. E. J. Ariens, Ed. Academic Press. New York/London, p. 16.
4. MILLS, A. & M. J. DUGGAN. 1993. Orphan seven transmembrane domain receptors: Reversing pharmacology. Trends Pharmacol. Sci. **14:** 394–396.
5. MEUNIER, J.-C., C. MOLLEREAU, L. TOLL, C. SUAUDEAU, C. MOISAND, P. ALVINERIE, J.-L. BUTOUR, J.-C. GUILLEMOT, P. FERRARA, B. MONSARRAT, H. MAZARGUIL, G. VASSART, M. PARMENTIER & J. COSTENTIN. 1995. Isolation and structure of the endogenous agonist of opioid receptor-like ORL_1 receptor. Nature **377:** 532–534.

6. STEPHENSON, R. P. 1975. Interaction of agonists and antagonists with their receptors. INSERM **50:** 15–28.
7. ROLE, L. W. 1992. Diversity in primary structure and function of neuronal nicotinic acetylcholine receptor channels. Curr. Opin. Neurobiol. **2:** 254–262.
8. MONYER, H., R. SPRENGEL, R. SCHOEPFER, A. HERB, M. HIGUCHI, H. LOMELI, N. BURNASHEV, B. SAKMANN & P. H. SEEBURG. 1992. Heteromeric NMDA receptors: molecular and functional distinction of subtypes. Science **256:** 1217–1221.
9. SCHWARTZ, T. W. 1994. Locating ligand-binding sites in 7TM receptors by protein engineering. Curr. Opin. Biotech. **5:** 434–444.
10. GIROS, B., P. SOKOLOFF, M.-P. MARTRES, J.-F. RIOU, L. J. EMORINE & J.-C. SCHWARTZ. 1989. Alternative splicing directs the expression of two D_2 dopamine receptor isoforms. Nature **342:** 923–926.
11. EPELBAUM, J., P. DOURNAUD, M. FODOR & C. VIOLLET. 1994. The neurobiology of somatostatin. Crit. Rev. Neurobiol. **8:** 25–44.
12. DOWNIE, D. L., A. G. HOPE, J. J. LAMBERT, J. A. PETERS, T. P. BLACKBURN & B. J. JONES. 1994. Pharmacological characterization of the apparent splice variants of the murine 5-HT_3 R-A subunit expressed in xenopus laevis oocytes. Neuropharmacology **33:** 473–482.
13. GERALD, C., N. ADHAM, H.-T. KAO, M. A. OLSEN, T. M. LAZ, L. E. SCHECHTER, J. A. BARD, P. J.-J. VAYSSE, P. R. HARTIG, T. A. BRANCHEK & R. L. WEINSHANK. 1995. The 5-HT_4 receptor: molecular cloning and pharmacological characterization of two splice variants. EMBO J. **14:** 2806–2815.
14. HOUSLEY, G. D., D. GREENWOOD, T. BENNETT & A. F. RYAN. 1995. Identification of a short form of the P2XR1-purinoceptor subunit produced by alternative splicing in the pituitary and cochlea. Biochem. Biophys. Res. Commun. **212:** 501–508.
15. REGOLI, D., A. BOUDON & J.-C. FAUCHERE. 1994. Receptors and antagonists for substance P and related peptides. Pharmacol. Rev. **46:** 551–599.
16. BIRNBAUMER, L., J. ABRAMOWITZ, A. YATANI, K. OKABE, R. MATTERA, R. GRAF, J. SANFORD, J. CODINA & A. M. BROWN. 1990. Roles of G proteins in coupling of receptors to ionic channels and other effector systems. Crit. Rev. Biochem. Mol. Biol. **25:** 225–244.
17. OKSENBERG, D., S. A. MARSTERS, B. F. O'DOWD, H. JIN, S. HAVLIK, S. J. PEROUTKA & A. ASHKENAZI. 1992. A single amino-acid difference confers major pharmacological variation between human and rodent 5-HT_{1B} receptors. Nature **360:** 161–163.
18. GETHER, U., T. E. JOHANSEN, R. M. SNIDER, J. A. LOWE, S. NAKANISHI & T. W. SCHWARTZ. 1993. Different binding epitopes on the NK_1 receptor for substance P and a non-peptide antagonist. Nature **362:** 345–347.
19. GINGRICH, J. A. & M. G. CARON. 1993. Recent advances in the molecular biology of dopamine receptors. Ann. Rev. Neurosci. **16:** 299–321.
20. SARGENT, P. B. 1993. The diversity of neuronal nicotinic receptors. Ann. Rev. Neurosci. **16:** 403–443.
21. AHLQUIST, R. P. 1948. A study of the adrenergic receptors. Am. J. Physiol. **153:** 586–600.
22. NICKERSON, M. 1949. The pharmacology of adrenergic blockade. Pharmacol. Rev. **1:** 27–101.
23. BLACK, J. W. & J. S. STEPHENSON. 1962. Pharmacology of a new adrenergic β-receptor blocking compound. Lancet **2:** 311–314.
24. DOXEY, J. C., C. F. C. SMITH & J. M. WALKER. 1977. Selectivity of blocking agents for pre- and postsynaptic α-adrenoceptors. Br. J. Pharmacol. **60:** 91–96.
25. BLACK, J. W. 1990. Development of ideas about noradrenaline receptors. *In* The Heyman Institute of Pharmacology 1890–1990. A. F. De Shaepdryver, Ed. Heymans Foundation. Ghent, Belgium, pp. 109–115.
26. LANDS, A. M., A. ARNOLD, J. P. MCAULIFF, F. P. LUDUENA & T. G. BROWN, JR. 1967. Differentiation of receptor systems activated by sympathomimetic amines. Nature **214:** 597–598.
27. FRIELLE, T., K. W. DANIEL, M. G. CARON & R. J. LEFKOWITZ. 1988. The structural basis of β-adrenergic receptor subtype specificity: studies with chimeric β_1/β_2-adrenergic receptor. Proc. Natl. Acad. Sci. USA **85:** 9494–9498.

28. ASH, A. S. F. & H. O. SCHILD. 1966. Receptors mediating some actions of histamine. Br. J. Pharmacol. **27:** 427–439.
29. BLACK, J. W., W. A. M. DUNCAN, C. J. DURANT, C. R. GANELLIN & E. M. PARSONS. 1972. Definition and antagonism of histamine H_2-receptors. Nature **236:** 385–390.
30. HOYER, D., D. E. CLARKE, J. R. FOZARD, P. R. HARTIG, G. R. MARTIN, E. J. MYLECHARANE, P. R. SAXENA & P. P. A. HUMPHREY. 1994. VII. International Union of Pharmacology Classification of Receptors for 5-Hydroxytryptamine (Serotonin). Pharmacol. Rev. **46:** 157–203.
31. COLEMAN, R. A., W. L. SMITH & S. NARUMIYA. 1994. VIII. International Union of Pharmacology Classification of Prostanoid Receptors: Properties, Distribution, and Structure of the Receptors and their Subtypes. Pharmacol. Rev. **46:** 205–229.
32. FREDHOLM, B. B., M. P. ABBRACCHIO, G. BURNSTOCK, J. W. DALY, T. K. HARDEN, K. A. JACOBSON, P. LEFF & M. WILLIAMS. 1994. VI. International Union of Pharmacology Nomenclature and Classification of Purinoceptors. Pharmacol. Rev. **46:** 143–156.
33. HUMPHREY, P. P. A. 1984. Peripheral 5-hydroxytryptamine receptors and their classification. Neuropharmacology **23:** 1503–1510.
34. BRADLEY, P. B., G. ENGEL, W. FENIUK, J. R. FOZARD, P. P. A. HUMPHREY, D. N. MIDDLEMISS, E. J. MYLECHARANE, B. PRICHARDSON & P. R. SAXENA. 1986. Proposals for the classification and nomenclature of functional receptors for 5-hydroxytryptamine. Neuropharmacology **25:** 563–576.
35. HUMPHREY, P. P. A., P. HARTIG & D. HOYER. 1993. A proposed new nomenclature for 5-HT receptors. Trends Pharmacol. Sci. **14:** 233–236.
36. MARTIN, G. R. & P. P. A. HUMPHREY. 1994. Receptors for 5-Hydroxytryptamine: Current perspectives on classification and nomenclature. Neuropharmacology **33:** 261–273.
37. HARTIG, P. R., D. HOYER, P. P. A. HUMPHREY & G. R. MARTIN. 1996. Alignment of receptor nomenclature with the human genome. Effects on classification of 5-HT_{1B}/5-HT_{1D} receptor subtypes. Trends Pharmacol. Sci. **17:** 103–105.
38. HUMPHREY, P. P. A., M. SPEDDING & P. M. VANHOUTTE. 1994. Receptor classification and nomenclature: The revolution and the resolution. Trends Pharmacol. Sci. **15:** 203–204.
39. SACHAIS, B. S., R. M. SNIDER, J. A. LOWE III & J. E. KRAUSE. 1993. Molecular basis for the species selectivity of the substance P antagonist CP-96,345. J. Biol. Chem. **268:** 2319–2323.
40. ADHAM, N., J. A. TAMM, J. A. SALON, P. J.-J. VAYSSE, R. L. WEINSHANK & T. A. BRANCHEK. 1994. A single point mutation increases the affinity of serotonin $5\text{-HT}_{1D\alpha}$, $5\text{-HT}_{1D\beta}$, 5-HT_{1E} and 5-HT_{1F} receptors for β-adrenergic antagonists. Neuropharmacology **33:** 387–391.
41. KENAKIN, T. 1995a. Agonist-receptor efficacy I: Mechanisms of efficacy and receptor promiscuity. Trends Pharmacol. Sci. **16:** 188–192.
42. KENAKIN, T. 1995b. Agonist-receptor efficacy II: Agonist trafficking of receptor signals. Trends Pharmacol. Sci. **16:** 232–238.
43. KENAKIN, T. P., R. A. BOND & T. I. BONNER. 1992. Definition of pharmacological receptors. Pharmacol. Rev. **44:** 351–362.
44. PROBST, W. C., L. A. SNYDER, D. I. SCHUSTER, J. BROSIUS & S. C. SEALFON. 1992. Sequence alignment of the G-protein coupled receptor superfamily. DNA Cell Biol. **11:** 1–20.
45. LEWIS, C., S. NEIDHART, C. HOLY, R. A. NORTH, G. BUELL & A. SURPRENANT. 1995. Coexpression of $P2X_2$ and $P2X_3$ receptor subunits can account for ATP-gated currents in sensory neurons. Nature **377:** 432–435.
46. DOHLMAN, H., M. CARON & R. J. LEFKOWITZ. 1987. A family of receptors coupled to guanine nucleotide regulating proteins. Biochem. **26:** 2657–2664.
47. HARTIG, P. R., T. A. BRANCHEK & R. L. WEINSHANK. 1992. A subfamily of 5-HT_{1D} receptor genes. Trends Pharmacol. Sci. **13:** 152–159.
48. HALL, J. M., M. P. CAULFIELD, S. P. WATSON & S. GUARD. 1993. Receptor subtypes or species homologues: relevance to drug discovery. Trends Pharmacol. Sci. **14:** 376–383.
49. HOYER, D., H. LÜBBERT & C. BRUNS. 1994. Molecular pharmacology of somatostatin receptors. Naunyn-Schmiedeberg's Arch. Pharmacol. **350:** 441–453.

50. Reisine, T. & G. I. Bell. 1995. Molecular biology of somatostatin receptors. Endocrine Rev. **16:** 427–442.
51. Jackson, T. 1991. Structure and function of G protein coupled receptors. Pharmac. Ther. **50:** 425–442.
52. Milligan, G. 1993. Mechanisms of multifunctional signalling by G protein-linked receptors. Trends Pharmacol. Sci. **14:** 239–244.
53. Watson, S. & S. Arkinstall. 1994. The G-protein linked receptor. Factsbook, Academic Press.
54. Kleuss, C., H. Scherubl, J. Hescheler, G. Schultz & B. Wittig. 1993. Selectivity in signal transduction determined by gamma subunits of heterotrimeric G proteins. Science **259:** 832–834.
55. Vallar, L., C. Muca, M. Magni, P. Albert, J. Bunzow, J. Meldolesi & O. Civelli. 1990. Differential coupling of dopaminergic D_2 receptors expressed in different cell types. J. Biol. Chem. **265:** 10320–10326.
56. Birnbaumer, L. & M. Birnbaumer. 1994. Signal transduction by G proteins. J. Rec. Sig. Trans. Res. **15:** 213–252.
57. Spengler, D., C. Waeber, C. Pantaloni, F. Holsboer, J. Bockaert, P. H. Seeburg & L. Journot. 1993. Differential signal transduction by five splice variants of the PACAP receptor. Nature **365:** 170–175.
58. Stephenson, R. P. 1956. A modification of receptor theory. Br. J. Pharmacol. **11:** 379–393.
59. Vanhoutte, P. M., P. P. A. Humphrey & M. Spedding. 1996. X. International Union of Pharmacology Recommendations for Nomenclature of New Receptor Subtypes. Pharmacol. Rev. **48:** 1–2.
60. Bylund, D. B., D. C. Eikenberg, J. P. Hieble, S. Z. Langer, R. J. Lefkowitz, K. P. Minneman, P. B. Molinoff, R. R. Ruffolo & U. Trendelenburg. 1994. IV. International Union of Pharmacology Nomenclature of Adrenoceptors. Pharmacol. Rev. **46:** 121–136.
61. Vanetti, M., G. Vogt & V. Hollt. 1993. The two isoforms of the mouse somatostatin receptor (mSSTR2A and mSSTR2B) differ in coupling efficiency to adenylate cyclase and in agonist-induced receptor desensitization. FEBS Lett. **331:** 260–266.
62. Offermanns, S., T. Wieland, D. Homann, J. Sandmann, E. Bombien, K. Spicher, G. Schultz & K. H. Jakobs. 1994. Transfected muscarinic acetylcholine receptors selectively couple to G_i-type G proteins and $G_{q/11}$. Mol. Pharmacol. **45:** 890–898.
63. Gilman, A. G. 1995. G. proteins and regulation of adenylyl cyclase. Bioscience Reports **15:** 65–97.

Protein Structures in Receptor Classification[a]

ERIC A. BARNARD[b]

Molecular Neurobiology Unit
Royal Free Hospital School of Medicine
London NW3 2PF, United Kingdom

INTRODUCTION

Broad and Narrow Meanings of "Receptor"

We ought firstly to consider the boundaries of the field which we seek to classify. If we do not use the term "receptor" rigorously and consistently, the range of structures involved becomes too heterogeneous for a classification which is useful in pharmacology. The first level of classification, therefore, is of "receptors" versus "non-receptors".

The term "receptor" is often used, even while keeping to the molecular level, in a very broad sense (as in "cell adhesion receptor", "LDL receptor", *etc.*), but the narrow sense definition is required here. For a pharmacological classification scheme the term "receptor" means "signal-transducing receptor" (as has been defined and discussed for IUPHAR classification purposes by Kenakin *et al.*[1]; see also P. P. A. Humphrey,[2] this volume) and the native agonists for the receptor activation are messenger molecules which specifically relay a signal into a cell or between compartments within a cell. Generally, one can distinguish binding and biological activity at these receptors *in sensu strictu* from such interactions at a surface protein which are not primarily transductive, as in interactions with extracellular matrix molecules or in transporter activity or in "receptor-mediated endocytosis"[3] for nutrient uptake or for recycling, catabolism or scavenging of certain proteins.[4] Sometimes, however this distinction is not so clear: as examples, there is evidence that a few transporters (*e.g.*, the cystic fibrosis transmembrane conductance regulator (CFTR)[5] or glutamate transporters[6]) can be also ligand-gated channels, that oxidized LDL can interact at a transductive macrophage receptor,[7] and that some adhesion/migration activities mediated by integrin or "Eph" receptors are linked to signaling pathways[8,9] to the cytoskeleton or to gene activation.

The receptors in the narrow sense may be in one of three locations: (a) in the plasma membrane (the receptors for neurotransmitters, neurotrophins, growth factors and cytokines, other immuno-mediators, morphogens, sensory stimulants and chemoattractants, and most circulating hormones); (b) in an organelle membrane (*e.g.*, those receptors where the transduction process is the release of Ca^{2+} from an intracellular store); (c) in cytosolic solution. Case (c) involves migration to the cell nucleus, and is employed for the receptors which belong to a particular superfamily of ligand-regulated transcription factors, whose ligands are steroid hormones and certain other fat-soluble hormones.

[a] Tables 1 and 2, and associated text, are modified from the paper by E. A. Barnard (1996) Trends Pharmacol. Sci. **17:** 305–309, by permission of Elsevier Science Ltd.

[b] The author is supported by the Wellcome Trust.

THE BASIS OF RECEPTOR CLASSIFICATION

The Receptor Classes

The application of the immensely powerful methods of molecular biology has, of course, wrought a revolution in our knowledge of the signal-transducing receptors. The deduced amino acid sequences of a large number of receptor subunit types are now known, and within most of these the multiple sub-types or isoforms have yielded further sequences. In all, well above a thousand receptor subunit amino acid sequences are now available. By analyzing this considerable database, some generalizations can be reached about their structures overall. It is found that each receptor molecule either is a monomeric polypeptide, or contains two or more identical subunits, *i.e.*, is homo-oligomeric, or is built using two to five types of non-identical subunits, *i.e.*, is hetero-oligomeric. Apart from the ligand-regulated transcription factor class, these are all integral membrane proteins and hence all of their subunits contain at least one hydrophobic trans-membrane domain (TM) (or at least a membrane anchor[10]: see CNTF receptor, FIG. 1) plus extracellular and intracellular domains. In fact, despite the enormous multiplicity of receptor types, only four basic structural designs have been revealed to account for all of the membrane receptors (FIG. 1).

It appears, therefore, logical to distinguish the broad classes of receptors by their main structural motifs in relation to the membrane. This yields four Classes, as shown in TABLE 1. However, each class of membrane topology is associated with, and obviously arises from, a different general type of transduction (FIG. 1), so that this higher-order classification of receptors is jointly structural and functional.

Several general conclusions emerge from a survey of receptor types and their allocation between the Classes 1–4:

(i) The number of *different* subunit types which can be present in one receptor molecule is not a fundamental distinction between Classes: this number is one in Class 2, is one or two in Class 4 and varies from one to five between receptor types in both Classes 1 and 3.

(ii) Different receptor types in Class 1 can have one of several different membrane topologies, this still being within the structural definition of a single Class, as will be discussed below.

(iii) Compared to the wide range of structural motifs employed across the range of all proteins, only a relatively small number of structural elements (FIG. 1) are used for transducing receptor function. These elements are not defined at the level of amino acid sequence but at the level of chain topology in relation to the membrane and three-dimensional structure. Thus, no requirement is observed in practice for general conservation of subunit sequence within a given Class even at a low homology level. Each Class, except Class 4, contains several protein superfamilies.

(iv) It is not possible, it emerges, to separate the receptors for neurotransmitters and hormones from the receptors which transduce signals specific to the immune system, although those two groups have usually been treated separately. The latter are built on the same structural principles as the former and they can all be accommodated in the Classes 2 and 3 erected for the former. (In addition, at least one immune system receptor is a transmitter-gated ion channel, namely the P_{2Z} receptor for ATP, which is important in the function of macrophages, mast cells and lymphocytes).[14] Some receptors of differentiated neurons (*e.g.*, the CNTF receptor) contain a subunit in common with an interleukin receptor of immune system cells[10,15] and have intracellular domains homologous to regions in receptors for interferons and some other cytokines.[16] Likewise, some receptors for hormones (such as growth

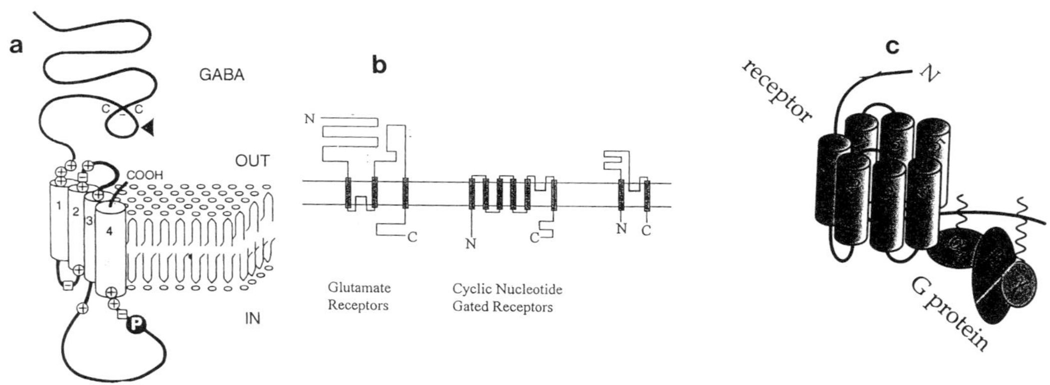

FIGURE 1. Structures of the membrane-bound receptors. The upper face is external to the cell or organelle membrane. **a:** A subunit of the $GABA_A$ receptor.[37] The TMs are numbered 1–4. The arrowhead points to the 15-residue Cys-loop. The structure typifies the subunits of all of the Cys-loop receptors. **b:** Other structures found in Class 1 receptors. The P-region is marked diagrammatically as a loop, forming the second MIS (re-entrant from IC side) in the glutamate receptor subunits and the sixth MIS (re-entrant from EC side) in the cyclic nucleotide receptors (diagram from Sun *et al.*[11]). The structure on the right represents subunits in the sub-Class 1.6 receptors; it can also represent sub-Classes 1.4 and 1.5, if the P-region is omitted (although the absence therein of any MIS in that region just before TM2 is as yet not certain[66]). The lengths of the EC segment, and of the N- and C-terminal extensions, are not to scale and vary considerably between those three sub-Classes. **c:** A 7-TM receptor, associated with its G-protein, showing the relative arrangement in the membrane of the TMs (based on the bovine rhodopsin structure). The lengths of the N- and C-terminal extensions vary considerably between families.

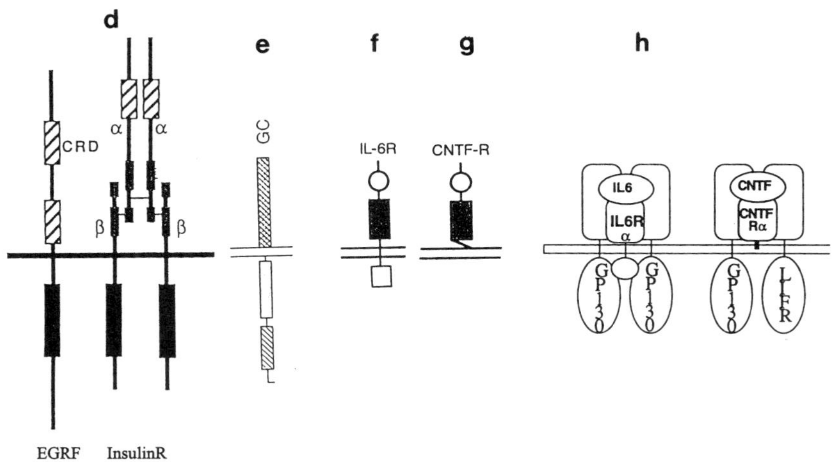

d–h, Different forms in the Class 3 receptor subunits. **d:** The epidermal growth factor receptor subunit and the insulin receptor subunits, typical of intrinsic TK receptors. The insulin receptor is in a sub-group of 3.1, of receptors containing a single subunit post-translationally cleaved to produce two chains (which may, as here, become linked by disulfides). The striped boxes show cysteine-rich domains (CRD) involved in the binding of the agonist polypeptide. The IC dark box shows the TK enzymic region. **e:** The ANP receptor.[13] The cross-hatching shows the agonist binding domain and the IC striped box shows the guanylate cyclase (GC) enzymic region. **f,g:** A cytokine receptor ligand-binding (α) subunit, here from the interleukin-6 receptor (IL-6R)[64] or from the CNTF receptor.[10] The EC domains of such α subunits contain some shared homologous features, as indicated by the corresponding symbols here. **h:** Heteromers form the active receptors in this series. The IC region of the IL-6Rα-subunit lacks a transduction-linked domain; the binding of IL-6 to it recruits two copies of a promiscuous structural (β) subunit, gp130. These then associate with the IC domain of the α subunit; the complex shown is activated to bind a diffusible TK (JAK kinase) and these initiate a transduction pathway. The same process occurs with the CNTF receptor, but using two β forms, gp130 and the LIFR (leukemia inhibitory factor, or cholinergic differentiation factor) β-subunit. (Panel **h** is from Ip & Yancopoulos.[68])

TABLE 1. Receptor Classes on the Basis of Structure

1	Transmitter-gated channels.[a] All have two or more MIS in each subunit, and four or more of these line a central channel in the oligomer.
2	7-TM proteins (G-protein-linked receptors)
3	Receptors with subunits each having one MIS, enzyme-associated[b]
4	Ligand-dependent transcriptional regulators[c] (Soluble proteins with DNA-binding domains; no membrane insertion)

TM, transmembrane segment; MIS, membrane-inserted segment(s). MIS include TMs, but also a P region where present (see text). MIS also includes, in Class 3, a glycolipid membrane anchor as employed in the recognition subunit of the ciliary neurotrophic factor (CNTF) receptor[10] and in certain receptors in the immune system.[12] Representative structures of Classes 1–3 are illustrated in FIGURE 1.

[a] Listed in TABLE 2.

[b] With intrinsic guanylate cyclase (ANP receptor)[13] or tyrosine kinase (TK) or Ser/Thr kinase activity, *or* associating with a diffusible TK or possibly another transduction-linked enzyme. This Class includes the receptors for growth factors, neurotrophins, cytokines (but not chemokines, which are in Class 2) and other mediators noted in the text.

[c] Steroid hormone, thyroid hormone, retinoic acid, and Vitamin D receptors.

hormone and prolactin) or for neurotrophins have extracellular sections (containing immunoglobulin (Ig)- or fibronectin-type domains) which are homologous in sequence to those in some cytokine receptors.[15–17]

(v) Further, we can draw the general conclusion that a limited range of structural features, and four structural Classes alone, are sufficient to cover all of the transducing receptors of the nervous, endocrine, sensory, immune, and hematopoietic systems. They encompass also all of the transductions specific for embryonic development, as well as the intracellular receptors present in all cells for signaling between their compartments. A natural economy of means is shown thus in accommodating a very wide range of molecules, small and large, as signaling messengers.

Definitions of Receptor Types

The Classes 1 to 4 can be divided into numbered sub-Classes primarily on the basis of their superfamilies, and secondarily on the basis of functional kinship, a procedure exemplified by the sub-Classes 1.1 to 1.8 of Class 1 (TABLE 2). Within each sub-Class a family of homologous sequences forms a receptor type. This would give three numerical divisions (which are each not limited to one digit), again as illustrated in TABLE 2.

Definitions of Receptor Subtypes

For the classification and nomenclature of the subtypes of a given receptor type, pharmacological criteria have in the past been far more practicable than others,[1] even when the molecular identity of all of the subtypes involved is known. Well-studied receptors were, initially, classified in terms of the relative activities of their agonists or their antagonists, *i.e.*, functionally. Thus, adrenoceptors were firstly recognized as distinct from other bioamine receptors, then α and β types, and then the subtypes thereof based on their ligand selectivities.[18] When the known subtypes

TABLE 2. Superfamilies in Class 1 (Transmitter-gated Channels)

Superfamily	Receptor Types	Membrane Topology	Transmitter Origin
1.1	Superfamily of Cys-loop receptors	4TM	EC
	1.1.1 anion channels ($GABA_A$)		
	1.1.2 anion channels (glycine)		
	1.1.3 anion channels (glutamate-gated)[a]		
	1.1.4 cation channels (nicotinic ACh)		
	1.1.5 cation channels ($5\text{-}HT_3$)		
1.2	Glutamate-gated cation channels	3TM + P	EC
	1.2.1 non-NMDA receptors		
	1.2.2 NMDA receptors		
1.3	Related to voltage-gated cation channels		
	1.3.1 cyclic nucleotide receptors	6TM + P[b]	IC
	1.3.2 IP_3 receptors	6TM + P[b]	IC
	1.3.3 cADP-ribose/Ca^{2+}-activated Ca^{2+} release channels ("ryanodine receptors")	4–6TM + P	IC
1.4	Related to epithelial Na^+ channels; non-peptide		
	1.4.1 (P2X) extracellular-ATP-gated cation channels	2TM[c]	EC
	1.4.2 ($P2X_7$/P2Z) ATP-gated dual channel	2TM[c]	EC
1.5	Related to epithelial Na^+ channels; peptide-gated channels		EC
	1.5.1 FMRFamide-gated Na^+ channel	2TM[c]	
1.6	Related to Kir channels		
	1.6.1 ATP-activated K^+ channel[d]	2TM + P	IC
	1.6.2 ATP-antagonized K^+ channel (K_{ATP})[d]	2TM + P	IC
1.7	Related to ABC transporters		
	1.7.1 CFTR (ATP-activated anion channel)[f]	12TM	IC
1.8	Related to amino acid transporters		
	1.8.1 glutamate-activated chloride channel/ EAA transporter[6]	10–12TM[e]	EC

For further details of each, see text. The membrane insertions presently deduced (in each subunit of the receptor) are indicated in the third column. EC = extracellular. IC = intracellular. P = pore region, in a re-entrant loop. EAA = excitatory amino acid. ABC = ATP-binding cassette (see text). Kir = inward rectifier K^+ channel. (Not classified are some potential members [see text]).

[a] Known from *C. elegans*[21] and *Drosophila.*[69]

[b] There are 6 deduced TMs and a P region in the membrane, in each subunit. In this case the channel in the assembled tetrameric molecule is believed to be built from 8 TMs (S5,S6 domains), with 4 P regions forming an iris-like central gate.[11] In the 1.3 set the agonist binds on the cytosolic, not the luminal, side of the receptor.

[c] There are two hydrophobic domains, presumed to be TMs; whether there is also an equivalent of a P region is unclear at present.[66] The 1.4 type has similarity in topology to the epithelial Na^+ channels but no sequence homology; 1.5.1 has both. Type 1.5.1 is known so far from *Helix.*[22] All of the P2X receptor subunits known so far[66] except $P2X_7$ can be regarded as in one receptor type, 1.4.1, but the $P2X_7$ receptor can logically be separated from those, because, although structurally related to them, its transduction differs. $P2X_7$ is also the P2Z receptor: in this type, a small cation channel is opened by ATP at physiological concentrations of Mg^{2+} but a very large non-selective channel is opened in the same homomeric receptor by ATP applied repetitively in low-divalent-cation medium.[14]

[d] The G-protein–activated Kir pairs, GIRK1,2 and GIRK2,4 (at least), in brain or heart respectively, can form K^+ channels which require the binding (non-hydrolytic) of IC ATP for activity.[23] The 1.6.2 type is the classical K_{ATP} channel *closed* by IC ATP,[24] which can be formed[25,26] from a related Kir (Kir 6.2) combined with the sulfonylurea binding protein (see text).

[e] The hydropathy plots are not very clear for the glutamate transporters and there is no independent evidence yet on their topology; 10 TMs has been thought to be a probable structure,[27] but other transporters have been deduced to have 12TMs.

[f] That the CFTR is an ATP-gated ion channel has been proposed, but at present the basis of its action is not fully clear: see text.

of a receptor became fairly numerous, structural information also influenced the classification, as in some of the subtypes of 5-HT receptors.[19] For example, the closer sequence similarity of the 5-HT_{1C} receptor (as defined pharmacologically) to the 5-HT_2 receptor led to the relevant IUPHAR Subcommittee re-classifying the former as the 5-HT_{2C} receptor, while the gene sequence dissimilarity to all the others of the provisionally numbered 5-ht_5 receptor led to confirmation of a separate 5-ht_5 category, which will become the 5-HT_5 receptor when the activity is identified *in vivo*.[19,20] However, since new receptor types and subtypes in the 7-TM Class are now more commonly discovered by DNA cloning (and more such can be expected from the recognition of the 7-TM motif in a set of the anonymous sequences being generated by the Genome Cloning programs), it can be predicted that the structural criteria will become those most readily applied in classifying new subtypes of the 7-TM receptors. This can be done while still weighing functional differences in making assignments, as is discussed by P. P. A. Humphrey in this volume.[2] The same general considerations apply also to Class 4 receptors. For some of the transmitter-gated ion channel receptors, on the other hand, there are special difficulties with this, to be noted where these are discussed.

TRANSMITTER-GATED ION CHANNELS

The term "transmitter-gated" is more exact than the previously used "ligand-gated" in describing these channels, since there are many ligands which can open or close ion channels (*e.g.*, voltage-gated types) without being natural transmitters, *e.g.*, a variety of toxins. "Transmitter" is a convenient term for all molecules signaling by direct binding to a receptor, whether they are extracellular or intracellular. Further, these receptors should not be specified as *ion* channels, since they include a channel that selects permeant molecules mainly by size (the $P2X_7$ receptor, discussed below).

These Class 1 receptor subunits have structures that can assemble in an oligomer surrounding a membrane pore. Transduction in this case is via the opening of a cation *or* an anion channel [or, in at least one potential case (the ATP-gated K^+ channel), the closing of a cation channel]. These receptors are adapted to perform fast signaling, in that their transduction is independent of any other component. So far, seven entirely distinct superfamilies of receptors are known in this class (TABLE 2) and there is a potential eighth (the CFTR). CFTR is a chloride channel but (exceptionally) by its sequence it is in the superfamily of ATP-binding cassette (ABC) transporters and contains two nucleotide binding domains.[28,29] Hydrolysis of ATP at one of those domains is coupled to the chloride channel opening, but there is also evidence to show that then a non-hydrolyzed ATP, bound at a second domain, maintains the open state of the channel.[30] If this is established, then the CFTR would be an ATP-gated channel as well as an active transporter.

Further evidence that the distinction between ion channels and transporters can in certain cases be blurred is afforded by certain glutamate transporters.[31] These form a separate superfamily,[27] 1.8, and combine a chloride channel with the transport of glutamate. This channel is, in fact, gated by bound glutamate.[6,31] It is unknown whether all of the TMs which determine the transport of glutamate are involved in the permeation of chloride, which is independently gated by glutamate.[31]

Even leaving aside those special cases of receptors associated with a transporter structure, the six sub-classes of "pure" transmitter-gated ion channels show very varied membrane topologies of their subunits and there is no relationship between these and the action of the agonist being at the external or the internal face of the receptor (TABLE 2). In the Class 1 receptors, most structural information has been available for the nicotinic acetylcholine (ACh) receptor of skeletal muscles or electric organs. In this case it has been definitively established, by a structural analysis at 9Å resolution, that all of the subunits span the membrane, forming a pentamer which encloses a central ion channel.[33] For the native $GABA_A$ receptors the same pentameric structure has also been established by electron microscopic image analysis[34] and there is evidence that the glycine, neuronal ACh and $5\text{-}HT_3$ receptors are also pentamers. All of the subunits of all of these receptor types are in one superfamily, 1.1. An invariant motif in those subunits is a pair of cysteines at a 15-residue spacing at a fixed position in the extracellular domain, which are known (at least in the case of the nicotinic ACh receptor[35]) to form a disulfide-bridged loop. From molecular modeling on the various subunit types of these receptors for this loop, a constant β-folded amphipathic structure is predicted.[36] This is a defining feature of this superfamily,[37] which can conveniently be termed the Cys-loop superfamily.[36,38]

In the Cys-loop superfamily the four hydrophobic domains deduced in each subunit have been shown to be truly transmembrane (extended recently to the $5\text{-}HT_3$ receptor, 1.1.5),[32] and they are at similar positions in each case. One of these, TM-2 (α-helical), is in the lining of the channel[33,38] and a segment of TM-1 is also implicated therein.[38] The 10 other TMs in the pentameric molecule would surround this channel structure. With this structural plan, the difference between anion- and cation-selective channels is secondary in the classification (TABLE 2). It is interesting that an anion channel in the Cys-loop superfamily is gated by glutamate (known as yet only from invertebrates: type 1.13), the only case in Class 1 where alternative anion and cation native channels are known that are gated by the same transmitter. The classification features in this superfamily are maintained from man down to the nematode.[21,39]

Cation Channels Gated by Glutamate or Cyclic Nucleotides

The glutamate or EAA receptors constitute a separate superfamily, with no significant homology to the Cys-loop receptors. The mammalian EAA receptor subunits are 2 to 3 times the size of the latter types. Although the hydropathy plots had initially suggested 4 or 5 TMs per subunit in the EAA receptor subunits,[40] recent mapping of the extracellular and intracellular access has led to a major revision of the assigned topology.[41–43] As indicated in FIGURE 1, only 3 domains are transmembrane, and the previous "TM-2" is apparently a re-entrant loop (both ends facing the cytoplasm), of the type of the re-entrant β-hairpin P (pore) region of the channel in the voltage-gated ion channels.[41–43] The evidence on this so far has, however, largely been obtained on the non-NMDA receptors.

A β-hairpin region also exists in the receptors for cGMP and cAMP.[44] These cyclic nucleotide-gated channels are found in photoreceptor cells and olfactory sensory neurons and in sperm.[45] Their subunits have some homology to voltage-gated K^+ channels and both those structures have been deduced to contain 6 TMs, S1–S6. The sequence between S5 and S6, known to control the pore in the K^+ channels, is homologous there in the cyclic nucleotide receptors.[46] Accessibility determination in cysteine-substituted mutants by Karlin and co-workers[11] across this region, in the retinal cGMP receptor in the open and closed channel states, have confirmed this for these receptors. The combined evidence from this and previous reports supports a model in which S5 and S6 TMs from each of the 4 subunits in one receptor molecule line the pore, with 4 P-region β-hairpin loops forming an iris-type shutter in its center.[11]

From this case and that of the EAA receptors, the term transmembrane domain (TM) is clearly inadequate to cover all receptor subunit regions lying in the membrane. The term MIS (membrane-inserted segment) is therefore preferable where it is at present unclear whether a hydrophobic segment present traverses the membrane (*i.e.*, is a TM) or is a P region or has some other association with the membrane. The case of the EAA receptors confirms that plots of hydropathy to identify hydrophobic domains of ~20 residues can no longer be used to define TMs or even MIS. Direct identification is needed of which segments are exposed on the extracellular side and which on the intracellular, by the various methods of accessibility determination which have been applied in the glutamate receptor and other cases discussed here, as well as mutagenesis studies for a P-region within the channel.

The two receptor types known to contain Ca^{2+} channels which operate in the release of Ca^{2+} from its intracellular stores also have homologies to voltage-gated K^+ channels but otherwise are not in the same superfamily as the cyclic nucleotide receptors (TABLE 2). One is gated by IP_3 and there is evidence that this receptor has the same arrangement of TMs and a P region as the latter type.[47] The other is the "ryanodine receptor" family, where it has been proposed that the channel is gated by both Ca^{2+} itself and by the second messenger cyclic ADP-ribose, although the roles of these in different ryanodine receptors are not yet entirely clear. This very large subunit assembles as a tetramer. The N-terminal region, ~80% of the subunit, of the ryanodine receptor is unrelated to others and is a cytoplasmic "foot" structure of the endoplasmic reticulum, but the remaining C-terminal region of the ryanodine receptor has strong homology to the IP_3 receptors, although here with possibly 4 TMs as well as the P region.[48] Hence the superfamily 1.3 is based on the employment of a P region apparently as in voltage-gated K^+ channels, but the number of associated TMs is not yet established to be the same in all.

Receptor Superfamilies with Similarities to Other Na^+ or K^+ Channels

Two groups of receptors are related in some way to the epithelial Na^+ channel. That structure is a non-receptor Na^+ channel which can contain 3 homologous subunit types, each with 2 hydrophobic domains which are presumed to be TMs.[49,50] The second TM in those subunits has been implicated in the channel structure.[51] A receptor (1.5.1) for the peptide FMRFamide (Phe-Met-Arg-Phe-NH_2) in a mollusc has been found[22] to be a peptide-gated form of a Na^+-selective channel protein of similar structure, with the channel-implicated region TM2 32% identical to TM2 in the mammalian[49,50] epithelial Na^+ channel. This may indicate that a family of peptide-gated channels of this type exists.[22]

The same topology of this channel type, with a long (~300 amino acids) extracellular loop between TM1 and TM2 (FIG. 1) is present in the P2X ATP-gated channels,[52,53] a family of which is now known. There is no sequence homology to the epithelial Na^+ channels, but the specific topology of the latter is shared only by the P2X type among known receptors. It has been suggested that the region immediately preceding TM2 in the P2X subunit is a P-region,[53] but it has clear sequence homology with K^+ channel P-regions only in the $P2X_2$ subtype. The existence of a P-region in the 1.4 and 1.5 receptor types is at present an open question. Of special interest is the recent discovery (see TABLE 2) in the $P2X_7$ receptor of a two-stage channel opening, either by Mg ATP or by $ATP4^-$, to produce the P2X or the P2Z type of channel, respectively.

The inward rectifier K^+ channels (Kir or IRKs) contain 2 TMs and a P-region which controls the channel.[54] An ATP-requiring channel is formed[23] by a heteromer containing 2 types of these channels, GIRK1 and 2, and another by GIRK2 and 4. The ATP action is not kinase-mediated and a non-hydrolyzable ATP analogue also maintains activity.[23] Hence, allowance must be made for a receptor superfamily, 1.6, based upon the Kirs. This superfamily extends to further types in which the K^+ channel is constitutively open but is closed by a transmitter: another Kir, Kir 6.2, has recently been shown[25,26] to be a subunit of such a channel, the K_{ATP} channel. This is a channel closed by intracellular ATP[24]; it is a hetero-oligomer which contains (so far as present evidence goes) Kir 6.2 subunits and subunits of the sulfonylurea-binding protein.[25,26] The latter is structurally of the ABC transporter superfamily and contains two nucleotide-binding sites and a sulfonylurea (channel-inhibitory) site per subunit (all of these being absent in Kir 6.2). Hence the K_{ATP} channel can be placed in sub-Class 1.6 for its K^+ channel structure but also in sub-Class 1.7 for its ligand-binding subunit.

The separate case of the ATP-activated chloride channel of the CFTR, superfamily 1.7, as well as that of the glutamate transporter/receptors (1.8) (TABLE 2), is described earlier here.

The sub-Classes shown in TABLE 2 do not necessarily exhaust the categories of the transmitter-gated ion channels. Various other channels have been proposed to be gated by a transmitter or intracellular messenger, of uncertain status at present because the physiological information is too incomplete and DNA clones are lacking. One example would be the putative messengers gating cell membrane channels repleting Ca^{2+} stores. Another type could arise from reports that arachidonic acid can activate a K^+ channel or that its metabolite, leukotriene C4, can activate a Ca^{2+} channel.[55] Other possibilities include claims of invertebrate ion channels activated directly by histamine or dopamine. In all of these cases, critical evidence for a direct action on the channel is awaited. If any of these potential members are verified, they may fit structurally into one of the sub-Classes of TABLE 2, but if not the classification would extend further to nine or more categories.

Defining the Subtypes of the Transmitter-gated Ion Channels

For certain receptors in this Class, it emerges that the subtypes cannot be defined in practice in the standard manner, with a large pharmacological input as described by P. P. A. Humphrey[2] (this volume). This occurs where the receptor is a hetero-oligomeric structure drawn from a variety of subunit types. This applies to the neuronal nicotinic ACh, $GABA_A$, non-NMDA, NMDA and perhaps the P2X receptors but not to the 5-HT_3 or glycine receptors.

A fundamental characteristic of that set of the transmitter-gated ion channels is their *combinatorial assembly*. This refers to the construction of a variety of subtypes of one receptor due to the existence of a considerable repertoire of subunit types, which is drawn upon in different combinations to the total, *e.g.*, of 5, subunits per receptor molecule. Thus the NMDA receptors can be built (on present knowledge) from combinations of 8 isoforms (alternatively spliced) of the NR1 subunit and 4 multiple types of the NR2 subunit, while the non-NMDA receptors are built from a repertoire of at least 9 subunit types.[40] Of the $GABA_A$ receptor subunits, the ρ types apparently assemble separately from the rest, but the others, including the dual splice variants of $\beta 2$ and $\gamma 2$ known so far, give a pool of 15 subunit types to form the receptors. The great majority of these receptors contain α, β and γ or δ subunits, to the total per molecule[34] of 5. Some restrictions have been observed upon the combinations detectable, and the probable theoretical maximum number of $GABA_A$ receptor subtypes is ~800, as discussed in full elsewhere.[56] The actual number *in vivo* is believed to be far less than this maximum, but nevertheless must be considerable.

The major element in the classification of these subtypes has to be the difference between the individual subunits which form their oligomers; this is necessarily structure-based, since they are in practice defined from cloning-derived sequences. However, the functional properties, if any, of a given subunit of the heteromeric receptors, when expressed as a recombinant alone, cannot be used in receptor classification. The functional characteristics obtained when a given subunit is co-expressed with proposed partners necessarily supplement the structural data in the classification. The same applies (to a lesser degree) to several sub-Classes of the receptors of Class 3 where multiple heteromeric receptors exist.

We cannot, with present techniques, know the subunit composition of a given subtype of such a multiplicitous receptor, nor necessarily can we recognize *in situ* a pharmacology which defines that composition alone. This problem is especially acute, obviously, with the brain receptors. It is therefore of little value to set up a classification of these subtypes based on subunit compositions alone. At present we can try to match the properties of a given combination formed *in vitro* with some observed native receptor. However, the number of pharmacologically recognized native $GABA_A$ or EAA receptors is, for example, far below the number of combinations which can be assembled successfully in co-transfected cultured cells.

Such combinations of subunits formed in stable cell lines will surely, by means of high-throughput screening programs, with time yield a much larger and superior range of pharmacological discriminants of those subtypes. New technologies of mapping *in situ* can also be anticipated to reduce the number of compositions known actually to occur. The definitions of the subtypes by function alone, at present very limited for these receptors, can be envisaged thus to become sufficiently powerful when the structural information can be assimilated into them. At present, we must recognize the gap between the two in these cases.

7-TM RECEPTORS

These Class 2 receptors fall into three structural sub-Classes (TABLE 3). There is virtually no sequence homology between these three but they all have the pre-

TABLE 3. 7-TM Receptors (Class 2)

Superfamily	Receptor Types	Note
2.1	2.1.1 Main (rhodopsin) superfamily: intact polypeptide chain	
	2.1.2 Main (rhodopsin) superfamily: cleaved polypeptide chain	(a) Thrombin receptor type[a] (b) Thyrotropin receptor type[a]
2.2	Secretin/VIP superfamily	Also contains parathyroid hormone and calcitonin receptors.
2.3	Metabotropic glutamate receptors	

VIP, vasoactive intestinal polypeptide.
[a] See text for the definition of these types.

dicted 7-TM regions in equivalent locations. Class 3.1 receptors are homologous to vertebrate rhodopsin and comprise by far the largest number of receptor types, the other two Classes being limited to those listed. In any of these sub-classes, very long terminal extensions of the chain outside the 7-TM core can occur, as in the luteinizing hormone receptor[57] (N-terminal domain, with a signal peptide, of 367 residues) in the 2.1 sub-Class, and all of the mGluR (1.3) sub-Class, where the C-terminus is exceptionally large. The former type is the extreme case of the receptors for protein hormones, which require such a domain. The extra C-terminal domain of the mGluRs is thought to play some additional role. Other structural features of the 7-TM receptors are reviewed by Strader *et al.*[58]

A further structural division which should be recognized in the classification is whether a single 7-TM receptor molecule has one or two chains. Hence, the sub-Class 2.1.2 contains the thrombin receptor and protease-activated receptor, in which the agonist specifically cleaves the receptor chain to liberate a new N-terminal segment and activate the receptor.[59,60] It also contains the thyrotropin receptor,[61] where a peptidase produces two extracellular chains to give the final receptor structure.

The 7-TM receptor Class includes by far the largest number of all receptor types known. Further, if the odorant and gustatory receptors, which are in Class 2.1, are each regarded as one type, then they contribute by far the largest number of subtypes. In man or rat, there is evidence for more than 500 genes encoding the set of 7-TM receptors needed as the basis of olfactory information coding.[62] Classification and nomenclature of these receptors when all are cloned will obviously require special measures.

CLASS 3 RECEPTORS

This Class of receptors is defined structurally, *i.e.*, those with subunits having a single TM (or MIS). They hence have an extracellular (EC) domain which can contain the ligand-binding site and an intracellular domain which is either specialized for a role in transduction or for association with other subunit types if in a heteromer (see FIG. 1). There is one structural exception to this definition, in that a very few of the ligand-binding subunits in receptors in this Class have no TM or intracellular domain, but have their EC domain anchored to the membrane by a glycolipid anchor (FIG. 1). These include[10,12] the ligand-binding subunit in the CNTF neurotrophin receptor, the Fcγ RIIIB receptor of neutrophils and the Fcμ receptor of B cells, all of which are linked to the tyrosine kinase transduction pathway. The Class 3.2 receptors are therefore divided into those with (3.2.1) or without (3.2.2) a TM peptide. The term 1-MIS (see above) must therefore be used to define the Class as a whole rather than 1-TM, since the membrane-inserted segment need not be peptidic.

Another possible structural subdivision was suggested by a proposal that a transducing receptor of platelets and monocytes, CD36, contains (from its hydropathy plot) 2 TMs, to form the EC domain as a long loop.[67] This would provide a new structural type for receptors, of theoretical interest. However, Pearce *et al.*[63] have truncated the chain just ahead of the second TM; the expressed protein was secreted, having lost its prior membrane attachment, but ligand binding was unaffected. Hence, the first apparent TM is not in fact in the membrane, and CD36 is a regular member of Class 3.2. The generality of the receptor structural patterns (FIG. 1) has therefore not yet found an exception.

TABLE 4. Class 3 Receptors

3.1	Receptors with intrinsic tyrosine kinase (TK) activity.
3.2	Non-TK-containing receptors associating with a cytoplasmic tyrosine kinase[a]
3.3	Receptors with intrinsic serine/threonine kinase activity [*e.g.*, for transforming growth factor β (TGFβ)]
3.4	Intrinsic cyclase receptors [*e.g.*, with guanylate cyclase activity (ANP receptors)]

ANP = atrialnatriuretic peptide.

[a] This sub-Class contains two structural subdivisions: see text. Class 3.2 may associate with a diffusible kinase directly, or indirectly through intermediary proteins.

Class 3 encompasses a very wide set of receptor functions, in the nervous, immune and other systems. The classification within this structural Class is, as can be seen from TABLE 4, logically made on a functional basis. A wide variety of sequence types is employed in the functional sub-Class 3.1, and similarly for 3.2. Most of these receptors are heteromers, and the sharing of one "structural" subunit within a group of receptors, each with its specific ligand-binding subunit, is a common device here. When single subunits are involved, the ligand causes their dimerization or clustering, as occurs in the growth factor, growth hormone, prolactin and similar receptors.[64] This process has been established definitively by crystallography of the complex.[65]

Other sub-Classes of the 1-MIS receptors may well, in view of the very wide field of signaling which they cover, yet be found. A more detailed classification of the Class 3 receptors will be presented elsewhere.

REFERENCES

1. KENAKIN, T. P., R. A. BOND & T.-I. BONNER. 1992. Pharmacol. Rev. **44:** 351–363.
2. HUMPHREY, P. P. A. 1997. Ann. N. Y. Acad. Sci. **812:** 1–xx.
3. WILEMAN, T., C. HARDING & P. STAHL. 1985. Biochem. J. **232:** 1–14.
4. STRICKLAND, D. K., M. Z. KOUNNAS & W. S. ARGRAVES. 1995. FASEB J. **9:** 890–898.
5. SHEPPARD, D. N., L. S. OSTEDGAARD, D. P. RICH & M. J. WELSH. 1994. Cell **76:** 1091–1098.
6. FAIRMAN, W. A., R. J. VANDENBERG, J. L. ARRIZA, M. P. KAVANAUGH & S. G. AMARA. 1995. Nature **375:** 599–603.
7. MATTHYS, K. E., C. E. VAN HOVE, P. G. JORENS, M. ROSSENEAU, B. MARESCAU, A. G. HERMAN & H. BULT. 1996. Eur. J. Pharmacol. **298:** 97–103.
8. CLARK, E. A. & J. S. BRUGGE. 1995. Science **268:** 233–239.
9. PANDEY, A., D. F. LAZAR, A. R. SALTIEL & V. M. DIXIT. 1994. J. Biol. Chem. **269:** 30154–30157.
10. DAVIS, S., D. S. ALDRICH, D. M. VALENZUELA, V. WONG, M. E. FURTH, S. P. SQUINTO & G. D. YANCOPOULOS. 1991. Science **253:** 59–63.
11. SUN, Z.-P., M. H. AKABAS, E. H. GOULDING, A. KARLIN & S. A. SIEGELBAUM. 1996. Neuron **16:** 141–149.
12. KINOSHITA, T., N. INOUE & J. AKEDA. 1996. Adv. Immunol. **60:** 57–103.
13. SCHULTZ, S., M. CHINKERS & D. L. GARBERS. 1989. FASEB J. **3:** 2026–2035.
14. SURPRENANT, S., F. RASSENDREN, E. KAWASHIMA, R. A. NORTH & G. BUELL. 1996. Science **272:** 735–738.
15. YAMAMORI, T. & A. SARA. 1992. Neurosci. Res. **15:** 151–161.
16. STAHL, N., T. J. FARRUGGELLA, T. G. BOULTON, Z. ZHONG, J. E. DARNELL, JR. & G. D. YANCOPOULOS. 1995. Science **267:** 1349–1353.
17. CHAPMAN, B. S. 1995. FEBS Lett. **374:** 216–220.
18. BYLUND, D. B., C. EIKENBERG, J. P. HIEBLE, S. Z. LANGER, R. J. LEFKOWITZ, P. MINNE-

MAN, P. B. MOLINOFF, R. R. RUFFOLO, JR. & U. TRENDELENBURG. 1994. Pharmacol. Rev. **46:** 121–136.
19. HOYER, D., D. E. CLARKE, J. R. FOZARD, P. R. HARTIG, G. R. MARTIN, E. J. MYLECHARANE, P. R. SAXENA & P. P. A. HUMPHREY. 1994. Pharmacol. Rev. **46:** 157–204.
20. MARTIN, G. R. & P. P. A. HUMPHREY. 1994. Neuropharmacology **33:** 261–273.
21. CULLY, D. F., D. K. VASSILATIS, K. K. LIU, P. S. PARESS, L. H. T. VAN DER PLOEG, J. M. SCHAEFFER & J. P. ARENA. 1994. Nature **371:** 707–710.
22. LINGUEGLIA, E., G. CHAMPIGNY, M. LAZDUNSKI & P. BARBRY. 1995. Nature **378:** 730–733.
23. LESAGE, F., E. GUILLEMARE, M. FINK, F. DUPRAT, C. HEURTEAUX, M. FOSSET, G. ROMEY, J. BARHANIN & M. LAZDUNSKI. 1995. J. Biol. Chem. **270:** 28660–28667.
24. QUAST, U. 1993. Trends Pharmacol. Sci. **14:** 332–337.
25. INAGAKI, N., T. GONOI, J. P. CLEMENT, C. Z. WANG, L. AGUILAR-BRYAN, J. BRYAN & S. SEIND. 1995. Neuron **16:** 1011–1017.
26. SAKURA, H., C. AMMALA, P. A. SMITH, F. M. GRIBBLE & F. M. ASHCROFT. 1995. FEBS. Lett. **337:** 338–344.
27. KANAI, Y., C. P. SMITH & M. A. HEDIGER. 1994. FASEB J. **8:** 1450–1459.
28. WELSH, M. J. & A. E. SMITH. 1993. Cell **73:** 1251–1254.
29. SHEPPARD, D. N., L. S. OSTEDGAARD, D. P. RICH & M. J. WELSH. 1994. Cell **76:** 1091–1098.
30. BAUKROWITZ, T., T. C. HWANG, A. C. NAIRN & D. C. GADSBY. 1994. Neuron **12:** 473–482.
31. WADICHE, J. I., S. G. AMARA & M. P. KAVANAUGH. 1995. Neuron **15:** 721–728.
32. MUKERJI, J., A. HAGHIGNI & P. SEGUELA. 1996. J. Neurochem. **66:** 1027–1032.
33. UNWIN, N. 1993. Cell **72:** 31–41.
34. NAYEEM, N., T. P. GREEN, I. L. MARTIN & E. A. BARNARD. 1994. J. Neurochem. **62:** 815–818.
35. KAO, P. N. & A. KARLIN. 1986. J. Biol. Chem. **261:** 8085–8088.
36. COCKCROFT, V. B., D. J. OSGUTHORPE, E. A. BARNARD & G. G. LUNT. 1990. Proteins **8:** 386–397.
37. BARNARD, E. A., M. G. DARLISON & P. SEEBURG. 1987. Trends Neurosci. **10:** 502–508.
38. KARLIN, A. & M. H. AKABAS. 1995. Neuron **15:** 1231–1244.
39. SQUIRE, M. D., C. TORNOE, G. A. BAYLIS, J. T. FLEMING, E. A. BARNARD & D. B. SATTELLE. 1995. Receptors & Channels **3:** 101–109.
40. HOLLMAN, M. & S. HEINEMANN. 1994. Ann. Rev. Neurosci. **17:** 31–108.
41. WO, Z. & R. E. OSWALD. 1994. Proc. Natl. Acad. Sci. USA **91:** 7154–7158.
42. HOLLMANN, M., C. MARON & S. HEINEMANN. 1994. Neuron **13:** 1331–1343.
43. BENNETT, J. A. & R. DINGLEDINE. 1995. Neuron **14:** 373–384.
44. GOULDING, E. H., G. R. TIBBS & S. A. SIEGELBAUM. 1994. Nature **372:** 369–374.
45. WEYAND, I., M. GODDE, S. FRINGS, J. WEINER, F. MÜLLER, W. ALTENHOFEN, H. HATT & B. KAUPP. 1994. Nature **368:** 859–863.
46. JAN, L. Y. & Y. N. JAN. 1990. Nature **345:** 672.
47. YAMAMOTO-HINO, M., T. SUGIYAMA, K. HIKICHI, M. G. MATTEI, K. HASEGAWA, S. SEKINE, K. SAKURADA, A. MIYAWAKI, T. FURUICHI, M. HASEGAWA & K. MIKOSHIBA. 1994. Receptors & Channels **2:** 9–22.
48. HAKAMATA, Y., J. NAKAI, H. TAKESHIMA & K. IMOTO. 1992. FEBS Lett. **312:** 229–235.
49. LINGUEGLIA, E., N. VOILLEY, R. WALDMANN, M. LAZDUNSKI & P. BARBRY. 1993. FEBS. Lett. **318:** 95–99.
50. CANESSA, C. M., L. SCHILD, G. BUELL, B. THORENS, I. GAUTHSCHL, J.-D. HORISBERGER & B. C. ROSSIER. 1994. Nature **367:** 463–467.
51. WALDMANN, R., G. CHAMPIGNY & M. LAZDUNSKI. 1995. J. Biol. Chem. **270:** 11735–11737.
52. VALERA, S., N. HUSSY, R. J. EVANS, N. ADAMI, R. A. NORTH, A. SURPRENANT & G. BUELL. 1994. Nature **371:** 516–519.
53. BRAKE, A. J., M. J. WAGENBACH & D. JULIUS. 1994. Nature **371:** 519–523.
54. FICKER, E., M. TAGLIALATELA, B. A. WIBLE, C. M. HENLEY & A. M. BROWN. 1994. Science **266:** 1068–1072.
55. PEPPELENBOSCH, M. P., L. G. J. TERTOOLEN, J. DEN HERTOG & S. W. DE LAAT. 1992. Cell **69:** 295–303.
56. BARNARD, E. A. 1996. The molecular biology of $GABA_A$ receptors and its applications.

In $GABA_A$ Receptors, Transporters and Metabolism. C. Tanaka & N. Bowery, Eds. Birkhauser (Switzerland), pp. 129–143.
57. McFarland, K. C., R. Sprengel, H. S. Phillips, M. Köhler, N. Rosemblit, K. Nikolics, D. L. Segaloff & P. H. Seeburg. 1989. Science **245:** 494–499.
58. Strader, C. D., T. M. Fong, M. P. Graziano & M. R. Tota. 1995. FASEB J. **9:** 745–754.
59. Vu, T. K., D. T. Hung, V. I. Wheaton & S. R. Coughlin. 1991. Cell **64:** 1057–1068.
60. Nystedt, S., A. K. Larsson, H. Åberg & J. Sunderlin. 1996. J. Biol. Chem. **270:** 5950–5955.
61. Misrahi, M., N. Ghinea, S. Sar, B. Saunier, A. Jolivet, H. Loosfelt, M. Cerutti, G. Devauchelle & E. Milgrom. 1994. Science **222:** 711–719.
62. Ressler, K. J., S. L. Sullivan & L. B. Buck. 1994. Curr. Opin. Neurobiol. **4:** 588–596.
63. Pearce, S. F., J. Wu & R. L. Silverstein. 1994. Blood **84:** 384–389.
64. Sprang, S. R. & J. F. Bazan. 1993. Curr. Opin. Struct. Biol. **3:** 815–827.
65. Somers, W., M. Ultsch, A. M. DeVos & A. A. Kossiakoff. 1994. Nature **372:** 478–481.
66. North, R. A. & E. A. Barnard. 1997. Curr. Opin. Neurobiol. **7**. In press.
67. Greenwalt, A. 1992. Blood **80:** 1105–1115.
68. Ip, N. Y. & G. D. Yancopoulos. 1996. Ann. Rev. Neurosci. **19:** 491–515.
69. Cully, D. F., P. S. Paress, K. K. Liu, J. M. Schaeffer & J. P. Arena. 1996. J. Biol. Chem. 20187–20191.

Transduction Is a Major Factor Influencing Receptor Characterization

MICHAEL SPEDDING AND CATHERINE DACQUET

Institut de Recherches Servier
Croissy Research Centre
125, Chemin de Ronde
Croissy 78290 France

INTRODUCTION

Structural, operational and transductional characteristics are required to characterize receptors,[1] and structural differences between subtypes have been clearly shown, for example in the muscarinic$_{1-5}$ subtypes, although in this example the rank order of potencies of antagonists are not very disparate, with the exception of pirenzapine. However, it is not always clear whether the operational characteristics of a receptor, such as rank orders of potencies of agonists and antagonists, are necessarily a primary function of the structure (*i.e.*, the receptor itself) or of the transduction system (coupling characteristics) used to activate the response. In this article we demonstrate that transductional effects may be major determinants of the operational characteristics. For the purposes of this article we consider transduction to refer only to direct coupling with the receptor; multiple messenger cascades downstream from receptor activation may contribute to different tissue responses but these are not differences ascribable to receptors.

Many of these aspects were initially predicted in a prescient article[2] on two, or more states, of the receptor. We also argue for multiple agonist states, which may be important physiologically.

Colquhoun[2] (see also Leff[3]) described the application of two state models to receptors in comparison with "classical" theory,

$$\begin{array}{ccc} R & \overset{L}{\rightleftharpoons} & R^{*} \\ \downarrow\uparrow K_{AR} & & \downarrow\uparrow K_{AR^{*}} \\ RA & & R^{*}A \end{array}$$

where R represents receptors in the resting non-coupled state, R* the active state, L the equilibrium constant determining the distribution between the two states, and an agonist A will have affinity (K, the dissociation equilibrum constant) for R (inverse agonist) or R* (agonist).

Several points are crucially relevant to receptor classification, even if they were not widely appreciated at the time. In a two-state cooperative model a competitive antagonist would have equal affinities for the R and R* states, but if the antagonist prefers R ($M_B > 1$), then the compound may act as an inverse agonist, or alternatively appear to have variable affinity for the receptor, depending on the interconversion of R and R*. Such differences may have been responsible for several problems in receptor classification in the past, and may still overemphasize differences between receptor subtypes. For example, inherent in Colquhoun's analysis of an antagonist preferring R to R*, the shift of the concentration-response curve to an agonist may not be *precisely* parallel. Precisely parallel concentration-response curves (or displacement curves) are difficult to define, and the very small changes predicted by Colquhoun[2] may well fall within the experimental error of many experimental systems. These differences become clear if pA2 values are calculated at 20% and 80% of maximum response, rather than at 50% as is usual practice. More seriously, the classical means used for many years to define receptor populations in isolated tissues, the Schild plot, would be modified by an antagonist preferring R in a way which caused maximum confusion for receptor classification, in that agents preferring R would still lead to linear Schild plots with unit slope, but the intercept (pA2 value, K value) would change. Thus Schild plots cannot be used to separate "competitive" interactions from tight allosteric coupling, as has been appreciated,[2,4] and furthermore, agents which prefer R or R* may yield different pA2 values, depending on the proportions of R and R* available. Differences between pA2 values in different tissues for antagonists have been previously ascribed to disparate receptor populations, but these differences may be due to local factors altering the populations of R and R*. The claims for subgroups of μ-opioid receptors ($\mu 1$, $\mu 2$[5]) would appear to be due to differences in G protein coupled and uncoupled receptors.[6] Also inherent in this analysis is that antagonists may not necessarily act at the same site as agonists, but compete for R and R* at different sites, which has been well established for ion channels; Schild analysis would not be capable of differentiating tight allosteric coupling of the sites.[2] Transduction, and the ability to modify the proportion of R and R*, may therefore be a prime aspect in the definition of agonist and antagonist potency. High affinity agonist binding (R*) is increased by increasing transfection levels of G proteins, with different G proteins being able to differentially modify the total proportion.[4] We consider that pA2 and K values may be modified by the receptor system involved, by local factors such as membrane characteristics, and by the transduction system coupled to the receptor.

The differences between Colquhoun's states can lead to major discrepancies because the difference between agonist and antagonist affinity states for many G protein-linked receptors is of several log orders, *i.e.*, *more* than the usual selectivity between subtypes. Drugs with variable affinity between these two states will therefore have very different affinity values (K_{app}) depending on their affinities for R and R*. This has been modeled by Leff.[3] Classically it was assumed that R* would activate several or many G proteins, of the same type, similarly to the way in which rhodopsin amplifies photon signals; where one absorbed photon in the rhodopsin receptor may activate 500 Gt molecules,[7] and rhodopsin diffuses through 1 μm^2 membrane in 600 msec.[8] There is now biochemical evidence for several states of R*, with G-protein linked receptors. Thus the G protein Gα passes through several conformational states: α.GDP (inactive); an empty state ($\alpha_e\beta,\gamma$) where GDP dissociates and the β,γ subunits associate, which binds to R*; and α.GTP which dissociates from R* and diffuses to its effectors. However, the G protein can also signal to the receptor to induce the high affinity state but it is not known whether there are substates depending on the G-protein conformation. In rhodopsin the C terminal of the G protein α_{t1} induces a conformational change in the receptor,[9] which is presumably different from the change induced by an agonist. Furthermore, palmitoylation of G proteins may be modified by agonists[10–15] and whether the G protein is palmitoylated or not will critically modify its interaction with the receptor and with the membrane.[16,17] Thus acylation of the G protein (or of the receptor) may be expected to induce several states of R* (see below).

COMPARISON WITH EFFECTS OF DRUGS AT ION CHANNELS

It is well established that the difference between K_{app}, the apparent equilibrium dissociation constant under a given experimental system, and K, the real equilibrium dissociation constant under optimized conditions, for drugs binding to sites on ion channels may differ by several log orders owing to:

- differential selectivity for the different states of the channel (resting, activated, inactivated),
- the proportion of time the channel is in a given state,
- the kinetics of the drug interacting with the given state as K_{app} in a given situation will depend on whether the drug has time to interact with its high affinity state,
- the accumulation of lipophilic drugs in the cell membrane.

Allosteric binding interactions have been clearly defined between the ligands binding to the different sites in L-type calcium channels, with drugs such as diltiazem and diclofurime increasing or decreasing dihydropyrine binding and modifying dihydropyridine dissociation rates;[18] some of these effects are markedly temperature dependent. Such allosteric interactions have now been described for muscarinic receptors where gallamine decreases and alcuronium increases [^{3}H]-N-methylscopolamine binding.[19] Can voltage-dependent channels, where the allosteric interactions have been clearly defined, be used as models for allosteric interactions at receptor systems? Rank orders of antagonist potency for a given channel subtype can only be obtained under highly optimized conditions where the channel is maintained in a particular state for complete association to occur; under more physiological conditions the rank orders are lost. For example, the high (>1000-fold) selectivity of dihydropyridines could be assessed in smooth muscle only under very precise conditions, but then nifedipine could be shown to interact with its site with high affinity, and a Schild plot revealed an apparent pA2 of 9.8 with a slope of 1.00.[20] These conditions were optimized (40 mM K^+), and changing the proportion of inactivated channels with potassium reduced K_{app} and reduced the apparently competitive nature of the effects of nifedipine. Resultant analysis was developed simultaneously for analysis of complex interactions at channels[21] and receptors.[22] Apparently competitive interactions of nifedipine with the channel activator, Bay K 8644, could be shown.[21] Thus "classical" receptor analysis can, if applied to voltage-dependent ion channels maintained in the high affinity state for the binding site of interest, give answers which are similar to those seen with receptor systems. However, as K_{app} is dependent on the ratios of the different channel states, the time the channel is in the high affinity state and the association/dissociation kinetics of the drug, under physiological conditions it is difficult to describe channel subtypes by rank order of potency of ligands. Furthermore, channels may not only exist in different states (closed, C, open, O, inactivated, I, with the inactivated state having high affinity for nifedipine), but also in different modes, with Bay K 8644 causing a mode shift, modifying the channel to increase activation, with long channel open times.[23] Clearly, Bay K 8644 can "competitively" interact with the mode blocked by nifedipine. However, other channel blockers (class III blockers, *e.g.*, fluspirilene, pimozide[24]) prevent the mode switch and completely block channel activation by Bay K 8644, at lower concentrations than they modify normal channel functioning and without modifying the inhibitory effects of nifedipine.[21,25] The model for this is proposed below:

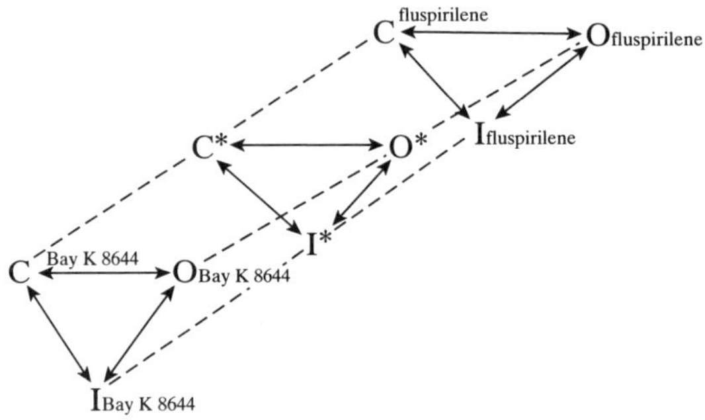

This model has some similarities to that proposed for receptors by Kenakin.[4] Do similar allosteric interactions occur for receptors?

ALLOSTERIC STATES AT RECEPTORS

It is evident that if the receptor is an ion channel, then major differences result from the local conditions (*e.g.*, differential binding at the glycine and the ion channel site labeled with [^{3}H]-MK801 for the NMDA receptor), but in this case the receptor is also the transducer. Can different states of the G protein-linked receptors give similar differences in K_{app}?

For "agonist-antagonist" states there is clear evidence for this to be the case. Furthermore, it is difficult to differentiate these states from G protein-coupled/uncoupled states. Receptors labeled with agonists show different pharmacology to receptors labeled with antagonists.[26,27] Such differences may well be important because we proposed that the agonist-labeled α_2-adrenoceptor showed little subtype-selective pharmacology whereas the subtype differences became apparent when the receptors were labeled with antagonists. Thus, there was no subtype-selective pharmacology between α_{2A}-adrenoceptors and α_{2B}-adrenoceptors, when the receptors were labeled with [^{3}H]-adrenaline, but the expected differential pharmacology was seen when the receptors were labeled with the antagonist [^{3}H]-RS-15385-197 (MacKinnon *et al.*,[26] FIG. 1). We conclude that antagonists (or synthetic partial agonists) may induce different states of the receptor which show subtype selectivity. Unfortunately, there are few binding studies where receptor subtyping has been compared with natural and synthetic agonists. However, synthetic agonists can induce high subtype selectivity.

Differential activation of coupling pathways has also been described for m1 receptors.[28] Thus synthetic agonists may favor differential coupling to G proteins, but via the same receptor. This situation has also been studied for muscarinic receptors where lipophilic side arms are essential to assure subtype selectivity for agonists. Alkylthiothiadozylazabicycles were synthesized by Sauerberg *et al.*[29] and the *n*-hexyl substituent was optimal for efficacy at m1 receptors, as assessed by a PI assay in BHK cells transfected with the human recombinant receptor. However, the functional efficacy was not correlated with measures of binding, because ratios

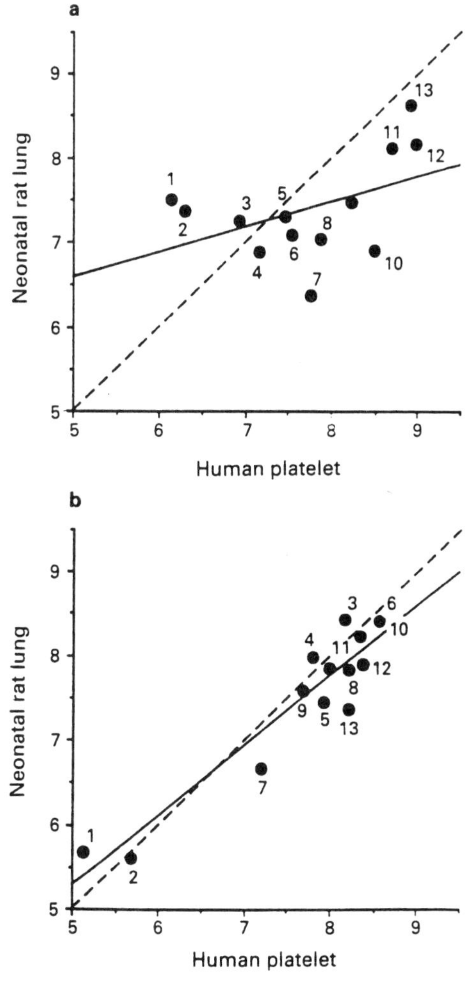

FIGURE 1. Correlation of affinities (pKi) of binding to α_{2A}-adrenoceptors (human platelets) and α_{2B}-adrenoceptors (neonatal rat lung), when the receptors were labeled with the antagonist [^{3}H]-RS-15385-197 **(a)** or the agonist [^{3}H]-adrenaline **(b)** (data from MacKinnon *et al.*[26]). The solid line represents the line of best fit and the dotted line represents the line of identity. **a:** [^{3}H]-RS-15385-197 binding in human platelet vs neonatal rat lung, slope = 0.29, r = 0.20; **b:** [^{3}H]-adrenaline binding in human platelet vs neonatal rat lung, slope = 0.83, r = 0.88. (1) Prazosin, (2) imiloxan, (3) noradrenaline, (4) BHT920, (5) clonidine, (6) adrenaline, (7) BRL 44409, (8) UK-14,304, (9) idazoxan, (10) oxymetazoline, (11) phentolamine, (12) rauwolscine, (13) yohimbine. Note that there was the expected divergence of affinities when the receptors were labeled with the antagonist, but not when labeled with the agonist.

of [^{3}H]-QNB (antagonist)/[^{3}H]-oxotremorine (agonist) binding were only sensitive o *n*-methyl, *n*-ethyl and *n*-propyl substitution which did not change efficacy; the *n*-hexyl substitution was not detected by binding, but this was by far the best substituent for PI turnover. As the sequences for the acetylcholine recognition site are essentially the same for m1–m5 receptors,[30] it is likely that the lipophilic side arms may interact with different TMs, modifying G protein activation. Thus the conformation of the receptor may be very different when activated by such agonists, which may resemble the mode switch induced by Bay K 8644 (FIG. 1); inverse agonists would occupy the third (antagonist) mode shift.

A different model of muscarinic receptor activation has been advanced by Wreggett and Wray,[31] whereby the receptors may cluster together in oligomeric cooperative arrays and mutually transmit conformational changes after interactions with the agonists or antagonists; in these experiments allosteric interactions occurred in the relative absence of G proteins, but G proteins could modify the changes.

Hamm *et al.*[9] have shown that G proteins and terminal peptides can induce conformational changes in receptor proteins. It is also evident that uncoupling G proteins from receptors profoundly changes the receptor state and mutation studies have shown that receptors may be made to couple differentially to G proteins[32–36] and the interactions have been extensively modeled.[35,37,38] Thus agonists at the μ-opioid receptor transfected into CHO cells couple have different EC_{50}s and maximal effects depending on whether the cells are transfected with Gi2α, Gi3α or Go2α.[39] The α_2-adrenoceptor is a receptor which has been claimed to be particularly "promiscuous" in its coupling to different G proteins.[40] Eason *et al.*[41–43] transfected the different α_2-adrenoceptor subtypes into CHO cells and then studied adenylyl cyclase activity via Gi or Gs inhibition or stimulation. The endogenous agonists noradrenaline and adrenaline acted as full agonists at the three receptor subtypes studied, in activating Gi or Gs. However, the imidazoline ligands UK 14304, oxymetazoline and BHT-920 and BHT-933 were full agonists at Gi, but only partial agonists or without effect at Gs-mediated responses. Thus entirely different agonist rank orders of potencies may be obtained at the same receptors, depending on the transduction system which is activated. Is this effect dependent on R and R* or on other factors? We showed that at the antagonist-labeled α_{2A}-adrenoceptor and the agonist-labeled α_{2B}-adrenoceptor, the affinities of adrenaline and noradrenaline were reduced by sodium, whereas this was not the case with oxymetazoline and UK14304 (MacKinnon *et al.*;[26] FIG. 2), indicating that these agonists interact differently at the α_2-adrenoceptor, but that this interaction can be classically differentiated by sodium and GTP. Thus this difference in G protein coupling between amine and imidazoline α_2-adrenoceptor ligands may be mediated by the sodium site. A similar difference between α_2-adrenoceptor coupling to G proteins has been shown to be changed by mutation of the transmembrane II (TMII) Asp responsible for sodium binding.[32] Thus as the TMII Asp is hypothesized to control the Arg switch at the base of TMIII, which couples the receptor to G proteins,[44] a model of differential activation by imidazolines and amine ligands can be made where the differential coupling depends on the signal supplied to the receptor coupling sequences.

Two further examples of receptor interactions are advanced, where the ability of a "side-arm" of the drug may induce differential pharmacology, which is incompatible with classical models. RS-30199 is an azepine with similar affinity for 5-HT_{1A} receptors compared with buspirone (pKis 7.9 and 8.0 respectively), and the affinity, as assessed by a range of binding assays in rat cerebral cortex, did not change when a buspirone side-arm was substituted (RS-64459, pKi 8.1, Small *et al.*;[45] FIG. 3). However, in a functional test of 5-HT_{1A} receptor activation (inhibition of twitch height in the guinea-pig ileum), RS-30199 was devoid of either agonist or antagonist pharmacology whereas RS-64459 behaved as a partial agonist. As there is no evidence for different subtypes of 5-HT_{1A} receptors, it is more likely that the side-arm is necessary for inducing functional coupling to the 5-HT_{1A} receptor in the guinea-pig ileum. Thus RS-30199 may not be able to stabilize the functional state which is induced by buspirone and RS-64459.

A second example at the 5-HT_{1A} receptor involves the most potent agonist known to date, S 14506. This compound is a full agonist in behavioral tests[46] and has a pKi at 5-HT_{1A} receptors of 9.0 (Hill slope 1.4). However, [^{3}H]-S 14506 binding was not reduced in the presence of GppNHp, although the binding of [^{3}H]-8-OH-DPAT was, implying a difference in the mode of coupling to the receptor. Furthermore S-14506 and 8-OH-DPAT binding to rat hippocampal membranes, or to membranes prepared from CHO cells transfected with the human recombinant 5-HT_{1A} receptor,[47] was differentially modified by cations.[48] We set up a simple method assaying whether protein acylation of either the receptor or the G protein

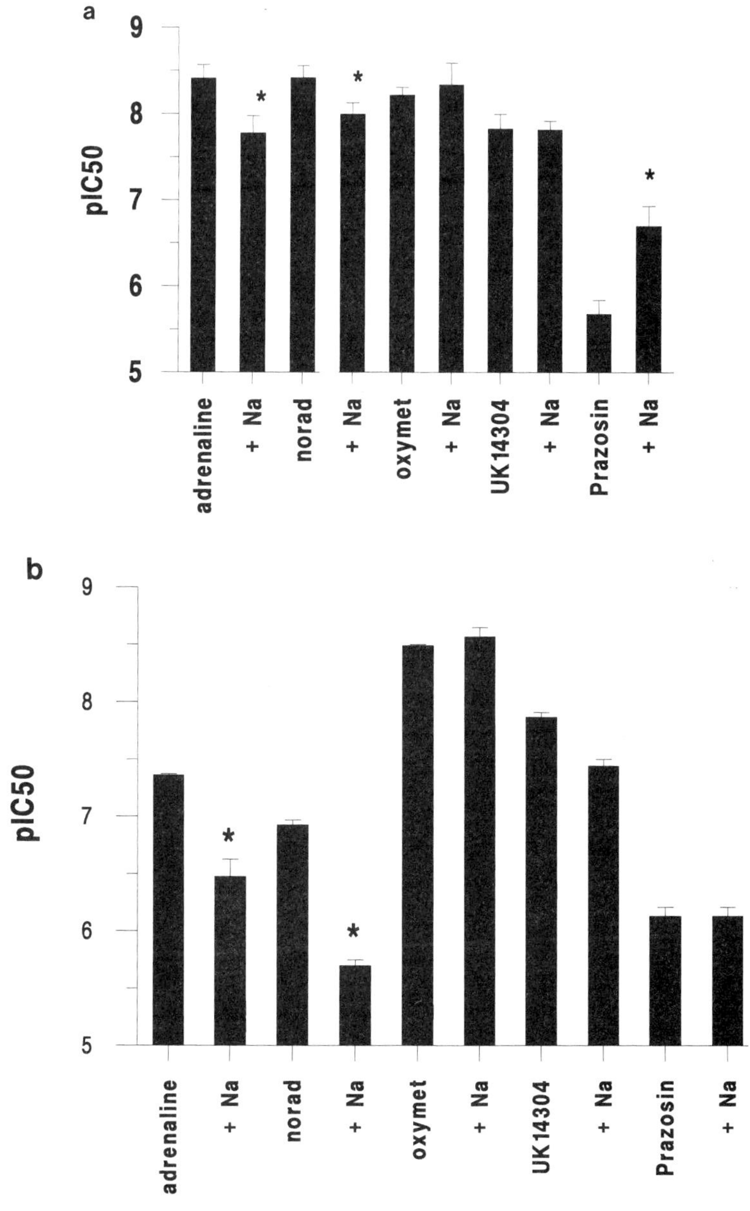

FIGURE 2. Differential modulation of binding to α_{2A}-adrenoceptors and α_{2B}-adrenoceptors, when the receptors were labeled with [^{3}H]-adrenaline (**a:** α_{2B}-adrenoceptors : neonatal rat lung), or with the antagonist [^{3}H]-RS-15385-197 (**b:** α_{2A}-adrenoceptors : human platelets) (data from MacKinnon *et al.*[26]), by the inclusion of 100 mM sodium chloride in the Tris buffer. Binding was measured in the presence of adrenaline, noradrenaline, oxymetazoline, UK 14304 or prazosin.

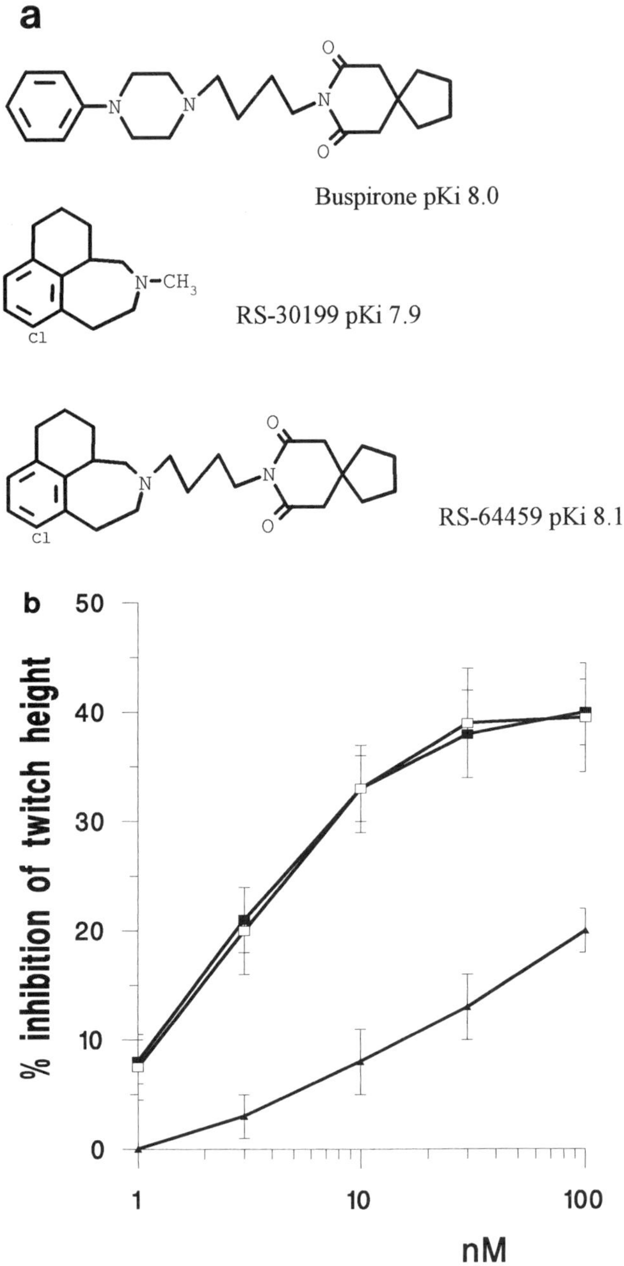
a
Buspirone pKi 8.0
RS-30199 pKi 7.9
RS-64459 pKi 8.1
b
% inhibition of twitch height
0
10
20
30
40
50
1
10
100
nM

←

FIGURE 3. a: Structure of ligands with affinity for 5-HT_{1A} receptors in rat cerebral cortex (pKi). **b:** Failure of RS-30199 (control ■, n = 50; RS-30199 ❑, 1 μM for 20 min, n = 7) to modify the inhibitory effects of the 5-HT1A agonist N,N-dipropyl-5-carboxamidotryptamine (DP-5-CT) on the contractile response of the guinea-pig ileum to electrical stimulation (0.1 Hz, 1 msec pulse duration) in the presence of low Ca^{2+} (0.9 mM). In contrast, RS-64459 behaved as a potent partial agonist at this preparation, causing an initial inhibition of the responses to field stimulation which waned over 20 min; after 20 min the responses to DP-5-CT were antagonized by RS-64459 (▲, 1 μM). Note that the addition of the buspirone side chain, which does not change affinity for the 5-HT1A receptor in rat cerebral cortex, is needed for functional effects of RS-30199 in the guinea-pig ileum. (Data from Small *et al.*[45])

by adding 1-palmitoyl coenzyme A (1 μM) to the membrane homogenates; 1-palmitoyl CoA differentially modified the effects of S 14506 and 8-OH-DPAT binding showing that G protein coupling is differentially affected.[48] S 14506 may therefore behave as an atypical agonist affecting one of the R*-G protein coupled states. Thus there may be several agonist states for receptor activation.

Nevertheless it is important to note that receptors have evolved to recognize agonists and not antagonists; but that synthetic agonists may favor certain states of the receptor which show subtype-selective pharmacology. The induced structure of the receptor will favor particular binding sites either for other drugs or for G protein interactions. The differences induced by these synthetic agonists or partial agonists may be greater than the differences due to receptor structure. Thus receptor "subtypes" may depend on the local environment and on transduction to a greater extent than the receptor aminoacid structure, although the structure must remain a prime index for classification (*e.g.*, Probst *et al.*[49]).

SUMMARY

Two state (agonist-antagonist) receptor systems may explain many discrepancies in receptor classification, but the role of transduction (G protein coupling) may be critical. We propose that in some instances synthetic agonists and antagonists may interact with the receptor in such a way as to modify coupling compared with endogenous agonists, and that the transduction system together with the local environment, may contribute more to the rank order of potency of agonists and antagonists than the receptor subtype as defined by structure. Allosteric interactions at ion channels and receptors require a modification of concepts of coupling. Imidazoline ligands have different efficacy in coupling α_2-adrenoceptors to G proteins, compared with adrenaline and noradrenaline, and do not show a marked sodium shift, implying that the sodium site, and by implication the arginine switch, is implicated in the differential coupling. The α_2-adrenoceptor labeled with a natural agonist does not show subtype selectivity whereas antagonist-labeled α_2-adrenoceptors show subtype selectivity. In the 5-HT_{1A} receptor, palmitoylation (of receptor or G proteins) allows the expression of different agonist states. Thus transduction and G protein coupling must be taken into account in receptor classification, even if the primary classification may be structural.

REFERENCES

1. Humphrey, P., M. Spedding & P. Vanhoutte. 1994. Receptor classification and nomenclature: The revolution and the resolution. TIPS **15:** 203–204.

2. COLQUHOUN, D. 1973. The relation between classical and cooperative models for drug action. *In* Drug Receptors. H. P. Rang, Ed.: 149–182. MacMillan.
3. LEFF, P. 1995. The two-state model of receptor activation. TIPS **16:** 89–97.
4. KENAKIN, T. 1995. Agonist-receptor efficacy I: Mechanisms of efficacy and receptor promiscuity. TIPS **16:** 188–193.
5. PASTERNAK, G. W. & P. J. WOOD. 1986. Multiple μ-opioid receptors. Life Sci. **38:** 1889–1890.
6. RICHARDSON, A., C. DEMOLIOU-MASON & E. A. BARNARD. 1992. Guanine nucleotide-binding protein-coupled and -uncoupled states of opioid receptors and their relevance to the determination of subtypes. Proc. Natl. Acad. Sci. USA **89:** 10198–10202.
7. STRYER, L. 1991. Visual excitation and recovery. J. Biol. Chem. **266:** 10711–10714.
8. POO, M. M. & R. A. CONE. 1974. Lateral diffusion of rhodopsin in the photoreceptor membrane. Nature **247:** 438–441.
9. HAMM, H. E., D. DERETIC, A. ARENDT, P. A. HARGRAVE, B. KOENIG & K. P. HOFMANN. 1988. Site of G protein binding mapped with synthetic peptides from the α subunit. Science **241:** 832–835.
10. LINDER, M. E., P. MIDDLETON, J. R. HEPLER, R. TAUSSIG, A. G. GILMAN & S. M. MUMBY. 1993. Lipid modifications of G proteins: α Subunits are palmitoylated. Proc. Natl. Acad. Sci. USA **90:** 3675–3679.
11. DEGTYAREV, M. Y., A. M. SPIEGEL & T. L. Z. JONES. 1993. Increased palmitoylation of the Gs protein α subunit after activation by the β-adrenergic receptor or cholera toxin. J. Biol. Chem. **268**(32), 23769–23772.
12. WEDEGAERTNER, P. B. & H. R. BOURNE. 1994. Activation and depalmitoylation of Gsα. Cell **77:** 1063–1070.
13. CASEY, P. J. 1994. Lipid modifications of G proteins. Curr. Biol. **6:** 219–225.
14. CASEY, P. J. 1995. Protein lipidation in cell signaling. Science **268:** 221–225.
15. MILLIGAN, G., M. PARENTI & A. I. MAGEE. 1995. The dynamic role of palmitoylation in signal transduction. TIPS **20:** 181–186.
16. PARENTI, M., M. A. VIGANO, M. H. NEWMAN, G. MILLIGAN & A. I. MAGEE. 1993. A novel N-terminal motif for palmitoylation of G-protein α-subunits. Biochem. J. **291:** 349–353.
17. GALBIATI, F., F. GUZZI, A. I. MAGEE, G. MILLIGAN & M. PARENTI. 1994. N-terminal fatty acylation of the α-subunit of the G protein Gi1: Only the myristoylated protein is a substrate for palmitoylation. Biochem. J. **300:** 697–700.
18. MIR, A. K. & M. SPEDDING. 1987. Calcium-antagonist properties of diclofurime isomers 2. Molecular aspects: Allosteric interactions with dihydropyridine recognition sites. J. Cardiovasc. Pharmac. **9:** 469–477.
19. TUCEK, S. & J. PROSKA, 1995. Allosteric modulation of muscarinic acetylcholine receptors. TIPS **16:** 205–212.
20. SPEDDING, M. 1982. Assessment of "Ca^{++}-antagonist" effects of drugs in K^+-depolarised smooth muscle. Differentiation of antagonist sub-groups. Naunyn-Schmiedeberg's Arch. Pharmacol. **318:** 234–240.
21. SPEDDING, M. 1985. Competitive interactions between Bay K 8644 and nifedipine in K^+-depolarized smooth muscle: A passive role for Ca^{++}. Naunyn Schmiedeberg's Arch. Pharmacol. **328:** 464–466.
22. BLACK, J. W., P. LEFF, N. P. SHANKLEY & J. WOOD. 1985. An operational model of pharmacological agonism: The effect of E/[A]curve shape on agonist dissociation constant estimation. Br. J. Pharmacol. **84:** 561–571.
23. HESS, P., J. B. LANSMAN & R. W. TSIEN. 1984. Different modes of Ca channel gating behaviour favoured by dihydropyridine Ca agonists and antagonists. Nature (Lond.) **311:** 538–544.
24. SPEDDING, M. & R. PAOLETTI. 1992. IUPHAR classification of calcium channels and drugs modifying channel function. Pharmac. Rev. **44:** 363–376.
25. SPEDDING, M. & C. BERG. 1984. Interactions between a "calcium agonist" Bay K 8644 and calcium antagonists differentiate calcium antagonist subgroups. Naunyn-Schmiedeberg's Arch. Pharmacol. **328:** 69–75.
26. MACKINNON, A. C., C. M. BROWN & M. SPEDDING. 1993. Sodium modulation of ^{3}H-

agonist and ^{3}H-antagonist binding to α_2-adrenoceptor subtypes. Br. J. Pharmacol. **109:** 371–378.

27. Westphal, R. S. & E. Sanders-Bush. 1994. Reciprocal binding properties of 5-hydroxytryptamine type 2C receptor agonists and inverse agonists. Mol. Pharmacol. **46:** 937–942.
28. Gurwitz, D., R. Haring, E. Heldman, C. M. Fraser, D. Manor & A. Fisher. 1994. Discrete activation of transduction pathways associated with acetylcholine m1 receptor by several muscarinic ligands. Eur. J. Pharmacol. **267:** 21–31.
29. Sauerberg, P., P. H. Olesen, M. J. Sheardown, P. D. Suzdak, H. E. Shannon, F. P. Bymaster, D. O. Calligaro, C. H. Mitch, J. S. Ward & M. D. B. Swedberg. 1995. Muscarinic agonists as analgesics. Antinociceptive activity versus M1 activity: SAR of alkylthio-TZTP'S and related 1,2,5-thiadiazole analogs. Life Sci. **56**(11/12): 807–814.
30. Wess, J., N. Blin, E. Mutschler & K. Bluml. 1995. Muscarinic acetylcholine receptors: Structural basis of ligand binding and G protein coupling. Life Sci. **56**(11/12): 915–922.
31. Wreggett, K. A. & J. W. Wray. 1995. Cooperativity manifest in the binding properties of purified muscarinic receptors. J. Biol. Chem. **270:** 1–12.
32. Surprenant, A., D. A. Horstman, H. Akbarali & L. E. Limbird. 1992. A point mutation of the α_2-adrenoceptor that blocks coupling to potassium but not calcium currents. Science **257:** 977–980.
33. Kosugi, S., F. Okajima, T. Ban, A. Hidaka, A. Shenker & L. D. Kohn. 1993. Substitutions of different regions of the third cytoplasmic loop of the thyrotropin (TSH) receptor have selective effects on constitutive, TSH-, and TSH receptor autoantibody-stimulated phosphoinositide and 3′,5′-cyclic adenosine monophosphate signal generation. Mol. Endocrinol. **7:** 1009–1020.
34. Parma, J., L. Duprez, J. Van Sande, P. Cochaux, C. Gervy, J. Mockel, J. Dumont & G. Vassart. 1993. Somatic mutations in the thyrotropin receptor gene cause hyperfunctioning thyroid adenomas. Nature **365:** 649–651.
35. Baldwin, J. M. 1994. Structure and function of receptors coupled to G proteins. Curr. Biol. **6:** 180–190.
36. Wu, D., H. Jiang & M. I. Simon. 1995. Different α_1-adrenergic receptor sequences required for activating different Gα subunits of Gq class of G proteins. J. Biol. Chem. **xx:** 9828–9830.
37. Hibert, M. F., J. Hoflack, S. Trumpp-Kallmeyer & A. Bruinvels. 1993. Modèles tridimensionnels des récepteurs couplés aux protéines G. Méd. Sci. **9:** 31–40.
38. Conklin, B. R. & H. R. Bourne. 1993. Structural elements of Gα subunits that interact with G$\beta\gamma$, receptors and effectors. Cell **73:** 631–641.
39. Chakrabarti, S., P. L. Prather, L. Yu, P. Y. Law & H. H. Loh. 1995. Expression of the μ-opioid receptor in CHO cells: Ability of μ-opioid ligands to promote α-azidoanilido [32P]GTP labeling of multiple G protein α subunits. J. Neurochem. **64**(6): 2534–2541.
40. Eason, M. G., H. Kurose, B. D. Holt, J. R. Raymond & S. B. Liggett. 1992. Simultaneous coupling of α_2-adrenergic receptors to two G-proteins with opposing effects. J. Biol. Chem. **267**(22): 15795–15801.
41. Eason, M. G. & S. B. Liggett. 1993. Functional α_2-adrenergic receptor-Gs coupling undergoes agonist-promoted desensitization in a subtype-selective manner. Biochem. Biophys. **193**(1): 318–323.
42. Eason, M. G., M. T. Jacinto, C. T. Theiss & S. B. Liggett. 1994. The palmitoylated cysteine of the cytoplasmic tail of α_{2A}-adrenergic receptors confers subtype-specific agonist-promoted downregulation. Proc. Natl. Acad. Sci. USA **91:** 11178–11182.
43. Eason, M. G., M. T. Jacinto & S. B. Liggett. 1994. Contribution of ligand structure to activation of α_2-adrenergic receptor subtype coupling to Gs. Mol. Pharmacol. **45:** 696–702.
44. Oliveira, L., A. C. M. Paiva, C. Sander & G. Vriend. 1994. A common step for signal transduction in G protein-coupled receptors. TIPS **15:** 170–172.
45. Small, C., C. M. Brown, W. S. Redfern & M. Spedding. 1990. The guinea pig ileum as a model for 5-HT_{1A} receptors: Anomolous effects with RS-30199. Br. J. Pharmacol. **104:** 519–525.
46. Schreiber, R., M. Brocco, A. Gobert, S. Veiga & M. J. Millan. 1994. The potent

activity of the 5-HT_{1A} receptor agonists, S 14506 and S 14671, in the rat forced swim test is blocked by novel 5-HT_{1A} receptor antagonists. Eur. J. Pharmacol. **271:** 537–541.
47. NEWMAN-TANCREDI, A., R. WOOTTON & P. G. STRANGE. 1992. High-level stable expression of recombinant 5-HT_{1A} 5-hydroxytryptamine receptors in Chinese hamster ovary cells. Biochem. J. **285:** 933–938.
48. DACQUET, C., C. MACIA & M. SPEDDING. 1995. Acylation differentiates two forms of agonist binding to rat 5-HT1A receptors: Palmitoylated G protein—agonist states? Ann. N. Y. Acad. Sci. **812:**. This volume.
49. PROBST, W. C., L. A. SNYDER, D. SCHUSTER, J. BROSIUS & S. C. SEALFON. 1992. Sequence alignment of the G-protein coupled receptor superfamily. DNA & Cell Biol. **11**(1): 1–20.

To What Extent Can Binding Studies Allow the Quantification of Affinity and Efficacy?

NIGEL J. M. BIRDSALL[a] AND SEBASTIAN LAZARENO[b]

[a]*National Institute for Medical Research*
[b]*MRC Collaborative Centre*
Mill Hill
London NW7 1AA United Kingdom

INTRODUCTION

Over twenty years ago it was found that whereas the antagonist binding properties of receptors in membrane preparations could be described by a single affinity constant, agonist binding was more complex. There were reports relating to the properties of both ion-channel coupled receptors[1,2] as well as other receptors such as muscarinic receptors[3–7] and opioid receptors[8] whose molecular mechanisms were not known at that time. This article concentrates on the binding properties of this latter class of receptors, now known as G-protein coupled receptors, and attempts to relate these properties to the affinity and efficacy of ligands, particularly agonists.

In the case of muscarinic receptors it was demonstrated that there were multiple states of the receptor and that the ratios of affinities of ligands for the different states were correlated with the 'efficacy' of these ligands, as measured in functional studies.[5–7] That is, the 'flatness' of the ^{3}H-antagonist-ligand competition curves was the greatest for high efficacy agonists such as acetylcholine or carbachol, intermediate for agonists of lower efficacy such as pilocarpine and for antagonists the inhibition curves were of the form predicted for a uniform population of receptor binding sites. This type of behavior has also been described subsequently for many receptor systems (*e.g.*, β-adrenergic receptors[9]).

The original binding studies also examined the effects of an irreversible antagonist on the binding of agonists to muscarinic receptors in rat cerebral cortex.[7] These results suggested that the high and low affinity agonist binding components were not interconvertible, at least under the conditions of the experiments, and were not resulting from homotropic negative cooperativity associated with interacting oligomeric arrays of receptors. It was suggested that the two agonist binding components were independent and represented receptors which were tightly coupled to and uncoupled from an effector, E. The postulated scheme [**(1)** and **(2)**] was a variant on the del Castillo/Katz model[10] **(1)**.

$$A + R \overset{K_1}{\rightleftharpoons} AR \overset{K_2}{\rightleftharpoons} AR^1 \qquad \textbf{(1)}$$

$$A + RE \overset{K_1}{\rightleftharpoons} ARE \overset{\alpha K_2}{\rightleftharpoons} AR^1E^* \qquad \textbf{(2)}$$

in which the mechanism of transduction of the agonist receptor interaction is by coupling of the $R \rightarrow R^1$ conformational change to the activation of the effector, $E \rightarrow E^*$. If the binding properties of the "ground states", R and RE, do not differ then the ratio of agonist affinity for coupled and uncoupled receptors is $(1 + \alpha K_2)/(1 + K_2)$, which is a function of K_2 or the 'intrinsic efficacy' of the agonist. This is true for α being greater or less than 1.

At the time the scheme was proposed, nothing was known about the nature or energetics of R.E coupling for muscarinic receptors and it was assumed that the energy of the agonist induced conformational change was utilized to drive the conformational change associated with the activation of the effector ($\alpha \ll 1$).[5-7] This implied (incorrectly as it turned out) that the coupled receptor had a lower agonist affinity than the uncoupled receptor. The assumption that receptor effector coupling perturbed agonist binding in an efficacy dependent manner was however correct.

GUANYL NUCLEOTIDES, G PROTEINS AND THE TERNARY COMPLEX MODEL

Subsequent investigations have revealed that the agonist binding properties of muscarinic receptors, β-adrenergic receptors *etc.* are perturbed by Mg^{2+} and guanyl nucleotides, *e.g.*, GDP and GTP: magnesium increases agonist binding and guanyl nucleotides decrease binding. In other words the states of the receptor can interconvert with the effector, E, being a guanine nucleotide binding protein, G, and the interconversion of states being facilitated by GDP/GTP binding to the G-protein. Instead of $R \rightarrow R^1$ driving $E \rightarrow E^*$ (SCHEME 2) just by the selective binding of the agonist, A, to R^1, there are (at least) two additional binding processes which drive the activation of G-protein coupled receptors. These are the positive heterotropic cooperativity associated with the binding of agonist and G-protein to the receptor (SCHEME 3) and the negative cooperativity between the binding of agonists and G-GTP (or G-GDP) complexes.

The agonist binding properties of these type of receptors have been analyzed according to a simple heterotropic cooperative model, the ternary complex model[11] shown below (SCHEME 3).

$$\begin{array}{ccccc} & & G & & G \\ & & + & K_1 & + \\ A & + & R & \rightleftharpoons & AR \\ & & K_G \downarrow\uparrow & \gamma K_1 & \downarrow\uparrow \gamma K_G \\ A & + & RG & \rightleftharpoons & ARG \end{array} \qquad (3)$$

$K_G[G]$ is generally much less than 1 and $\gamma > 1$ for agonists and $\gamma \leqslant 1$ for antagonists. This model describes a highly simplified subset of the mechanistic scheme for agonist-receptor-G-protein interactions. In the absence of GTP, agonist binding affinity is $K_1(1 + \gamma K_G[G])/(1 + K_G[G])$: in the presence of a high concentration of GTP (when it is assumed that [RG] and [ARG] $\rightarrow 0$) agonist affinity is K_1. The ratio of measured affinities for the coupled and uncoupled receptor is $(1 + \gamma K_G[G])/(1 + K_G[G])$ and is therefore, for different agonists, a function of γ which in itself is a measure of an agonist's ability to generate ARG complexes, *i.e.*, its efficacy. It should be noted that this concept of efficacy is different from that intrinsic to the del Castillo/Katz model. This will be discussed later.

Complications arise because, except under specified conditions, SCHEME 3 predicts agonist binding curves which have Hill coefficients of 1 in the absence and presence of GTP. This is almost always not observed, especially for agonist binding studies performed in the absence of guanyl nucleotide. In order to simulate heterogeneous agonist binding curves it has been necessary to postulate that $[R] \geqslant [G]$. However, the calculations and data of Wells and coworkers[12] suggest that SCHEME 3 cannot fit muscarinic agonist binding curves with $[R]/[G]$ constrained to be constant for different agonists. In addition, both measurements of G-protein levels and the fact that one receptor molecule can catalytically activate several G-protein molecules suggest that in the membrane, $[G] > [R]$.

A more complex model has been proposed which considers in addition the guanyl nucleotide, the α and $\beta\gamma$ subunits of the G-protein and the multiple additional positive and negative cooperative interactions between the five components.[13] However, this model does not appear to help with the understanding of the agonist binding curves although it can simulate a number of functional consequences of agonist receptor interactions.

R-G COUPLING IN THE ABSENCE OF AGONIST

A further complication has arisen in these systems in which receptor-G protein coupling in the absence of agonist can be detected. This was first manifest as a positive heterotropic cooperative interaction of guanyl nucleotide with ^{3}H-N-methylscopolamine binding to M_2 receptors.[14] The enhancement of antagonist binding by GTP appeared to be favored by low ionic strength, low temperature and magnesium ions. It was suggested that a major fraction of the receptors could exist under certain conditions as RG complexes and that a number of muscarinic antagonists had an approximately 20-fold lower affinity for RG than for R. In terms of SCHEME 3 this suggests $K_G[G]$ can exceed unity and that γ is less than one for those antagonists tested. It should be noted that RG complexes can be most readily detected in binding studies by this approach if the receptor concentrations are *low* ($[R] < [G]$) and the concentration of relevant G-protein is high because detection is dependent on a significant *proportion* of the total receptors existing as RG complexes.

CONSTITUTIVE ACTIVITY OF RECEPTORS

There is also an increasing body of evidence that receptors can bind to G proteins in the absence of agonists and generate *active* RG complexes. This is more often found in cell membrane preparations than in whole cells. The phenomenon is normally characterized as the ability of certain antagonists to decrease the 'basal' level of activity in an assay of receptor function. This was first shown as an inhibitory action of atropine on muscarinic receptor linked K^+ channels[15] and subsequently at δ-opioid receptors[16] as a small inhibitory action of the peptide antagonist ICI 174864 on basal GTPase activity of NG108-15 cell membranes, which was considerably enhanced by cyclohexylammonium ions. This latter phenomenon was not found in whole cells.[16,17]

Antagonists therefore have the potential of exerting a biological action in the absence of agonist by uncoupling pre-existing RG complexes. This action has been termed 'negative antagonism' or 'inverse agonism'. The study of Costa *et al.*[16] also indicated that the magnitude of the maximal decrease in activity was dependent on antagonist structure. Some antagonists produced a substantial decrease in basal activity; others exhibited smaller effects or had no action on basal activity. Antagonists therefore can possess 'negative efficacy' or 'negative intrinsic activity'.

The existence of constitutive activity in G-protein coupled receptors has been described for many systems (see *e.g.*, refs. 18, 19 for reviews), although caution has to be taken to ensure that the constitutive activity is not due to low levels of endogenous hormone or to the presence of long-lived agonist-receptor G-protein complexes. In fact constitutive activity has been most readily detected when receptors are overexpressed[20,21] or when the receptors have been mutated in appropriate positions.[18–20] The latter mutations are invariably associated with an increase in agonist binding. The implications of this are discussed below in terms of SCHEME 4. This scheme is basically the combination of SCHEMES 1–3 with the addition of two extra states, R^1 and R^1G^*. It also represents a modification of the 'extended version of the allosteric ternary complex model'[20] in which there is the possibility of formation of inactive RG and ARG complexes. It recognizes the existence of constitutive activity, associated with R^1G^*, the level of this activity being dependent on [R], [G], J, β and K_G.

$$\begin{array}{ccccccc} R^1 & \overset{J}{\rightleftharpoons} & R & \overset{K_1}{\rightleftharpoons} & AR & \overset{\alpha J}{\rightleftharpoons} & AR^1 \\ \beta K_G \upharpoonleft\downharpoonright & & K_G \upharpoonleft\downharpoonright & & \gamma K_G \upharpoonleft\downharpoonright & & \upharpoonleft\downharpoonright \gamma\delta K_G \\ R^1G^* & \underset{\beta J}{\rightleftharpoons} & RG & \underset{\gamma K_1}{\rightleftharpoons} & ARG & \underset{J\alpha\delta}{\rightleftharpoons} & AR^1G^* \end{array} \qquad (4)$$

In the simplest terms, high affinity agonist binding (measured in the absence of GTP and assuming $[R] \ll [G]$) is given by $K_1[1 + \gamma K_G[G] + \alpha J(1 + \gamma\delta K_G[G])]/[1 + J + K_G[G](1 + \beta J)]$ and the low affinity binding constant [in the presence of GTP and assuming no binary or ternary receptor-G-protein complexes exist] is $K_1(1 + \alpha J)/(1 + J)$.

It should be noted that this more complex model still has agonist affinity constants which contain the term αJ (equivalent to K_2 efficacy term in the del Castillo/Katz model). Therefore the intrinsic affinity and efficacy of an agonist, parameters relating to the binding of agonist to the receptor alone, still cannot be separated and measured by equilibrium binding studies.

As noted earlier, constitutively activated receptors generated by mutations have higher agonist affinities than the wild-type receptor for the uncoupled and coupled forms of the receptor. In fact the increase in affinity is efficacy related being largest for agonists of high efficacy whereas antagonist affinities are relatively unchanged.[18,20] The characteristic pattern can be explained by an increase in the value of J being a minimal requirement in constitutively activated receptors.[18,20] This is because the value of the affinity constant of a ligand for the uncoupled receptor is proportional to $(1 + \alpha J)/(1 + J)$ and any increase in affinity associated with an increase in J will be correlated with α.

AN AGONIST-INDUCED CONFORMATIONAL CHANGE

Agonists dramatically enhance the phosphorylation of β_2-adrenergic receptors[22] and M_2 muscarinic receptors[23] by the receptor kinase, βARK. The extent of phosphorylation is related to the efficacy of the agonist[22] and is not found in the unliganded and the antagonist bound receptor. The phosphorylation occurs in reconstitution studies in the absence of G-protein and must reflect a substantially changed receptor conformation which is induced or selected by agonists but not antagonists. This may well represent the manifestation of a component of agonist efficacy.

MULTIPLE G-PROTEINS, MULTIPLE CONFORMATIONAL STATES

There is now abundant evidence that some receptor species can bind to and activate more than one G-protein.[24,25] One example is the ability of m1 muscarinic receptors to activate both pertussis toxin sensitive and insensitive G-proteins. Some agonists activate both pathways, others only activate the pertussis toxin insensitive pathway.[26]

These and other findings raise the question of whether there is a unitary conformational change which is undergone by a receptor independent of the nature of the agonist bound. A recent study on the α_{1B} adrenergic receptor indicates that a mutation in the third transmembrane domain generates a receptor that is constitutively active for a phospholipase C but not phospholipase A_2 pathway.[27] In addition there is a selective enhancement in the binding of phenylethylamines but not imidazolines or antagonists and a selective increase in potency of phenylethylamines for the phospholipase C response. These results suggest that different classes of agonist can bind in different modes to α_{1B} receptors to generate two or more different active conformational states which have a specificity for different G-protein-effector pathways.[27] Therefore agonist efficacy is not necessarily only a property of the receptor itself but also depends on the energetics of specific conformational changes which lead to activation of specific effector pathways.

DISCUSSION

Historically, agonist 'efficacy' and 'affinity' were parameters, defined or measured in the context of whole tissue pharmacological studies. The aim was to be able to generate two independent parameters that would characterize the action of an agonist at a receptor. However, as Colquhoun[28] has pointed out, agonist 'occupancy' as suggested by the theory of Stephenson[29] relates in fact to a microscopic affinity constant which cannot be measured in an equilibrium experiment. All equilibrium agonist affinity measurements contain an efficacy component.

The del Castillo/Katz model[10] provided the first *mechanistic* scheme for understanding efficacy in terms of agonists binding to discrete molecular states of the receptor. In the case of the frog endplate nicotinic receptor it has been possible, using single channel recording techniques, to estimate agonist efficacies and microscopic affinity constants using a more complex variant of the del Castillo/Katz model.[30] For other ion channel coupled receptors, the complexity of the mechanisms (*e.g.*, channel block, desensitization, subconductance states, multiple closed states) and receptor heterogeneity makes any simple interpretation very difficult, even with the application of sophisticated methods of kinetic analysis of single channel openings.

At the present time the field of G-protein coupled receptors lack the equivalent tools and methods of detecting discrete receptor states and measuring rapid interconversions between these states. Even SCHEME 4, which is relatively simple, has eight states of the receptor. There is the additional difficulty of a multicomponent system in which there is a reversible and (sometimes) catalytic interaction between the components. Despite this complexity, the measurement of the ratio, r_{+G}, of the affinities of an agonist for the coupled and uncoupled forms of a G-protein coupled receptor in a defined system provides an *ad hoc* measure of agonist efficacy. This is not efficacy as proposed by Stephenson as it only relates to the ability of an agonist to enhance the coupling of the receptor to a G-protein. It is not an absolute scale as the measured ratio depends on the experimental conditions and the concentration of G-protein. Nevertheless the rank order of r_{+G} values for different agonists, Z_{+G}, when measured under a defined set of conditions, can be informative by providing one relative measure of efficacy.

A second measure of efficacy may be provided by the ratio, r_{act}, of the increased affinity of binding of agonists to the uncoupled state of a constitutively active receptor relative to that of the wild-type receptor. This provides a measure of the 'efficacy' of agonists *in the absence of G proteins*, but does assume that only J (SCHEME 4) is changed by the mutation which generates the constitutively active receptor. The ratio r_{act} values for different agonists also provide another rank order of efficacy of agonists, Z_{act}. This rank order should be independent of the nature of the mutation that generates the constitutively active receptor (if only J is changed).

These two measures of efficacy, r_{act} and r_{+G}, are related to α and $\gamma\delta$ (SCHEME 4), respectively. However, there are questions regarding assumptions implicit in this scheme.

(i) Does AR^1 only bind to and activate one G-protein species?

(ii) Is there only one AR^1 state or can different agonists induce or select different conformational states of the receptor?

(iii) Is the R^1 state the same as the AR^1 (or AR^1G^*) state?

From the evidence presented earlier, it is clear that some receptors can activate more than one G-protein species. Therefore β and (more importantly) γ, δ, K_G and [G] will be different for agonist activation of the different G-proteins. If the AR^1 state is unitary then the rank orders (but not affinities) for the binding of agonists to different receptor G-protein complexes will be the same and Z_{+G} will be independent of the G-protein activated.

However there is evidence from a number of receptor systems that the AR^1 state is not unitary.[31–33] Some agonists will activate one G-protein, different agonists will activate selectively another G-protein. In these case there is one Z_{+G} for each G-protein species activated.

If there are different AR^1 states, it may not be possible to assume that the R^1 state (which results in constitutive activity) is the same as the 'active' conformational states when agonist and G-protein are bound. This could well affect the usefulness of Z_{act} as a measure of efficacy.

CONCLUSIONS

For G-protein coupled receptors it is not possible at present to determine agonist affinity constants which are independent of an efficacy component. Nevertheless the agonist affinity constant at the uncoupled receptor provides a good base measure of 'agonist affinity', which can be determined readily.

There are two ratios, r_{act} and r_{+G}, which provide different measures of efficacy of an agonist. r_{act} is related to the ability of agonists to induce or select an 'active' receptor conformation in the absence of G-protein whereas r_{+G} estimates the increased ability of an agonist-receptor complex (relative to the unoccupied receptor) to couple to a G-protein. The ratio r_{act}, for a given agonist, is dependent on the nature of the mutation used to generate the constitutively active receptor whereas r_{+G} is sensitive to the experimental conditions and to the nature of the G-protein(s) it activates.

The rank orders of r_{act} and r_{+G} values, Z_{act} and Z_{+G} respectively, possibly provide more robust practical indices of efficacy. Z_{+G} will be dependent on the nature of the activated G-protein in receptor systems in which certain selective agonists can determine the G-proteins which will couple to the receptor.

The potential of these parameters for understanding agonist-receptor mechanisms and in drug design needs to be validated by careful measurements of receptor binding and G-protein activation in well-defined systems.

REFERENCES

1. Michaelson, D., R. Vandlen, J. Bode, T. Moody, J. Schmidt & M. A. Raftery. 1974. Arch. Biochem. Biophys. **165:** 796–804.
2. Maelicke, A., B. W. Fulpius, R. P. Klett & E. Reich. 1977. J. Biol. Chem. **252:** 4811–4830.
3. Burgen, A. S. V. & C. R. Hiley. 1974. Brit. J. Pharmac. **51:** 127 P.
4. Birdsall, N. J. M., A. S. V. Burgen, C. R. Hiley & E. C. Hulme. 1976. J. Supramol. Struct. **4:** 367–371.
5. Birdsall, N. J. M. & E. C. Hulme. 1976. J. Neurochem. **27:** 7–16.
6. Hulme, E. C., A. S. V. Burgen & N. J. M. Birdsall. 1976. INSERM **50:** 49–70.
7. Birdsall, N. J. M., A. S. V. Burgen & E. C. Hulme. 1978. Mol. Pharmacol. **14:** 723–736.
8. Bradbury, A. F., D. G. Smyth, C. R. Snell, N. J. M. Birdsall & E. C. Hulme. 1976. Nature (Lond.) **260:** 793–795.
9. Kent, R. S., A. De Lean & R. J. Lefkowitz. 1980. Mol. Pharmacol. **17:** 14–23.
10. del Castillo & B. Katz. 1957. Proc. R. Soc. Lond. **146:** 369–381.
11. De Lean, A., J. M. Stadel & R. J. Lefkowitz. 1980. J. Biol. Chem. **255:** 7108–7117.
12. Wong, H-M. S., M. J. Sole & J. W. Wells. 1986. Biochemistry **25:** 6995–7008.
13. Onaran, H. O., T. Costa & D. Rodbard. 1993. Mol. Pharmacol. **43:** 245–256.
14. Hulme, E. C., C. P. Berrie, N. J. M. Birdsall & A. S. V. Burgen. 1981. Eur. J. Pharmacol. **73:** 137–142.
15. Soejima, M. & A. Noma. 1984. Pflügers Arch. **400:** 424–431.
16. Costa, T., J. Lang, C. Gless & A. Herz. 1990. Mol. Pharmacol. **37:** 383–394.
17. Costa, T., Y. Ogino, P. J. Munson, H. O. Onaran & D. Rodbard. 1992. Mol. Pharmacol. **41:** 549–560.
18. Lefkowitz, R. J., S. Cotecchia, P. Samama & T. Costa. 1993. Trends Pharmacol. Sci. **14:** 303–307.
19. Milligan, G., R. A. Bond & M. Lee. 1995. Trends Pharmacol. Sci. **16:** 10–13.
20. Samama, P., S. Cotecchia, T. Costa & R. J. Lefkowitz. 1993. J. Biol. Chem. **268:** 4625–4638.
21. Milano, C. A., L. F. Allen, H. A. Rockman, P. C. Dolber, T. R. McMinn, K. R. Chien, T. D. Johnson, R. A. Bond & R. J. Lefkowitz. 1994. Science **264:** 582–586.
22. Benovic, J. L., C. Staniszewski, F. Mayor, M. G. Caron & R. J. Lefkowitz. 1988. J. Biol. Chem. **263:** 3893–3897.
23. Kwatra, M. M., J. Ptasienski & M. M. Hosey. 1989. Mol. Pharmacol. **35:** 553–558.
24. Kenakin, T. 1995. Trends Pharmacol. Sci. **16:** 188–192.
25. Kenakin, T. 1995. Trends Pharmacol. Sci. **16:** 232–238.
26. Lazareno, S., T. Farries & N. J. M. Birdsall. 1993. Life Sci. **52:** 449–456.
27. Perez, D. M., J. Hiva, R. Gaivin, M. Mathar, F. Brown & R. M. Graham. 1996. Mol. Pharmacol. **49:** 112–122.
28. Colquhoun, D. 1987. *In* Perspectives on Receptor Classification. J. W. Black, D. H. Jenkinson & V. P. Gerskowitch. Eds.: 103–114. A. R. Liss. New York.
29. Stephenson, R. P. 1956. Brit. J. Pharmacol. **11:** 379–393.
30. Marshall, C. G., D. Ogden & D. Colquhoun. 1991. J. Physiol. **433:** 73–93.
31. Meller, E., T. Puza, J. Diamond, H-S. Lieu & K. Bohmaker. 1992. J. Pharm. Exp. Therap. **263:** 462–469.
32. Robb, S., T. T. Cheek, F. L. Hannan, L. M. Hall, J. M. Midgeley & P. D. Evans. 1994. EMBO J. **13:** 1325–1330.
33. Spengler, D., C. Waeber, C. Pantaloni, F. Holsboer, J. Bockaert, P. H. Seeburg & L. Journot. 1993. Nature **365:** 170–175.

Criteria for the Classification of Hormone Receptors with Implications for Nomenclature

An Impossible Task?

JON LOMASNEY

Department of Pathology
Department of Molecular Pharmacology
and Biological Chemistry
Feinberg Cardiovascular Research Institute
Tarry 12-703
Northwestern University Medical School
303 E. Chicago Avenue
Chicago, Illinois 60611-3008

In recent years it has become apparent that though there is a great number of primary messengers, there is an even greater number of distinct receptors that elicit the specific cellular responses. This realization has come about in large part through the molecular cloning of a large number of cDNA/genes encoding for these receptors, some of which were not previously identified by radioligand binding or functional studies. With this virtual explosion of distinct molecular species and the information gleaned from studies of the expressed gene products in homologous and heterologous systems, has come the challenge of making sense of this extraordinary diversity.

Pharmacological, functional, biochemical, and molecular studies have suggested heterogeneity of receptors. While many different criteria from these studies have been used to classify receptors, often only criteria from one type of study is utilized by a single investigator. Therefore, in large part the criteria used for receptor classification and hence nomenclature have relied upon the inherent bias of the observer. If the investigator is a pharmacologist, then perhaps the rank order of affinities for agonists or more likely antagonists would be used; if a molecular biologist perhaps the human chromosomal localization or deduced primary structure, *etc.* This article will examine the classification and nomenclature of the α_2 adrenergic receptor family as an example of issues in hormone receptor classification and nomenclature. I hope to demonstrate in this article that a true multidisciplinary approach should be used to classify hormone receptors and hence should also be used to develop a practical and accurate nomenclature. I have used this principal in my own work and support the forthcoming recommendation by the IUPHAR Technical Working Party that receptors be classified according to operational (function and binding), structure (molecular), and transductional (receptor-effect coupling) characteristics (TABLE 1).

TABLE 1. Features of Proposed α_2-Adrenergic Receptor Subtypes

Basis of Classification	Nomenclature	Characteristics	Tissue
Radioligand Binding	α_{2A}	Yoh > Oxy ≫ Praz	Platelet, brain
	α_{2B}	Rau > Yoh > Praz > Oxy	Brain, rat neonatal lung NG108
	α_{2C}	Rau > Yoh > Praz = Oxy	Opposum (OK) cell line
	α_{2D}	High affinity phentolamine	Bovine pineal, rat submaxillary
Function	α_2 Presynaptic	Low affinity SKF 104078	Guinea pig atrium
	α_2 Postsynaptic	High affinity SKF 104078	Dog saphenous vein
Molecular Biology	α_2C10		
	α_2C4		
	α_2C2		
Biochemistry	α_2 Glycosylated	M_r = 64,000, 67,000	Platelet
	α_2 Nonglycosylated	M_r = 44,000 & 49,000	Neonatal rat lung

PHARMACOLOGIC CHARACTERIZATION

Pharmacological, functional, biochemical, and molecular studies have suggested heterogeneity of α_2 adrenergic receptors.[1] Different rank orders of potency for adrenergic ligands at α_2 adrenergic binding sites in various tissues have been described. The compounds prazosin, oxymetazoline, ARC 239 and chlorpromazine discriminate between two different subtypes termed α_{2A} and α_{2B}. The selectivity of the drugs varies from a high of 100-fold for prazosin and ARC 239 to a relative low of 18-fold for chlorpromazine. Competition radioligand binding experiments performed with these compounds have identified the human colonic carcinoma cell line HT29 and human platelet as containing only the α_{2A} subtype, while the neuroblastoma myeloma hybrid cell line NG108-15 and neonatal rat lung tissue contain only the α_{2B} subtype. A third pharmacologically defined α_2 adrenergic receptor subtype has been described in the opossum kidney derived OK cell line. This receptor has previously been thought to be more similar to the α_{2B} subtype as it has a relatively high affinity for prazosin, BAM 1303 and WB 4101 and low affinity for oxymetazoline; however, it clearly has a distinct pharmacology and has been designated as the α_{2C}. Finally a fourth subtype has been described in rat submaxillary gland which appears similar to the α_2 adrenergic receptor found in the bovine pineal gland. This receptor is characterized by having a higher affinity for phentolamine and lower affinity for rauwolsine, yohimbine and (−)-mianserin than the other three subtypes. The correlation between the PKI values of various subtype selective drugs from the bovine pineal with those of the α_{2A} (HG29 0.80), the α_{2B} (NG108-15 0.46) and α_{2C} (OK 0.61) have suggested that the bovine pineal and rat submaxillary gland α_2 adrenergic receptor be designated as a fourth subtype: the α_{2D}. Therefore, based on the affinities of agonists and antagonists at α_2 adrenergic ligand binding sites, it is proposed that four subtypes exist and they are designated as the α_{2A}, α_{2B}, α_{2C}, and α_{2D}. Functional studies support the pharmacologic classification of α_{2A} and α_{2B} receptors. In fact, functional studies describing differences in potency of adrenergic agents in rat vas deferens and anococcygeus represented some of the earliest data to suggest heterogeneity of α_2 adrenergic ligand binding sites. In addition to these pharmacological differences, biochemical differences were also noted with the observation that in contrast to the α_{2A} receptor, the α_{2B} receptor of the neonatal rat lung and the NG108 cell line was not glycosylated.

MOLECULAR GENETICS

The first α_2AR gene to be isolated was that encoding for the human platelet receptor. Somatic cell hybrid analysis revealed that this gene resided on human chromosome 10. Therefore, this receptor was given the designation α_2C10. A human genomic Southern blot probed with a 0.95 kb fragment of the α_2C10 exon revealed the presence of three related genes, each localizing to a different chromosome: 10, 4, and 2. Because of the relatively stringent conditions under which these experiments were performed these results did not prove but constituted highly suggestive molecular evidence for the existence of multiple α_2AR. This same 0.95 kb fragment of α_2C10 was used to isolate, from a human kidney cDNA library, a distinct α_2AR whose gene resides on chromosome 4, and is termed α_2C4. Lastly, a third human α_2AR gene located on chromosome 2 (α_2C2) was isolated using approaches based on the polymerase chain reaction and low stringency hybridization. cDNA clones encoding homologous receptors have been isolated from other species including pig, rat and mouse. The molecular studies have shown conclusively that three (and only three) distinct α_2AR exist in any one species. The receptors have distinct primary structures (see below) and are encoded by separate genes. Since the genes could be localized to different human chromosomes, a "molecular" or "genetic" nomenclature was proposed (α_2C2, α_2C4, and α_2C10) that was based on the human chromosomal localizations as determined by somatic cell hybridization. This nomenclature was disliked by pharmacologists and physiologists because of its lack of correlation with subtypes described by rank orders of affinity or potency, a major weakness. The nomenclature did, however, define in a practical sense the receptors into three distinct and easily recognized entities. It was used until the relationship between the cloned and pharmacologically defined receptors could be determined.

PRIMARY STRUCTURE

Hydropathy analysis of the α_2AR deduced amino acid squences indicate that there are seven hydrophobic clusters of 20–25 residues, each separated by stretches of hydrophilic residues; a pattern characteristic of all G-protein coupled receptors isolated to date. These hydrophobic transmembrane spanning domains (MSD) are the most conserved regions of the different members of the G-protein coupled receptors. The amino terminus, carboxyl terminus and third cytoplasmic loop are the regions which differ most in length and amino acid composition. Since all adrenergic receptors bind the same endogenous ligands (epinephrine, norepinephrine), it has been postulated that the conserved MSD may be involved in ligand binding while the more divergent domains, especially the third cytoplasmic loop, may mediate coupling to the heterogeneous G-proteins which couple to varied signaling pathways. In fact, analysis of adrenergic and other G-protein coupled receptors whose primary structures have been modified by either deletion, insertion or exchange of various receptor domains has indicated that the MSD do make up the ligand binding site while the third intracellular loop and portions of the carboxyl terminus predominately mediate G-protein interactions. Among adrenergic receptors, amino acid identities in the MSD range from 36 to 73%. Between members of the same subtype group (α_{2A}, α_{2B}, or α_{2C}), the MSD identity is usually 70–80%, while between different subtype groups (α_1 vs. α_2, α_2 vs. β, *etc.*) the MSD identities are characteristically lower, 30–40%. Therefore, by comparing the primary structure of receptors it is possible to predict with some accuracy the subfamily to which the receptor belongs. Of the three α_2AR, only one α_2C2 lacked potential sites for N-linked glycosylation. This structural analysis indicated that α_2C2 might encode the non-glycosylated α_{2B} subtype in rat neonatal lung.

MELDING PHARMACOLOGY AND MOLECULAR GENETICS

The relationship among the pharmacologically defined α_2 adrenergic receptor subtypes and those identified by molecular cloning was not firmly established for some time. This lack of a clear correlation between the molecular and pharmacological classifications led to much confusion in the literature. For example, the human α_2C4 and its rat homologues have been classified as an α_{2B}, an α_{2C}, and a non-α_{2A}, -α_{2B}, or -α_{2C} receptor! Thus the goal of one of our studies was to compare the pharmacological characteristics (as determined from radioligand binding) of the pharmacologically defined α_2 adrenergic receptor subtypes (as identified in the HT29 cell line, the neonatal rat lung, and the OK cell line) with the human α_2 adrenergic receptor subtypes isolated by molecular cloning (α_2C2, α_2C4, and α_2C10) and expressed on COS-7 cells.[2] The major conclusion of this work was that the cloned human α_2 adrenergic receptor subtypes designated as α_2C10, α_2C2 and α_2C4 have identical or very similar antagonist binding characteristics to the pharmacologically defined α_2A, α_2B, and α_2C receptor subtypes, respectively. This conclusion was based on the correlation of affinity values determined simultaneously at the three cloned receptors and at the three model tissues for twelve adrenergic antagonists that have been shown previously to be subtype discriminatory. The correlation coefficient for HT29 vs. α_2C10 (α_2A) was 0.98, for neonatal rat lung vs. α_2C2 (α_2B) was 0.95, and for OK cells vs. α_2C4 (α_2C) was 0.95. The importance of using many subtype selective compounds and determining the affinities and ultimately correlation coefficients at all the receptors simultaneously, cannot be overemphasized in studies attempting to determine relationships between receptors.

THE PROBLEM OF SPECIES HOMOLOGUES

The extent to which the pharmacological characteristics of receptors are conserved among species is somewhat variable and has given rise in the literature to a number of "novel" or "new" receptors. This has been true of the α_2AR. Even in the last year novel α_2 subtypes are being proposed. Perhaps the most is known about the proposed α_{2D} subtype. The α_{2D} serves as an example of the problem sometimes encountered with species homologues, or perhaps more accurately stated using genetic terms as species orthologues. The term homologue is used rather loosely to mean "similar in structure". For hormone receptors there is no clear scientific definition of what "similar" means (does it refer to conservation of amino acids or strict identity of amino acids at corresponding positions in the primary structure) nor of the magnitude of "similarity" that must be obtained in order to call two receptors from different species homologues. From our analysis of the adrenergic receptors we would expect true homologues to have overall amino acid identities in the >80% range, and >90% amino acid identities in the predicted seven membrane spanning domains. The term "orthologous" is defined as two genes (or proteins) that have diverged at the same time that the species harboring them diverged. Orthology is determined by estimating the time of divergence using rates of substitution in the two genes. To my knowledge this analysis has not been performed for the adrenergic receptor gene family.

A priori, one might expect that receptors having very similar amino acid sequences to also have very similar ligand binding characteristics. The rat homologue of the α_2C2 adrenergic receptor is the α_2 RNG (rat non-glycosylated) receptor. When comparing corresponding amino acid residues of the deduced primary sequences, these two receptors have an 82% overall amino acid sequence identity. These two receptors have an even greater amino acid squence identity 98% in the seven membrane spanning domains (MSD), which are the regions of the receptor protein thought to be most important in determining the specificity and affinity of ligand/receptor interaction. The correlation of the affinities of the rat lung receptor (α_2B) with the human COS-C2 (α_2) gives a high correlation coefficient (0.95) and the slope of the correlation line is close to unity (0.99).[3] Thus, on the basis of the highly conserved primary structures and the nearly identical ligand binding characteristics, these two receptors can be safely regarded as true species homologues.

An example of species homologues with very conserved primary sequences but demonstrating subtle differences in ligand binding can be found with the presumed rat homologue of the α_2C10 adrenergic receptor which has been designated as RG20. RG20 has an 89% overall amino acid sequence identity and a 98% amino acid sequence identity in the MSD to the human α_2C10 receptor. Therefore, based on the highly conserved primary sequences these two receptors should be species homologues or orthologues. However, while RG20 has a ligand binding profile very similar to α_2C10, there are three compounds which can discriminate the two receptors: rauwolscine, phentolamine and yohimbine. This distinct pharmacology has led to the classification of RG20 as the α_2D receptor subtype. Aside from the teleological argument over the significance of differences in affinity or potency for a few synthetic compounds at species homologues, there is the practical problem of how to denote these species homologues in the nomenclature. To give each homologue a distinct subtype assignment would be ludicrous. Perhaps to denote receptors by the subtype and species might be useful (*i.e.*, α_2D or RG20 would be denoted as the rat α_2A receptor). A weakness of this nomenclature is that it fails to indicate that there are differences in the pharmacology of these two homologues. The strength is that it uses operational, structural, and transductional characteristics as criteria for classification and nomenclature.

The previous example of α_2A species homologues has revealed that subtle differences in ligand binding between species homologues are to be expected since amino acid substitutions may slightly alter the ligand binding site. A structural analysis of α_2A receptors from human, porcine, mouse, and rat suggests that the structural determinants of these differences in ligand binding may be as small as a single amino acid! The human and porcine receptors have relatively high affinity for yohimbine and rauwolscine (α_2A pharmacology) while the rat, mouse, and bovine receptors have low affinity (α_2D pharmacology). When a serine residue found at position 201 in the rat and mouse receptors is changed via mutagenesis to a cysteine residue (the amino acid found in the human and porcine receptors at the corresponding position 201), the mouse receptor mutant cysteine 201 binds yohimbine with relatively high affinity.[4] Evidence for a single amino acid substitution causing pharmacological differences between species homologues of G protein-coupled receptors also exists in the serotonin family (5HT1B) and tachykinin family (NK1). Therefore, currently it would be folly to assume that we could accurately predict a receptor's complete pharmacology solely based on the primary structure. There is hope, however, that such predictions may be made in the future based on tertiary structure.

TRANSDUCTIONAL CHARACTERISTICS

To date in virtually every system examined the activation of α_2AR inhibits adenylyl cyclase. However, in many cases α_2AR-mediated physiological effects canot be solely explained by a decrease of intracellular cAMP. The activation of several signaling pathways has been reported for the α_2AR, including the activation of K^+ channels, the inhibition of Ca^{+2} channels, increased Na^+/H^+ exchange and mobilization of intracellular Ca^{+2}.

Activation of α_2C10, α_2C4, and α_2C2 expressed in different fibroblast cell lines decreased both basal and forskolin-stimulated adenylyl cyclase activity via a PTX-sensitive G protein. Interestingly, stimulation of α_2C10 showed a twofold greater inhibition of basal adenylyl cyclase activity than α_2C4. this result may reflect differences in the molecular interactions of α_2C10 and α_2C4 with the inhibitory G proteins, though reconstitution studies with the purified α_2C4 and α_2C10 receptors and different recombinant $G_{\alpha i}$ proteins have shown similar rank orders of coupling efficiency ($G_{i3} > G_{i1} > G_{i2}$) for each receptor. In fact, recent studies using transfected NIH 3T3 cells have shown preferential coupling of the rat α_2C4 (α_2C) subtype to $G_{\alpha o}$, while the rat α_2C10 homologue (α_2A) couples to G_{i3} and/or G_{i2}. This indicates that receptor subtypes may couple preferentially to different G proteins and perhaps to different effectors, offering the opportunity to classify receptors based on the primary coupling mechanisms. Unfortunately, promiscuity of receptor-effector coupling has also been demonstrated. The human α_2A can couple simultaneously to both the inhibition of adenylyl cyclase through G_i, and the stimulation of adenylyl cyclase via G_S. Therefore, at this time using receptor-effector coupling to classify receptors is difficult, since so many different permutations exist.

CONCLUSIONS

I hope that the aforementioned has shown the pitfalls and folly of using only one type of criterion (pharmacology, structure) to classify or name receptors. Even if one uses information gleaned from operational, structural and transductional studies, a simple clearcut classification is not always apparent[5] (for instance, what should we do with those species homologues that demonstrate pharmacological differences? How does one incorporate transductional criteria when there is evidence for the ability of a single receptor to couple to more than one second messenger pathway via distinct G proteins? How does one incorporate functional data when the same physiological process is controlled by distinct receptor subtypes in different species?). Though the IUPHAR is rightly striving for accurate and permanent nomenclatures, I would suggest that it is impossible to devise such a thing. To the credit of those investigators working in the various receptor fields, a vast amount of new exciting information is being generated to elucidate the mechanisms by which hormones, receptors and linked effectors are activated and mediate cellular responses. If we knew everything there is to know about this process in normal and abnormal states, perhaps we could attempt to devise "perfect" nomenclatures, though there would be (and is) great debate over the relative importance of structural characteristics vs. operational vs. transductional, *etc.* Obviously, there is a lot we don't still understand. Therefore, the IUPHAR should be congratulated and encouraged to finish the task of developing a consensus nomenclature for hormone receptors. The nomenclature developed and published by TIPS these past few years has perhaps not been reached in such a formal manner, but has nevertheless been accurate and very useful. It has perhaps been more dynamic than one would like to see, but I argue that this is primarily due to the nature of science and scientists. The TIPS receptor nomenclature supplement has demonstrated the need and utility for a unified nomenclature.

I argue that information gleaned from operational, structural and transductional studies must be used to classify and name receptors. The explosion in the literature of what I consider to be "false" receptor subtypes clearly points to the need for some sort of systematic approach to receptor nomenclature. My bias would be to perhaps put more emphasis on structure than on receptor operation or transduction, partly because structure is easily determined and mostly invariant among different investigators, but mostly because structure determines function. Proteins that have similar structures are bound to have more in common with one another in pharmacology and function than with others. Certainly, data from the study of receptors and enzymes supports this view. While small differences in receptor biology are noteworthy (*i.e.*, among receptor species homologues), I would argue for moderation in the number of receptor subtypes. One must realize that the assignment of a novel non-cloned receptor often sends many laboratories off a pointless fishing expedition. The nomenclatures developed should represent current best guesses and be allowed to evolve as their respective fields do. Hopefully, investigators will find the efforts of so many to classify these proteins useful and truthful.

REFERENCES

1. Lomasney, J. W., S. Cotecchia, R. J. Lefkowitz & M. G. Caron. 1991. Molecular biology of alpha-adrenergic receptors: implications for receptor classification and for structure-function relationships. [Review]. Biochim. Biophys. Acta **1095:** 127–139.
2. Bylund, D. B., H. S. Blaxall, L. J. Iversen, M. G. Caron, R. J. Lefkowitz & J. W. Lomasney. 1992. Pharmacological characteristics of alpha 2-adrenergic receptors: Comparison of pharmacologically defined subtypes with subtypes identified by molecular cloning. Mol. Pharmacol. **42:** 1–5.
3. O'Rourke, M. F., L. J. Iversen, J. W. Lomasney & D. B. Bylund. 1994. Species orthologs of the alpha-2A adrenergic receptor: The pharmacological properties of the bovine and rat receptors differ from the human and the porcine receptors. J. Pharmacol. Exp. Therap. **271:** 735–740.
4. Link, R., D. Daunt, G. Barsh, A. Chruscinski & B. Kobilka. 1992. Cloning of two mouse genes encoding alpha 2-adrenergic receptor subtypes and identification of a single amino acid in the mouse alpha 2-C10 homolog responsible for an interspecies variation in antagonist binding. Mol. Pharmacol. **42:** 16–27.
5. Kenakin, T. P., R. A. Bond & T. I. Bonner. 1992. Definition of pharmacological receptors. [Review]. Pharmacol. Rev. **44:** 351–362.

Do Recombinant Receptor Assays Provide Affinity and Potency Estimates?

JOËL BOCKAERT,[a] CÉLINE BRAND, AND LAURENT JOURNOT

CNRS, UPR 9023, CCIPE
Rue de la Cardonille
34094 Montpellier Cedex 5, France

Since the origin of the receptor concept, drug-receptor interaction has been characterized extensively using only two variables. The main variable was the drug itself, and the pharmacologist had the possibility of modifying its structure and concentration. The second variable was, in some specific systems, the possibility of decreasing (but not increasing) the number of receptors with irreversible ligands. With the recent cloning of receptors and the advent of recombinant DNA technology, the pharmacologist can now directly alter the receptor part of the drug-receptor interaction. The receptor can be manipulated in many ways. For the first time, the concentration of a given receptor can be adjusted according to the desire of the pharmacologist. In particular, receptor concentrations higher than that physiologically expressed can be obtained in transfected cell lines. As we will further discuss, this possibility has revealed, in some systems, non-classical drug receptor interaction behaviors. The receptor can also be expressed in cells containing different G proteins or enzymes (kinases, phosphatases, *etc.*) able to posttranslationally alter its activity. Finally, the receptor can be structurally modified directly by site directed mutagenesis.

In this review, we will address the question of the putative influence of the nature of the cell and of the number of receptors expressed in those cells on the affinity and potency of the drugs. The affinity is defined as the inverse of the equilibrium dissociation constant for binding (K_D or K_A) ($K_{affinity} = K_A^{-1}$) and the potency as the EC_{50} of the biological response. This review will only consider the G-protein-coupled receptors (GPCR).

In General, There is a Rather Good Similarity between Ligand Affinities Measured in Tissues and Those Determined in Cells Expressing the Equivalent Cloned Receptors

The cloning of the β2-adrenergic receptor, a GPCR different from rhodopsin, was first published one decade ago.[1] Since then, the number of published sequences of members of this receptor family has increased rapidly. If one includes the putative olfactory receptors, the total number certainly approaches half a thousand.[2,3] Generally, there was no important differences between the affinities of a series of

[a] Tel: 033 4 67142930; Fax: 033 4 67542432.

	Drug	HeLa cells binding pKD	Pig cortex binding pKD
1	5-HT	7.90 ± 0.11	8.51
2	5-OT	9.07 ± 0.16	9.53
3	8-OH-DPAT	8.34 ± 0.10	8.74
4	Flesinoxan	8.11 ± 8.08	8.32
5	Sumatriptan	5.99 ± 0.19	6.13
6	Ipsapirone	7.62 ± 0.12	7.73
7	Buspirone	7.11 ± 0.07	7.58
8	Spiroxatrine	8.13 ± 0.13	8.05
9	MDL 73005	8.06 ± 0.07	8.54
10	NAN 190	8.47 ± 0.08	8.90
11	Pindolol	7.43 ± 0.16	7.63
12	SDZ 216-525	9.38 ± 0.08	9.18

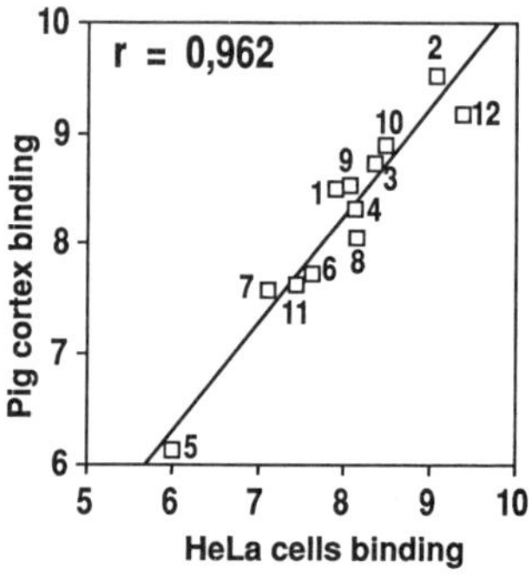

FIGURE 1. Comparison between receptor affinity values derived from binding studies performed in membranes of HeLa cells transfected with human 5-HT_{1A} receptors and in pig cortex membranes. (Modified from Boddeke *et al.*[4])

ligands for cloned receptors expressed in heterologous cell lines and their affinity for the corresponding receptors expressed in tissues. FIGURE 1 illustrates the comparison between the 5-HT_{1A} receptor affinity values derived from binding studies performed in membranes of HeLa cells transfected with the human 5-HT_{1A} receptor and those derived from binding studies in pig cortex membranes.[4] A remarkable correlation was obtained whatever the nature of the ligand (full agonists, partial agonists or antagonists). Notice that for this receptor there is no species difference in affinities which is, of course, not always the case. The now classical difference between the affinity of rodent and opposum 5-HT_{1B} receptor and its human, dog, guinea-pig, cow and pig equivalent 5-$HT_{1D\beta}$ for aryloxyalkylamine β-adrenergic antagonists, is the best example of species differences in receptor pharmacology (for a review see ref. 5). The rodent 5-HT_{1B} receptor (like the β-adrenergic receptors) has an asparagine in the 7th transmembrane region essential for high affinity binding of these aryloxyalkylamines, whereas the human 5-$HT_{1D\beta}$ contains a threonine at the equivalent position.[6]

However, despite this general agreement between ligand affinities for heterologously expressed receptors and for naturally expressed counterparts, several problems arose. Some were expectable: potential influence of the nature and the concentration of associated G proteins, GTP concentration, phospholipid environment *etc.*, some were not. For example, there are several reports which indicate a nonclassical behavior (when classical pharmacological models are considered) of the receptor density on receptor affinity and potency.

The Existence of High and Low Affinity GPCRs as a Potential Pitfall in Heterologous Receptor Expression

Before examining the experimental observations describing the influence of G protein coupling on receptor affinities, one has to recall the mechanisms by which GPCRs interact with these proteins. The agonist-receptor complex [AR] once formed will interact with the inactive trimeric G protein (α-GDP-$\beta\gamma$) and acts as a GDP/GTP exchanger. The first step is the decrease in the affinity of the nucleotide site for GDP and the formation of a ternary complex (AR-α empty $\beta\gamma$) (FIG. 2A; 2). It has been clearly demonstrated by Kühn[7] in rhodopsin-transducin (G-protein specific for the rhodopsin receptor) interaction, that if GTP is absent, the rhodopsin activated receptor is unable to dissociate from the transducin from which GDP has been released (FIG. 1; 2). FIGURE 2B symbolizes the rhodopsin-transducin complex as it is frozen if GTP is absent. The two proteins are locked together. The empty G protein increases the stability of the retinal (the agonist) on the receptor and therefore increases its affinity.[8] This is the high affinity state of the GPCRs. Notice that as soon as GTP is present, the transient state (FIG. 2A, 2) disappears to give two active components α-GTP and $\beta\gamma$ (FIG. 2A, 3) and of course [AR].

The ligand receptor-G protein interaction is described by equation 1:

$$A + R \underset{}{\overset{K_A}{\rightleftharpoons}} AR + G \overset{K_2}{\rightleftharpoons} ARG \quad \underset{K_{obs}}{\longleftrightarrow} \tag{1}$$

• When the G protein concentration is high, compared to the concentration of ARG, the observed equilibrium dissociation constant (K_{obs}) is equal to:

$$K_{obs} = \frac{K_A}{1 + [G]/K_2} \tag{2}$$

The greater the potency of the agonist in producing [ARG] (K_2 small; GTP is absent) the smaller is K_{obs} compared to K_A (FIG. 2C). When K_2 increases, for example when GTP is present, [ARG] does not accumulate, then $K_{obs} = K_A$ (FIG. 2D). On isolated membranes, K_{obs} is equal to the high (K_H) and low (K_L) affinity states of the receptors (defined in binding experiments) done in the absence of GTP and the presence of GTP respectively.

• When the G protein concentration is limiting or when some receptors cannot associate to G proteins (diffusion problems, receptors localized in intracellular compartments in which G is not present or unknown reasons) then a mixture of AR and ARG species can accumulate to give a heterogeneous binding curve (FIG. 2E). This behavior is generally observed for agonist binding on isolated membranes in the absence of GTP. FIGURE 3 gives the example of displacement of the QNB antagonist binding on rat heart muscarinic receptors by two agonists. In the absence of GTP, high and low affinity components are observed, whereas in the presence of GTP, only one low affinity component is observed.[9]

Influence of Receptor and G Protein Densities on Affinity of Heterologously Expressed Recombinant Receptors

On the basis of the above discussion, several differences in agonist affinities are expected depending on the relative concentrations of receptors and G protein present in a given cell line.

The theoretical evolution of binding curves in systems with varying receptor/G protein levels calculated by T. Kenakin[10] is described in FIGURE 4. This fits well with the decrease in the percentage of the high affinity state (K_H) of muscarinic receptors for carbachol as a function of receptor density expressed in fibroblasts B 82 cells[11] (TABLE 1). A more dramatic situation occurs when the radioligand is

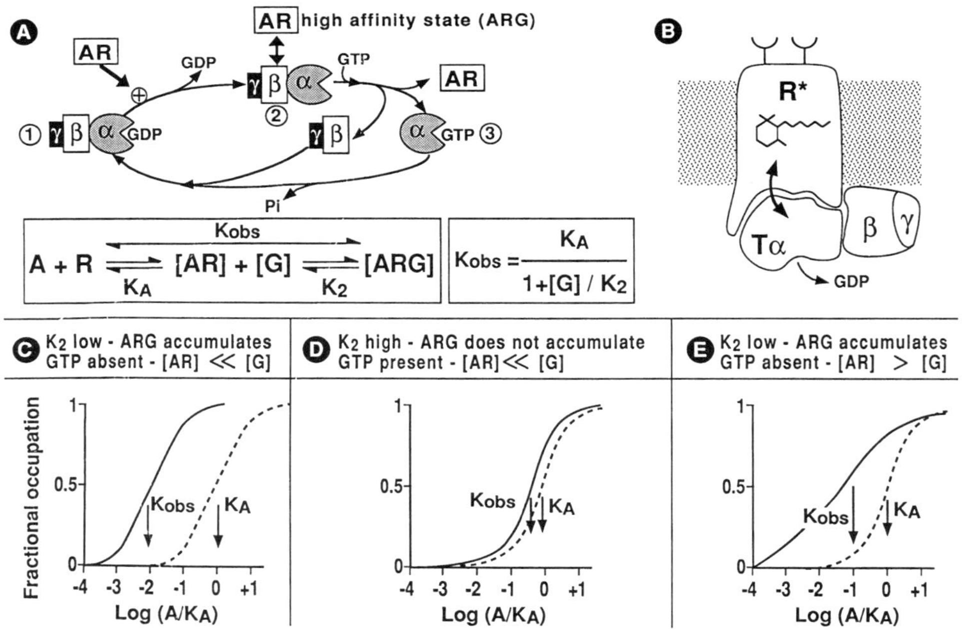

FIGURE 2. Receptor affinity states determined *in vitro* in the absence and the presence of GTP: **A:** schematic representation of the interaction of AR with the heterotrimeric G protein. The step 2 (ARG complex) corresponds to the high affinity state (KH) in the absence of GTP. **B:** schematic representation of the reciprocal interactions between the activated rhodopsin (R*) and the empty (no nucleotide) heterotrimeric G protein. **C, D, E:** Evolution of K_{obs} (equivalent to K_H) under different experimental conditions. K_A and K_2 are the equilibrium dissociation constants of the drug receptor and the ternary complex, respectively. C) $K_A = 1$; $K_2 = 0.01$; $[G] = 0.1$. D) $K_A = 1$; $K_2 = 100$; $[G] = 0.1$. E) only 50% of R is susceptible to form an ARG complex (constant values as in C) with G (G limiting or a fraction of receptors do not interact with G).

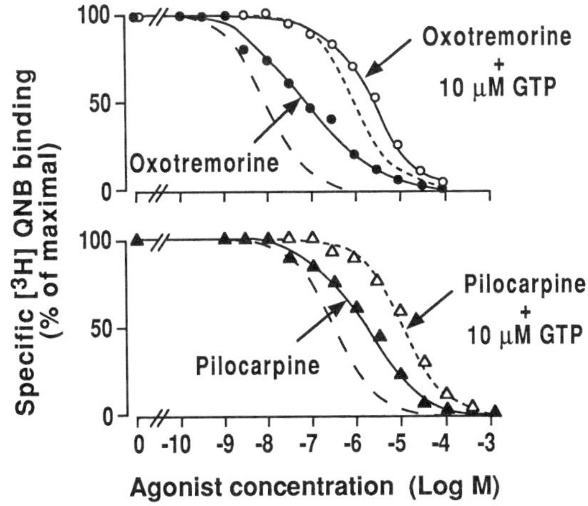

FIGURE 3. Displacement of [^{3}H]-L-[benzili4,4′-3H]quinuclidinyl benzilate from rat muscarinic receptors in heart membranes by cholinergic agonists. *Closed symbols* and *open symbols* represent the binding in the absence and the presence of agonist, respectively. *Dashed line*: binding to K_H; *dotted line*: binding to K_L. (Modified from Waelbroeck *et al.*[9])

TABLE 1. Binding Parameters for M_1 Muscarinic Receptors in a Series of Transfected B82 Cells

Clone	B_{max} (fmol 10^6 cells)	[^{3}H]QNB K_d (pM)	Carbachol[a] K_H (μM)	% H[b]	K_L (μM)
LK3-1	12 ± 0.98	100	7.7	100	
LK3-6	30 ± 4.4	97	3.8	68 ± 10	34
LK3-7	96 ± 15	150	5.8	62 ± 12	41
LK7-2	140 ± 19	150	7.1	47 ± 10	53
LK3-3	240 ± 15	160	4.8	40 ± 7.6	45
LK3-8	260 ± 48	190	2.4	42 ± 5.9	53

Modified from Mei *et al.*[11]

[a] Displacement of [^{3}H]QNB by carbachol.

[b] Percentage of sites in the high-affinity state (K_H).

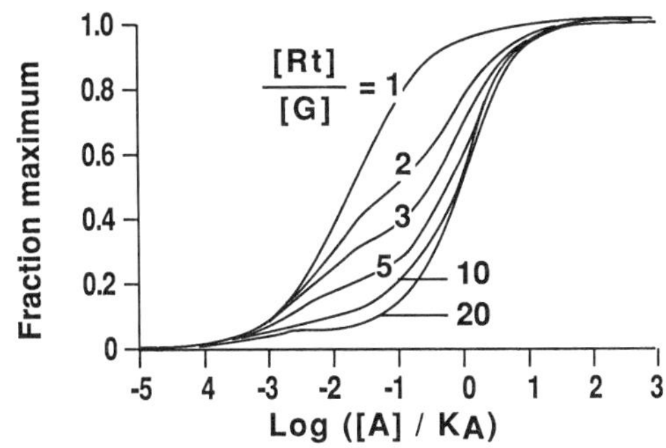

FIGURE 4. Saturation binding curves for ligands in systems with varying receptor/G-protein levels. $K_A = K_2 = 0.01$; [G] = 1 and [RT] = 1, 2, 3, 5, 10, 20. Ordinate normalized binding. (Modified from Kenakin.[10])

an agonist. In this situation only the high affinity state is easily detected because of the difficulty of having a good ratio between the specific and non-specific binding for the low affinity state when the receptor concentration is low. This is the case illustrated in FIGURE 5A when the binding of [^{3}H]-8-OH-DPAT is determined in fibroblasts transfected with human 5-HT_{1A} receptors at low density. Only high affinity binding sites are detected (B_{max} = 500 fmol/mg; K_H = 0.8 nM).[12] However, when the same receptor is expressed at a very high density (19000 fmol/mg) both a high and a low affinity component appear[13] (FIG. 5B). In addition, under these conditions, the determination of binding constants become problematic since in this experiment K_H was computed to be equal to 0.06 nM which is 10 times lower than determined by several authors expressing this receptor at low density in different cell lines.[14]

No data are available to discuss the influence of G protein concentration on K_H and K_L in transfected cells. Of course, some differences between the affinities for agonists on the same receptor transfected at low density in different cell lines can be interpreted along this line. For example, the 10-fold difference in 5-HT affinity for [^{3}H]-8-OH-DPAT binding in Swiss 3T3 and HeLa cells.[14] However, since in the same experiment this difference was not found for 8-OH-DPAT itself, this probably reflects laboratory data differences of unknown origin rather than a true relevant difference between cell lines. Similar conclusions can be drawn for the 10-fold difference in the affinity (K_H) for carbachol of transfected muscarinic m3 receptors in HEK and RAT-1 cells.[15] In some cases, these differences reported by some laboratories may be related to a difference in the contamination by GTP of the membrane preparations.

However, if the G protein content of cells is dramatically different, the influence of G protein concentration on K_H (but not on K_L) should be observed. In a reconstituted system consisting of purified muscarinic receptors and Go proteins, the influence of Go concentration was clearly measurable[16] (TABLE 2).

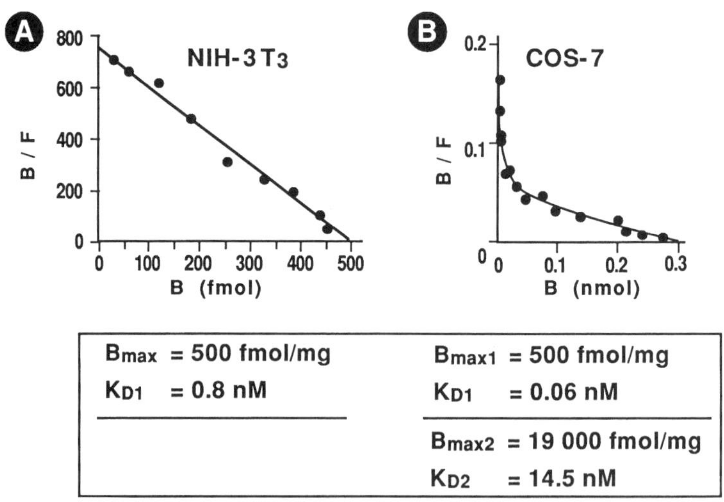

FIGURE 5. Scatchard plots of the saturation binding curves for [^{3}H]8-OH-DPAT in two cell lines at different densities. **A**: Modified from Varrault *et al.*[12] **B**: Modified from Fargin *et al.*[13]

TABLE 2. Effect of the Coupling Protein Go on Oxotremorine Potency

RT[a] (pmol)	Go[b] (pmol)	IC_{50} (−GTP)[c] μM	IC_{50} (+GTP)[d] μM
5.7	36	4.0	9.0
5.7	180	1.5	7.0
5.7	920	0.25	8.0
5.7	4600	0.02	15

Modified from Florio & Sternweis.[16]

[a] Purified muscarinic receptor.

[b] Purified Go.

[c] Concentration of oxotremorine to inhibit 50% of QNB binding (−GTP, a condition under which ARG accumulates).

[d] Concentration of oxotremorine to inhibit 50% of QNB binding (+GTP, a condition under which ARG does not accumulate).

The affinity of antagonists is not affected by the presence or absence of G proteins since the [AR] complex does not interact with G. However, there are intriguing GTP effects on antagonist binding, suggesting that some antagonists discriminate between receptors which may be associated or not with G proteins even in the absence of agonist (see the two-state model).

Does the High and Low Affinity State of the Receptor Have Any Significance on Intact Cells?

As we have already mentioned, the high affinity state is not observed in the presence of GTP. Therefore, in intact cells containing GTP all the receptors are under the low affinity state. This has been demonstrated by examining the binding of an adenosine agonist (N6-cyclohexyladenosine) on adenosine receptors labeled with an antagonist either on membranes or on intact DDT1 MF-2 cells.[17] The binding is heterogeneous in membranes (high and low affinity state) but homogeneous in intact cells (homogeneous low affinity state). Therefore, in intact cells it is not possible, using binding data, to distinguish between receptors which can interact with G proteins from receptors which cannot. All are under a low affinity state because of the presence of GTP. It is not evident whether or not the fraction of receptors coupled to G proteins differs in intact cells and in isolated membranes. It has been proposed that the ratio between K_H and K_L could be related to the intrinsic activity of the agonist (for a discussion see ref. 10).

Transfection of Recombinant Receptors Can Lead to Non-physiological Coupling to G Proteins

It is common that a single ligand elicits changes in more than one second messenger pathway. Very often, this is mediated by multiple receptor types, each of them being able to couple to multiple G proteins. However, through the expression of receptors in heterologous systems, accumulating evidence has shown that some receptors, when expressed in heterologous systems, can activate G protein and second messenger productions which are not found physiologically. Just to give two examples:

1) it has been shown that the β-adrenergic receptor, a receptor which has never been found coupled to Gq and thus to stimulate the phospholipase C (PLC) in normal tissues was found to do so when 300,000 receptors are expressed per fibroblast L cell.[18] This is also the case for all Gs coupled receptors transfected at high levels in these cells.[18]

2) 5-HT_{1A} receptors can stimulate PLC when expressed at reasonable density (490 fmol/mg) in HeLa cells[19] whereas such a coupling has not been reported in tissues. However, significantly higher EC_{50} values (at least two order of magnitude) were reported for agonists in the PLC coupling compared to those involving inhibition of adenylyl cyclase,[19] a coupling observed in cell expressing this receptor under physiological conditions.[20]

Theoretical Considerations About the Two-State Model in Relation to the Affinity of Ligand

Thanks to mutations in cloned receptors, it has been shown that receptors can have intrinsic activities (formation of an active [RG] complex in the absence of agonist)[21] (FIG. 6). In addition, some drugs have been found to reverse the formation of [RG] which are called inverse agonists. A useful model (FIG. 6) has been proposed by Costa *et al.*[22] to take into account these observations. In this two-state model,

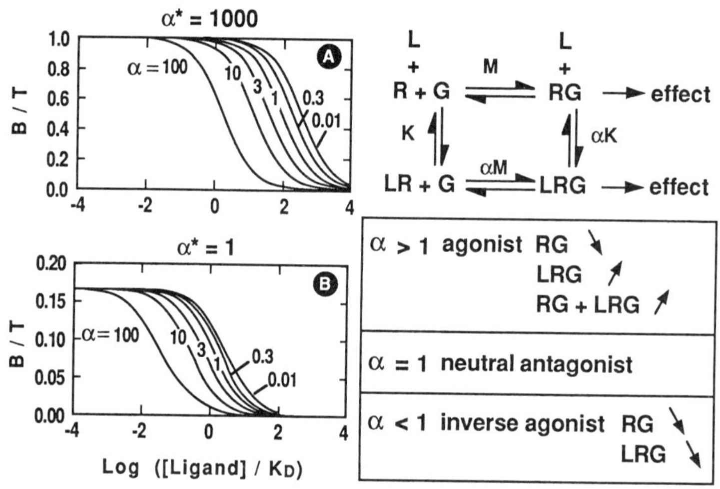

FIGURE 6. Influence of the two-state model on the agonist potencies. The two-state model is represented on the right. K and M are equilibrium affinity constants describing the formation of the binary species LR and RG. αK and αM are the equilibrium affinity constants for the formation of the ternary complex LRG. Effect of α on the competition isotherms of ligands in the two-state model. **A** and **B**: the competing ligands have the same K_A = 100 nM. The labeled ligand has a K_D of 1 nM and is present at 1 pM concentration. [RT] = [G] = 0.2 nM; M = 10^{10} M^{-1}. In **A** the α^* constant for the labeled compound is 1000 whereas in **B** the α^* is equal to 1. (Modified from Costa *et al.*[22])

the receptor can interact with G in the absence of agonist to form an active RG complex. The affinity of L for R and RG are different, equal to K and αK, respectively. The affinity of R and RL for G are different, equal to M and αM, respectively. In the presence of an agonist if its affinity for RG is higher than its affinity for R ($\alpha > 1$). The formation of LRG is facilitated by the presence of L and the sum of the active complexes [RG] + [LRG] increases. In the presence of a neutral antagonist ($\alpha = 1$), the sum of [RG] + [LRG] is constant. In the presence of an inverse agonist ($\alpha < 1$), the RG complex is destabilized toward [R] and [G]. According to this model, Costa *et al.*[22] have computed that the IC_{50} of ligands (in binding displacement experiments) having the same K_A, will depend on the α constant of the radioligand used (α^* in FIG. 6) as well as on the α constant of the displacing ligand (FIG. 6). Since the "coupling" constant α is both G protein- and ligand-related, the nature of the G proteins expressed in a given cell line can theoretically affect the IC_{50} and therefore the affinity of ligands computed from those experiments. It can be noted that for some values of α the displacement curves have a Hill coefficient below one (FIG. 6) which is indicative of the presence of several affinity states in classical pharmacological models. Similarly, as the affinity of L for R and RG are different the affinity of the ligand is dependent on M (which determines the ratio between R and RG). As M is receptor- and G protein-dependent, the nature of the G proteins expressed in a given cell line can theoretically affect the IC_{50} and therefore the affinity of ligands computed from those experiments. Whether or not the two-state model (precoupling of [R] to [G] in the absence of ligand) is a general model for non-mutated receptors expressed as physiological densities, remains to be determined.

The Classical Pharmacological Models Such as the Operational Model Does Not Describe Some Observations Made Using Recombinant Receptor Technology

In classical pharmacological models, such as the operational models, it is assumed that:

$$A + R \overset{K_A}{\rightleftharpoons} AR \tag{3}$$

A chain of hyperbolic coupling steps (including the interaction of [AR] with [G])

$$AR + G \rightleftharpoons ARG \rightleftharpoons \text{response} \quad (\underset{Kc}{\longleftrightarrow})$$

occurs between the formation of [AR] and the final response. Then, the final response is expected to be a hyperbolic function of the ligand concentration as generally experimentally observed.

Therefore, one can write that the final response E is a hyperbolic function of [AR]:

$$E = ET\frac{[AR]}{Kc + [AR]} \tag{4}$$

Kc being the coupling constant. ET is the theoretical maximal response which could be obtained if AR $\gg$ Kc. RT is the total number of receptors.

From equations 3 and 4, we can obtain the relationship between the final response (E) and the agonist concentration (A):

$$E = \frac{[ET]}{1 + Kc/[RT]} \times \frac{A}{\frac{KA}{1 + [RT]/Kc} + A} \quad \textbf{(5)}$$

K_A the binding dissociation constant of the agonist.

$$\frac{K_A}{1 + [RT]/Kc} = \text{potency} = EC_{50}$$

$$\frac{[ET]}{1 + Kc/[RT]} = \text{efficacy}$$

Note that theoretically, the EC_{50} of the response is shifted to the left as RT increases. There is no limit in the intensity of the shift which can be obtained within the classical pharmacological models. Although some shift of the dose response curves to the left has been observed in transfected cells when the concentration of receptor is increased, there are two cloned cell lines (NIH-3T3) expressing 45 and 500 fmol of 5-HT_{1A} receptor per mg of proteins (FIG. 7A), the EC_{50} of 5-HT in inhibiting the adenylyl cyclase was not different (110 ± 10 nM and 80 ± 20 nM, respectively) (FIG. 7B). Another unexpected finding was that the EC_{50} and also the relative efficacy of the partial agonist 8-OH-DPAT (compared to the full agonist 5-HT) was unchanged when the 5-HT_{1A} receptor concentration was increased more than 15-fold (FIG. 7B and C).[12] In those experiments, the efficacies of the agonists increased as the function of receptor density (FIG. 7B).[12] Similar results have been published following the transfection of the delta opioid receptor in CHO cells. Increasing the receptor density from 18000 to 1540000 receptors per cell, does not change the EC_{50} of the agonist DADLE for inhibiting the adenylyl cyclase.[24]

We have recently studied the evolution of the EC_{50} for two agonists PACAP38 and PACAP27 stimulating PLC via a single cloned receptor[25] expressed in the LLC PK1 cell line when the receptor density is increased from 0.1 to 2.5 pmol/mg of protein (FIG. 8). Surprisingly, the EC_{50} for PACAP38 remains constant whereas the EC_{50} for PACAP27 decreases, reaching the EC_{50} value for PACAP38 at high receptor density (FIG. 8). Note that the K_A was the same for PACAP38 and PACAP27 (data not shown). Such an observation can explain why in PC12 cells, PACAP38 is more potent than PACAP27 in stimulating PLC,[26] whereas in the pituitary cell line αT3-1 the two agonists have the same potency.[27]

As already stated, it is only with the introduction of receptor cDNA technology

FIGURE 7. Transfection of 5-HT_{1A} receptors in NIH-3T3 fibroblasts. Effects of receptor density on the affinity and potency of agonists and a partial agonist. **A**: Scatchard plot analysis of the saturation binding curve for [^{3}H]8-OH-DPAT on membranes of two clones which expressed 45 (clone 6) and 500 (clone 2B) fmol of receptor per mg of protein. **B** and **C**: effects of the receptor densities on the potencies of 5-HT and ipsapirone (a partial agonist) to inhibit adenylyl cyclase. *Dashed line* represents the theoretical dose-response curve for 5-HT in clone 2B calculated according to the classical pharmacological model described in the text. **D**: effect of increasing 5-HT_{1A} receptor densities on the efficacy of ipsapirone, relative to 5-HT in different cell models: polyclonal cells, monoclonal cells and monoclonal cells treated with dexamethasone (DXM).[23]

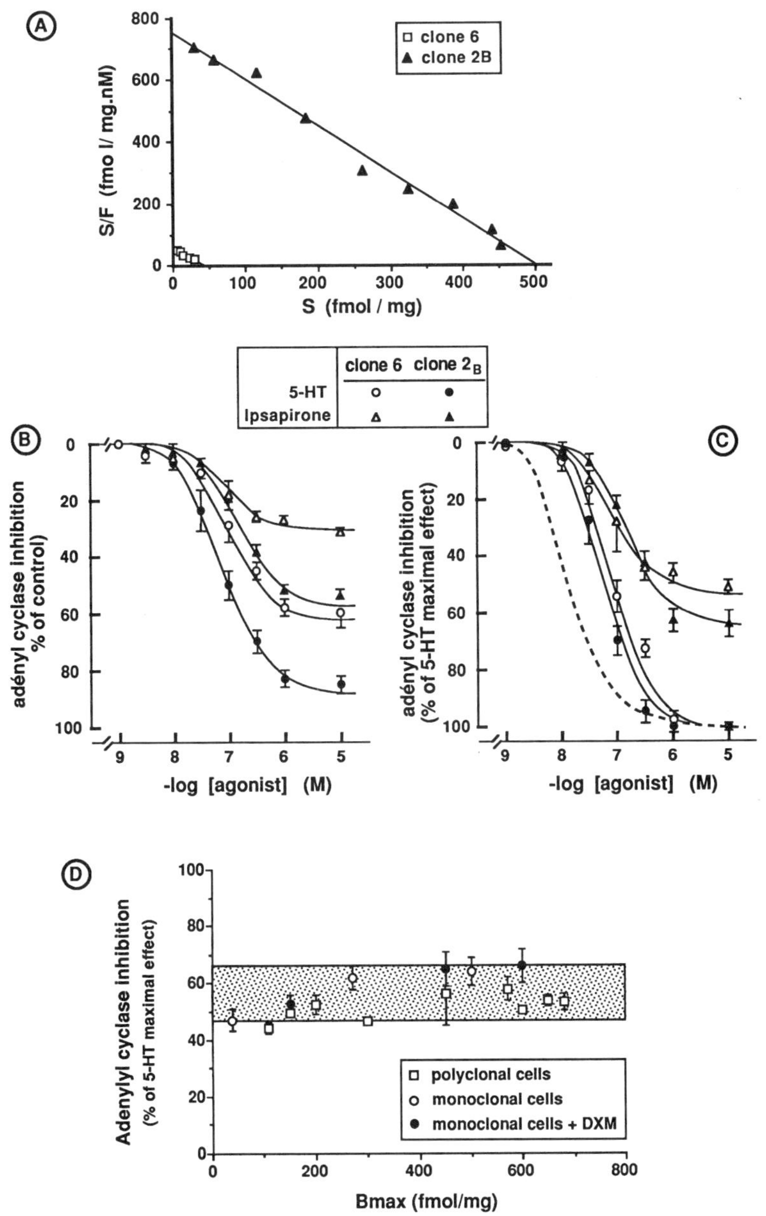
A
S/F (fmo l/ mg.nM)
S (fmol / mg)
clone 6
clone 2B
clone 6 clone 2B
5-HT
Ipsapirone
B
C
adényl cyclase inhibition % of control)
adényl cyclase inhibition (% of 5-HT maximal effect)
-log [agonist] (M)
D
Adenylyl cyclase inhibition (% of 5-HT maximal effect)
Bmax (fmol/mg)
polyclonal cells
monoclonal cells
monoclonal cells + DXM

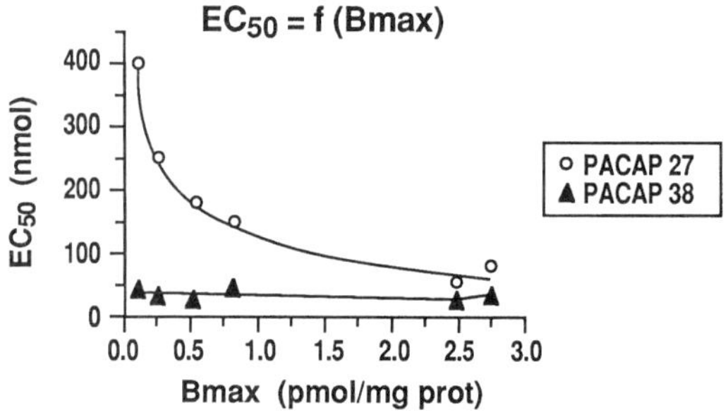

FIGURE 8. Evolution of the potencies (EC_{50}) of two agonists, PACAP27 and PACAP38 to stimulate PLC in LLCP-K1 cells transfected with a single clone of the PACAP receptor as the function of receptor concentration (B_{max}). Note that PACAP27 and PACAP38 have the same K_A (not shown).

that we have been able to increase the receptor density above the physiological values. Until recently, the pharmacologists were only able to reduce the number of receptors (using irreversible ligands) which is of course quite different. Experiments increasing the number of receptors indicate that there is a limiting factor in the intensity of the shift to the left which can be observed when the receptor density increases, at least in some cell systems. In addition, the PACAP receptor experiments indicate that this limit could be agonist-specific. We have no definitive explanation for these observations. Of course, a trivial explanation could be that when overexpressed, the receptors are not targeted to the functional compartment, *i.e.*, the plasmic membrane. This is certainly not the only explanation for the experiments reported by our laboratory. In the case of $5\text{-}HT_{1A}$ receptors, reported in FIGURE 7 increasing the number of receptors increases the E_{max}, indicating that functional receptors are introduced. In the case of experiments with PACAP receptors, the differences observed between the two agonists indicates that the explanation of non-functional receptors is not valid.

One can certainly find numerous other explanations. One possible one is the following. We may recall that the classical pharmacological models only apply if the ligand-receptor complexes, as well as the G proteins are free to diffuse in the cell membrane which is considered as a homogeneous system. However, it is possible that the receptor and the G proteins are distributed in restricted domains. Within these domains, free diffusion of receptors and G proteins occurs and the classical models apply. Increasing the number of receptors could simply result in an increase in the number of domains; the efficacy will increase but not the EC_{50}. This accounts for the results obtained with $5\text{-}HT_{1A}$ receptors. If one considers the results obtained with PACAP receptors, the difference in the evolution of EC_{50} for PACAP38 and PACAP27, as the function of receptor density, the following explanation can be given (FIG. 9). Within the restricted domain, the occupation of one PACAP receptor with PACAP38 will activate all the G protein molecules, whereas the occupation of the same receptors by PACAP27 (the receptor occupation is the same because the two peptides have the same K_A) will activate only a fraction of these G protein molecules. One reason could be that the receptor diffuses less when occupied by PACAP27 than when occupied by PACAP38 and therefore activates fewer G protein molecules. Another reason could be that the equilibrium dissociation of AR complex for G (K2, eq. 1) is higher when PACAP27 is the agonist than when PACAP38 is the agonist. When the concentration of PACAP27 increases, occupation of the remaining receptors of the domain increases as well as the stimulation of the remaining G proteins of the domain. Of course, under these conditions, the EC_{50} of the PACAP27 response is shifted to the right compared to the EC_{50} of the

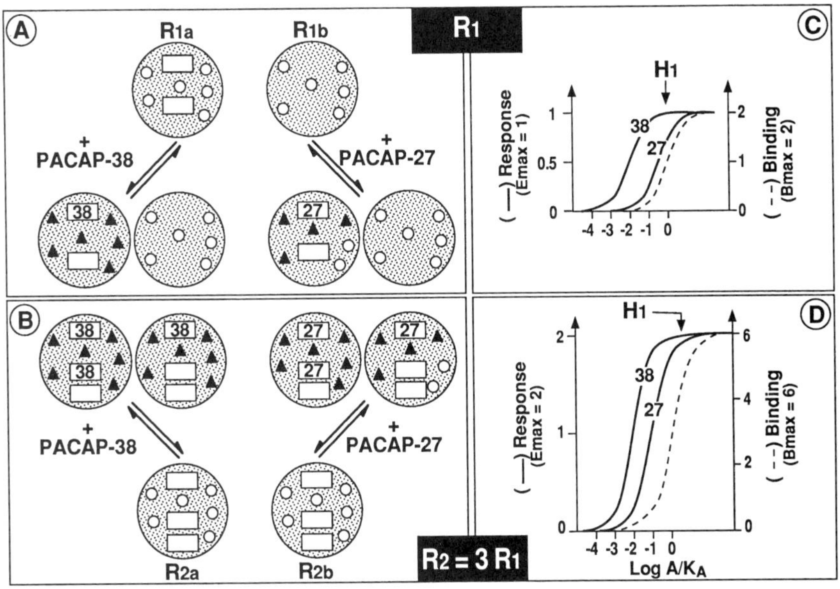

FIGURE 9. Microdomains explaining the non-identical evolution of EC_{50} for PACAP38 and PACAP27 as the function of B_{max}. The membrane is divided into microdomains containing a defined number of receptors and G proteins. In the example shown, 2, 0, 3, and 3 receptors (*open square*) in domains R1a, R1b, R2a, R2b, respectively. Each domain contains 6 G proteins (*open circles*). At a receptor concentration R1 (2 in R1a and 0 in R1b) and a hormone concentration H1 (= K_A), PACAP38 and PACAP27 occupy 50% of receptors. One receptor of the domain R1a is occupied by PACAP38, all the G proteins of that domain are activated (*open triangles*) **(A)**. In contrast, when 1 receptor of the domain is occupied by PACAP27, only 4 G proteins are activated **(A)**. Therefore, at this concentration, the E_{max} is reached with PACAP38 whereas it is not with PACAP27. Increasing the PACAP27 concentration will increase the occupation of receptors in the domain, all G proteins will be activated, and the response will reach E_{max} **(C)**. When the receptor concentration is increased [R2 = 3R1] **(B)**, 3 receptors are present in R2a and 3 in R2b, and the situation is different. At the same hormonal concentration (H1) 50% of total receptors are occupied, as in **A**. Let us suppose that 2 receptors are occupied in R2a and 1 in R2b (this should statistically occur). Therefore, in the presence of PACAP38, all the G proteins are activated in the two domains (12 G proteins), and E_{max} is reached, which is two times higher **(D)** than the E_{max} reached with the receptor concentration R1 (6 G proteins in **A**). In the presence of PACAP27, the occupation of 2 receptors in R2a activates 6 G proteins and the occupation of 1 receptor in R2b activates 4 G proteins. E_{max} is not reached, but the fractional response (10 G activated proteins for an E_{max} obtained with 12 activated G proteins) **(B)** is higher than with the receptor concentration R1 **(A)** (4 G proteins activated for an E_{max} obtained with 6 activated G proteins). Increasing the receptor concentration will not shift the PACAP38 dose-response curve to the left but will shift the PACAP27 dose response curve to the left **(D)**. (○) non-activated G protein; (▲) activated G protein; (□) receptor.

PACAP38 response. Increasing the number of receptors in the membrane will either (a) increase the number of receptors in the existing domains and therefore this will not shift the dose response curve for PACAP38 to the left (all the G proteins are activated by the occupation of one receptor), but will shift the dose response for PACAP27 to the left since the classical operational model applies within the domain, or (b) increase the number of independent domains containing receptors. Under these conditions the maximal response increases but the situation is similar to the previous one for the EC_{50} evolution (FIG. 9).

Conclusion

Expression of recombinant receptors in cell lines to estimate the affinity of ligands for a given receptor is obviously a great progress in pharmacology. Considering human receptors from tissues which are difficult to obtain surgically or postmortem, it is the only possibility to screen for the affinity of an important number of molecules. This is particularly important when species variation in the pharmacology of the receptor considered do exist. In general, the affinities determined in transfected cell lines are in accordance with those previously determined in cells in which they are normally expressed. However, owing to the formation of ternary complex between GPCR and G proteins which influences the affinity for the agonists (and in some cases for the antagonists) some care has to be taken, especially when affinities are determined *in vitro*, in the absence of GTP. Therefore, it is particularly important to keep the density of expressed receptor at a level comparable to physiological density when one wants to measure parameters as close as possible to the physiological ones. The influence of the nature and density of G protein on GPCR affinity has not be studied in detail experimentally, although in theory such an influence exists and could be experimentally demonstrated *in vitro* in the absence of GTP. The role of GPCR precoupling as proposed in the so-called "two-state model" on the determination of ligand affinity is also a source of possible variations depending on the nature of the G protein present in the cell.

Finally, thanks to the possibility of studying, for the first time, the influence of receptor densities higher than normal on biological responses, it is possible to find examples of paradoxical behaviors of recombinant receptors which were unexpected on the basis of classical pharmacological receptor theories.

Acknowledgments

The authors thank Drs. S. Jard and J.-P. Pin for stimulating discussions and interesting suggestions.

REFERENCES

1. DIXON, R. A. F., B. KOBILKA, D. STRADER, J. BENOVIC, H. DOHLMAN, T. FRIELLE, M. BOLANOWSKI, C. BENNET, E. RANDS, R. DIEHL, R. MUMFORD, E. SLATER, I. SIGAL, M. CARON, R. LEFKOWITZ & C. STRADER. 1986. Cloning of the gene and cDNA for mammalian β-adrenergic receptor and homology with rhodopsin. Nature **321:** 75–79.
2. BOCKAERT, J. 1995. Les récepteurs à sept domaines transmembranaires: Physiologie et pathologie de la transduction. Méd./Sci. **11:** 382–394.
3. BALDWIN, J. M. 1993. The probable arrangement of the helices in G protein-coupled receptors. EMBO J. **12:** 1693–1703.

4. BODDEKE, H. W. M., A. FARGIN, J. R. RAYMOND, P. SCHOEFFTER & D. HOYER. 1992. Agonist/antagonist interactions with cloned human 5-HT_{1A} receptors: Variations in intrinsic activity studied in transfected HeLa cells. Naunyn-Schmiedeberg's Arch. Pharmacol. **345:** 257–263.
5. SAUDOU, F. & R. HEN. 1994. Serotonin receptor subtypes in vertebrates and invertebrates. Neurochem. Inter. **25:** 503–532.
6. OKSENBERG, D., S. A. MARSTERS, B. F. O'DOWD, H. JIN, S. HAVLIK, S. J. PEROUTKA & A. ASHKENAZI. 1992. A single amino-acid difference confers major pharmacological variation between human and rodent 5-HT_{1B} receptors. Nature **360:** 161–163.
7. KÜHN, H. 1980. Light and GTP regulated interaction of GTPase and other proteins with bovine photoreceptor membranes. Nature **283:** 587–589.
8. BORNACIN, F., C. PFISTER & M. CHABRE. 1989. The transitory complex between photoexcited rhodopsin and transducin. Reciprocal interaction between the retinal site in rhodopsin and nucleotide site in transducin. Eur. J. Biochem. **184:** 687–698.
9. WAELBROECK, M., P. ROBBERECHT, P. CHATELAIN & J. CHRISTOPHE. 1981. Rat cardiac muscarinic receptors. Effects of guanine nucleotides on high and low affinity binding sites. Mol. Pharmacol. **21:** 581–588.
10. KENAKIN, T. 1993. Pharmacologic analysis of drug-receptor interaction. Raven Press. New York.
11. MEI, L., J. LAI, H. I. YAMAMURA & W. R. ROESKE. 1989. The relationship between agonist states of the M1 muscarinic receptor and the hydrolysis of inositol lipids in transfected murine fibroblast cells (B82) expressing different receptor densities. J. Pharmacol. Exp. Ther. **251:** 90–97.
12. VARRAULT, A., L. JOURNOT, Y. AUDIGIER & J. BOCKAERT. 1992. Transfection of human 5-HT_{1A} receptors in NIH-3T3 fibroblasts: Effects of increasing receptor density on the coupling of 5-HT_{1A} receptors to adenylyl cyclase. Mol. Pharmacol. **41:** 999–1007.
13. FARGIN, A., J. R. RAYMOND, M. J. LOHSE, B. K. KOBILKA, M. G. CARON & R. J. LEFKOWITZ. 1988. The genomic clone G-21 which resembles a β-adrenergic receptor sequence encodes the 5-HT_{1A} receptor. Nature **335:** 358–360.
14. BOESS, F. G. & I. L. MARTIN. 1994. Molecular biology of 5-HT receptors. Neuropharmacology **33:** 275–317.
15. RICHARDS, M. H. 1991. Pharmacology and second messenger interactions of cloned muscarinic receptors. Biochem. Pharmacol. **42:** 1645–1653.
16. FLORIO, V. A. & P. C. STERNWEIS. 1989. Mechanisms of muscarinic receptor action and Go in reconstituted phospholipid vesicles. J. Biol. Chem. **264:** 3909–3915.
17. GERWINS, P., C. NORDSTEDT & B. B. FREDHOLM. 1990. Characterization of adenosine A1 receptors in intact DDT1-MF2 smooth muscle cells. Mol. Pharmacol. **38:** 660–666.
18. ZHU, S., S. GILBERT, M. BIRNBAUMER & L. BIRNBAUMER. 1994. Dual signaling potential is common among Gs-coupled receptors and dependent on receptor density. Mol. Pharmacol. **46:** 460–469.
19. FARGIN, A., R. J. RAYMOND, J. W. REAGAN, S. COTECCHIA, R. J. LEFKOWITZ & M. G. CARON. 1989. Effector coupling mechanisms of the cloned 5-HT_{1A} receptor. J. Biol. Chem. **264:** 14848–14852.
20. DUMUIS, A., M. SEBBEN & J. BOCKAERT. 1988. Pharmacology of $5\text{-hydroxytryptamine}_{1A}$ receptors which inhibit cAMP production in hippocampal and cortical neurons in primary culture. Mol. Pharmacol. **33:** 178–186.
21. LEFKOWITZ, R. J., S. COTECCHIA, P. SAMANA & T. COSTA. 1993. Constitutive activity of receptors coupled to guanine nucleotide regulatory proteins. Trends Pharmacol. Sci. **14:** 303–304.
22. COSTA, T., Y. OGINO, P. J. MUNSON, H. O. ONARAN & D. RODBARD. 1992. Drug efficacy at guanine nucleotide-binding regulatory protein-linked receptors; thermodynamic interpretation of negative antagonism and of a receptor activity in the absence of ligand. Mol. Pharmacol. **41:** 549–560.
23. VARRAULT, A. & J. BOCKAERT. 1992. Differential coupling of 5-HT_{1A} receptors occupied by 5-HT or 8-OH-DPAT to adenylyl cyclase. Naunyn-Schmiedeberg's Arch. Pharmacol. **346:** 367–374.
24. PRATHER, P. L., T. M. MCGINN, L. J. ERICKSON, C. J. EVANS, H. H. LOH & P. LAW. 1994.

Ability of delta-opioid receptors to interact with multiple G-proteins is independent of receptor density. J. Biol. Chem. **269:** 21293–21302.
25. SPENGLER, D., C. WAEBER, C. PANTALONI, F. HOLSBOER, J. BOCKAERT, P. H. SEEBURG & L. JOURNOT. 1993. Differential signal transduction by five splice variants of the PACAP receptor. Nature **365:** 170–175.
26. DEUTSCH, P. J. & Y. SUN. 1992. The 38-amino acid form of pituitary adenylate cyclase-activating polypeptide stimulates dual signaling cascades in PC12 cells and promotes neurite outgrowth. J. Biol. Chem. **267:** 5108–5113.
27. RAWLINGS, S. R., I. PIUZ, W. SCHLEGEL, J. BOCKAERT & L. JOURNOT. 1995. Differential expression of pituitary adenylate cyclase-activating polypeptide/Vasoactive intestinal polypeptide receptor subtypes in clonal pituitary somatotrophs and gonadotrophs. Endocrinology **136:** 2088–2098.

How Receptor Mutagenesis May Confirm or Confuse Receptor Classification

THUE W. SCHWARTZ,[a–c] SIGNE PERLMAN,[a]
METTE M. ROSENKILDE,[a] AND SIV A. HJORTH[a]

[a]*Laboratory for Molecular Pharmacology*
The Laboratory Center, Rigshospitalet 6321
Blegdamsvej 9
Copenhagen DK-2100 Denmark

[b]*Department of Protein Chemistry*
Institute of Molecular Biology
University of Copenhagen
Copenhagen, Denmark

INTRODUCTION

To the knowledge of the authors there is no good example where mutagenesis has directly aided receptor classification as such. However, receptor mutagenesis has been used extensively to map receptor domains and single residues involved in ligand binding and G-protein coupling, which are important cornerstones in receptor classification. On the other hand, it could be argued that receptor mutagenesis in several cases has caused more confusion than clarification in this field. For example, some peptide systems have refused to present an agonist binding site which corresponds to the monoamine agonist binding site, and in some of these systems it has not been possible to obtain a nice overlap between binding sites for what is believed to be competitive ligands.[1] Thus, either the mutational approach—which is in no way flawless—is at fault in these cases, or some of our current dogmas of receptor activation and inactivation are not correct. In the following these issues will be addressed from a 7TM receptor point of view and with a bias for mutational analysis applied to characterizations of ligand binding sites.

A NOMENCLATURE AND A GENERIC NUMBERING SYSTEM FOR RECEPTOR RESIDUES IS GREATLY NEEDED

Many papers on receptor mutagenesis appear to be written with the assumption that all readers have an intimate knowledge of the particular model system in question. However, the specific number of a residue in an individual receptor does not hold much information about its location in the receptor structure. Thus,

[c] Author to whom correspondence should be addressed. Tel: +45 3545 6321; Fax: +45 3135 2995.

communication in the receptor field would benefit greatly by the introduction of a generic nomenclature or rather a generic numbering system for receptor residues.

Such a generic numbering system is especially relevant in the 7TM receptor system since the seven helical bundle probably is oriented or arranged rather similarly, at least among rhodopsin-like 7TM receptors. This assumption was recently supported by the direct transfer of a tridendate zinc-site located between the outer segments of TM-V and TM-VI from the tachykinin NK-1 to the rather distantly related κ-opioid receptor (only 31% identity) without loss of metal-ion affinity.[2] Thus, a residue located at a particular position in the primary structure of one 7TM receptor is probably also located at a particular position in the folded three-dimensional structure. The conserved "fingerprint" residues found in all of the seven TMs can help to identify the generic location of a particular residue, for example "seven residues after the conserved Asp residue in T-II" (FIG. 1). The main problem is, which generic number should be given to the particular conserved residue and what type of typography should be used (a classical nomenclature problem!).

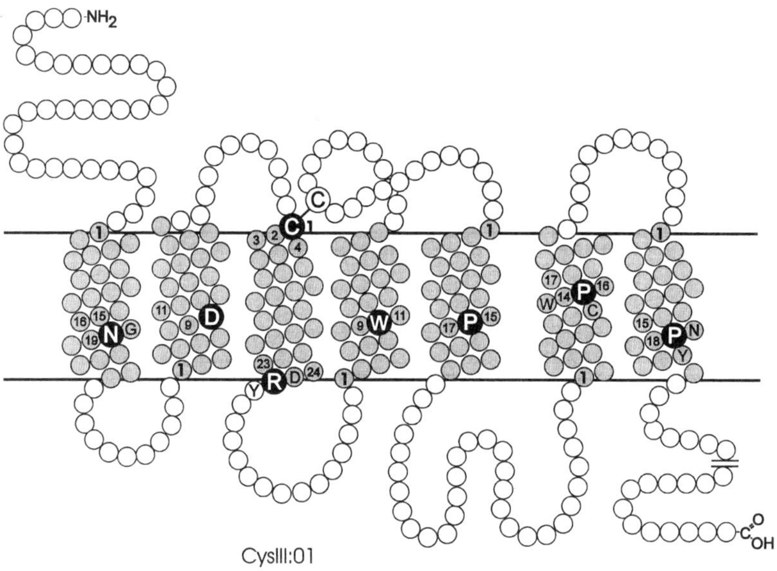

FIGURE 1. A serpentine diagram of a rhodopsin-like 7TM receptor with a proposed generic numbering scheme and nomenclature for transmembrane residues. Residues have been conserved especially in the transmembrane segments of 7TM receptors. However, because of large differences in especially the N-terminal segment, the specific numbering of residues in the individual receptors is of little use. Based on well conserved. "finger-print" residues found in each transmembrane segment, several generic numeric systems have been suggested.[3-6] These "finger-print" residues are shown in the serpentine model in white on black. Alignments of multiple rhodopsin-like receptors give an indication of the likely position where each helix enters the hydrophobic core of the membrane.[8] On this basis each of the "finger-print" residues can be assigned a number—as indicated below the serpentine diagram—and accordingly any residue in any TM can then also be assigned a number determined by its position in relation to the highly conserved "finger-print" residue in its helix.

Numbering

A couple of generic numbering schemes have been suggested over the last few years especially addressing the transmembrane segments, where this is a particularly relevant issue owing to the relatively high degree of sequence conservation.[3–6] In all generic numbering schemes the first digit identifies the number of the TM segment. However, there is still some controversy concerning the specific numbering of residues within a helix. Vriend and coworkers have, from a computational point-of-view suggested, that in each TM the most conserved residue be given a high, rather arbitrary number.[4] The other groups have all made an attempt to identify the actual start or end points of the transmembrane helices. Without reliable experimental evidence this is obviously a difficult task in the individual receptor. Moreover it is likely that some of the transmembrane segments will continue as helices at least into the interface of the membrane, which biophysically is characterized both by the presence of the polar headgroups of the phospholipids and, owing to the dynamic, fluid character of the lipid bilayer, also by more apolar molecular components.[7] Nevertheless, electron paramagnetic resonance studies with spin-labeling of rhodopsin, has clearly identified the position where TM-III and TM-V leaves the apolar hydrocarbon part of the lipid bylayer and where TM-IV and TM-VI enters the 30 Å hydrophobic core of the membrane (Wayne Hubbell, personal communication). Importantly, these demarcation points for transmembrane segments are similar to those suggested in generic terms by Probst and coworkers on the basis of multiple sequence alignments[8] and those used in most generic numeric schemes.[5,6] The main exception is TM-VII, which in systems based on the bacteriorhopsin scaffold usually is shifted "upwards".[3] Thus, the otherwise rather general agreement with respect to the generic demarcations of TMs indicates that a numbering scheme based on these segments can serve as a useful starting point for a generic numbering system.

Typography

It is most confusing to use an arabic number for identification of the helix number because the distinction between a specific and a generic, three-digit number is not possible. For example, is Asp^{308} in fact residue number 308 in this particular receptor sequence or is it residue number eight in TM-III? To avoid this potential confusion, roman numerals should be preferred for helix identification. In the given example the residue would be designated AspIII:08 to indicate that it is residue number eight in TM-III. As discriminator between the helix identification and the residue number ":" is preferred to "-" for computational reasons.

MUTAGENESIS CAN AID IN LOCATING LIGAND BINDING SITES

In the monoamine systems receptor mutagenesis has played a major role in the characterization of ligand binding sites. Here, a convincing picture has been presented with agonist ligand binding being located mainly between TM-III, -V, and -VI and with the interaction between the amino group of the ligand and the carboxylic acid function of an aspartate residue on TM-III as being the crucial element.[9,10]

However, rather few other cases have yet been presented where ligand binding sites have been thoroughly investigated by an extensive and systematic mutational analysis of all possible residues in a given area. In numerous cases in the literature

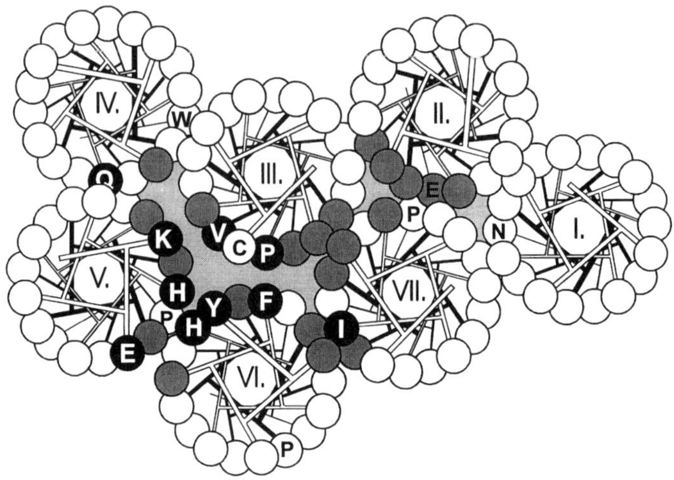

FIGURE 2. Mutational map of the presumed binding site for the proto-type non-peptide antagonists CP96,345 shown in a helical wheel diagram of the exterior part of the seven helical bundle of the tachykinin NK-1 receptor. Mutational "hits" for CP96,345 are shown in white on black, whereas residues which can be substituted without affecting the binding of the antagonist are shown in grey. These constitute important, "negative" or back-ground results.[6,11–17] The well-conserved "finger-print" residues in each TM (see FIG. 1) are indicated in black letters.

a few positive mutational "hits" are presented for one or more ligands in a receptor. At best this may only represent a small part of a binding site which possibly could involve side chains from a total of 6–12 or even more residues, besides possible backbone interactions which cannot be addressed by receptor mutagenesis. Importantly, a large number of negative results is in fact an essential part of the whole picture as these constitute the background from which the positive results should stand out. This point is illustrated by the mutational analysis of the presumed binding site for the prototype non-peptide antagonists CP96,345 in the tachykinin NK-1 receptor where a rather complete picture has been compiled over the last 3–4 years[10–15] (FIG. 2).

IT IS IMPORTANT TO DIFFERENTIATE BETWEEN REAL-, INDIRECT-, AND PSEUDO-"HITS" IN MUTATIONAL ANALYSIS

Interpretation of results from a mutational mapping of binding sites in a receptor is a difficult task. The major problem is, whether a substitution of a residue which affects the binding of a ligand does so directly or indirectly. Is the residue, which is mutated, directly in contact with the ligand, or is it a "second row" residue, which acts as support for a residue that directly touches the ligand? Or, is the substituted residue even further removed in the receptor structure but still indirectly important for the stability of the ligand binding conformation of the receptor? Unfortunately, there is no well-established mutational approach to differentiate between these possibilities. The overall picture of the location of positive "hits" and negative substitutions may, however, suggest whether a given substitution affect the binding indirectly simply by being located far removed from the other positive "hits". However, this is still guessing!

Although several types of control experiments can be performed to test whether, for example, the receptor is expressed appropriately on the surface of the cell, these are not always sufficient controls. Thus, a normal binding of an antibody or even an antagonist does not preclude that the agonist binding conformation of the receptor has been disturbed in some indirect manner. One simple example could be a mutation which impairs G-protein coupling. Such a substitution will selectively affect agonist binding and spare antagonist and receptor antibody binding. Thus, the differentiation between direct or real "hits" and indirect "hits" is far from simple in mutational mapping of binding sites.

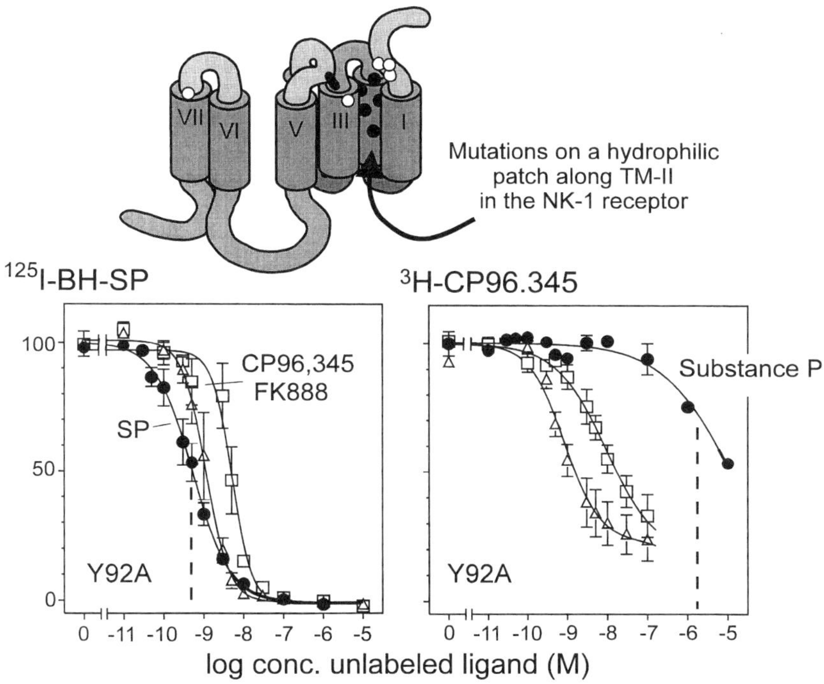

FIGURE 3. "Pseudo-hits" for substance P in the tachykinin NK-1 receptor affecting the ability of the peptide to compete for antagonist binding without impairing the actual high affinity binding of the agonist. Results from one, [Y92A]hNK1, out of a series of point mutations on the hydrophilic, inner face of TM-II are shown. Competition binding experiments are shown for the agonist substance P (*filled circles*) and for the antagonists, CP96,345 (*open triangles*) and FK888 (*open squares*). In the panel to the right ^{3}H-CP96,345 is used as radioligand and in the panel to the right parallel experiments performed with ^{125}I-substance P as radioligand. Note, that in these mutants substance P competes almost normally against itself but is severely impaired in its competition with the radiolabeled antagonist.[17] In the vignette at the top residues which upon substitution actually do affect the "real" binding affinity of substance P are indicated in white, thus being either "real" or "indirect" hits (see text).

More complicated to understand, but also more accessible to experimental analysis are the "pseudo-hits", *i.e.*, mutations that affect the ability of a ligand—usually an agonist—to compete for receptor binding with a chemically distinct radioligand (FIG. 3). This phenomenon was first described in the tachykinin NK-1 receptor, where a series of mutations down along a hydrophilic patch of TM-II dramatically (> 1000-fold) affected the ability of substance P to compete for binding with a radiolabeled version of the the prototype antagonist CP96,345.[16,17] Neverthe-less, radiolabeled substance P bound with normal or only slightly reduced affinity to all these mutant receptors—indicating, that the high affinity binding site for the agonist was in fact not disturbed by the substitutions.[17] This notion was supported by the observation that the ability of the agonist to stimulate inositolphosphate turnover was also only slightly impaired.[16] It was suggested that this type of mutation affects mainly the ability of the receptor to interchange between receptor conforma-tions that each bind agonist and antagonists with high affinity.[17] The same phenome-non was recently observed also in the κ-opioid receptor both in relation to the corresponding substitutions in TM-II but also in a series of chimeric constructs between the κ- and the μ-opioid receptor (Hjorth, Thirstrup & Schwartz, submitted for publication). Thus it appears to be a general phenomenon that mutations can affect the ability of agonists to compete for antagonist binding without really affect-ing the high affinity binding of the agonists themselves. This problem, which is particularly relevant when binding sites for agonists are being characterized, can be addressed by using a battery of different radiolabeled ligands and especially by performing homologous binding experiments with the agonist in question.

BINDING-SITES FOR ENDOGENOUS AGONISTS ARE NOT NECESSARILY CONSERVED AMONG RECEPTOR SUBTYPES?

It is generally presumed that the binding site for an agonist has to be almost identical in different subtypes of receptors. One reason is that receptor subtypes usually bind the endogenous agonist with similar affinity, and it is reasoned that the binding site for the endogenous ligand has been conserved during evolution. This certainly appears to be the case for small ligands like monoamines,[10] but apparently not necessarily for larger ligands, such as peptides.

AT_1 versus AT_2 Receptors

A number of residues have been identified as presumed interaction points for angiotensin in the AT_1 receptor.[18–21] One initial approach was in fact specifically to search among residues which were conserved between mammalian and *Xenopus laevis* AT_1 receptors.[18] As shown in FIGURE 4, this approach led to the identification of a series of residues which upon Ala substitution affected angiotensin binding seriously, in several cases 100 to 1000-fold. Surprisingly, as the angiotensin AT_2 receptor subsequently was cloned it turned out that several of the presumed interac-tion points for angiotensin were not conserved in this receptor subtype (FIG. 4). Among the mutational "hits" for angiotensin in the AT_1 receptor, only residues Tyr^{92}, Lys^{199}, and Asp^{281} are preserved in the AT_2 receptor.

Thus, either the binding site for the peptide is significantly different in the two subtypes of angiotensin receptors or several of the residues which have been presumed to be direct interaction points for angiotensin on the basis of receptor mutagenesis are perhaps instead indirectly involved in the binding of the ligand.

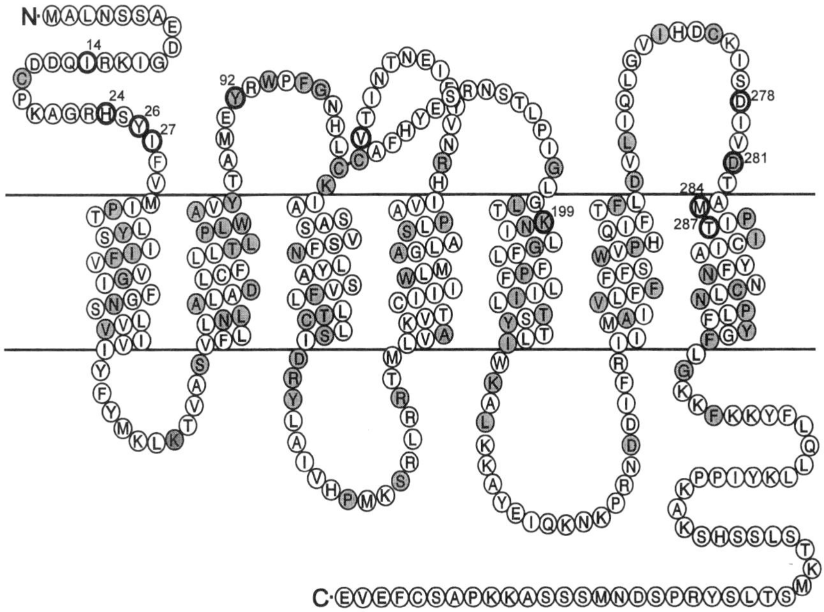

FIGURE 4. Apparent lack of conservation of presumed contact residues for angiotensin among receptor subtypes. In a serpentine diagram of the angiotensin AT_1 receptor bold circles and numbers indicate residues which upon mutational substitution affect the binding of angiotensin II seriously.[18–21] Residues are indicated in grey which are conserved between the different cloned AT_1 and AT_2 receptors, which both bind angiotensin II with similar high affinity. Note that only three of the presumed contact residues for angiotensin are conserved, Tyr^{92}, Lys^{199}, and Asp^{281}.

Tachykinin Receptors

Initially a series of residues in the N-terminal extension and at the top of TM-III and VII were identified as potential points of interaction for substance P in the NK-1 receptor.[11,16] The overall location of these NK-1 residues corresponded well to the residues identified for angiotensin in the AT_1 receptor discussed above. On the other hand, mutational analysis of the NK-2 receptor led to the identification of residues in the outer segments of TM-V and VI as being important for the binding of NKA.[22,23] This was surprising since substance P and NKA are homologous peptides and their respective receptors are also homologous. And, substitution of residues which corresponded to those which in the NK-2 receptor appeared to be crucial for NKA binding[22,23] did not at all affect substance P binding in the NK-1 receptor even though two of these were conserved residues (HisV:05 and HisVI:17).[15,24,25] Moreover, as shown in TABLE 1, substitution of these residues in the NK-1 receptor did not affect the binding of NKA to this receptor, although these substitutions in the NK-2 receptor did eliminate the binding of NKA.[22,23]

Thus, the mutational analysis tells us that tachykinin peptides are affected very differently by substitution of conserved residues in different receptor subtypes.

TABLE 1. Affinity for Substance P and NKA Determined in Wildtype NK-1 Receptors and in NK-1 Receptors with Mutations Corresponding to Those Which in the NK-2 Receptor Impair NKA Binding

		Substance P K_d (nM)	NKA K_d (nM)
WT hNK-1		0.074 ± 0.014 (8)	8.4 ± 2.0 (7)
[H197A]hNK-1	(~H198 in NK-2)	0.098 ± 0.034 (6)	7.7 ± 1.7 (4)
[H265A]hNK-1	(~H267 in NK-2)	0.127 ± 0.025 (4)	9.0 ± 4.5 (4)

The affinity of the ligands was determined in competition binding experiments against radiolabeled substance P in transiently transfected COS cells. Note that the binding of NKA (and substance P) is not affected by the substitution of the two conserved His residues in TM-V and TM-VI, respectively, which upon substitution in the NK-2 receptor seriously impair or eliminate NKA binding (they correspond to His[198] and His[267], respectively in the NK-2 receptor),[22,23] indicating that the tachykinin peptides may bind rather differently to the different tachykinin receptors.[23]

NPY Receptors

The information available in this system is still rather limited. However, a number of residues in the Y1 receptor, which was cloned several years ago, have been suggested to be involved in the binding of NPY.[26,27] Recently, two new members of this family of receptors were cloned, *i.e.*, the Y2 receptor and the PP1 or Y4 receptor, which preferentially binds the hormone pancreatic polypeptide but also has a reasonably high affinity for NPY.[26–29] As shown in FIGURE 5, the three presently

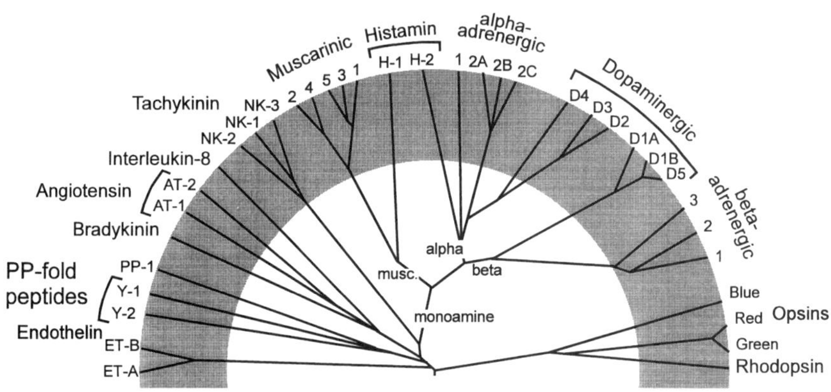

FIGURE 5. A small part of the sequence identity three for the family of rhodopsin-like receptors. The monoamine "branch" is shown together with an assortment of receptors for peptide ligands as well as rhodopsin and the opsins. The "root" indicate around 20% identity and the grey area more than 70% identity in amino acid sequence. Note that within certain families subtypes of receptors are relatively closely related (*e.g.*, muscarinic receptors and endothelin receptors), whereas receptor subtypes binding, for example, neuropeptide Y (Y-1 and Y-2) or angiotensin (AT-1 and AT-2) have surprisingly little sequence identity. As shown in FIGURE 4, many of the presumed interaction points for the endogenous ligand may not even be conserved. Note also the apparent "convergent evolution" among histamine and dopamine receptors.

cloned receptor subtypes, which all bind NPY with high affinity, are surprisingly different in sequence as compared to subtypes among, for example, muscarinic receptors. In fact, of the yet identified 9–10 presumed contact residues for NPY in the Y1 receptor (Philip Walker & Annette Beck, personal communication) only two are conserved among all three receptor subtypes, *i.e.*, a Tyr at the border between TM-II and the first extracellular loop and an Asp just outside TM-VI.[28–31] Thus, it would appear that also NPY binds rather differently to different Y-receptor subtypes.

With regards to identification of "contact residues" between receptors and ligands in general, it should be noted, that only a small fraction of the actual interface between growth hormone and its receptor turned out to be important for obtaining the high affinity binding of this large ligand. Basically all residues of the crystallographically characterized interface of the hormone-receptor complex had been subjected to Ala-substitution. Surprisingly, only a few centrally located residues on both the ligand and the receptor part of the interface accounted for the vast majority of binding energy.[32] If this principle also applies to peptide 7TM receptors, then a few crucial residues—like for example Tyr^{92} at the top of TM-II, Asp^{281} at the top of TM-VII, and Lys^{199} in TM-V of the angiotensin receptor—could be the decisive elements which have been conserved during evolution, whereas other parts of the binding "pocket" are less critical.

BINDING SITES FOR 'DUALISTS' AND SOME INSURMOUNTABLE ANTAGONISTS ARE SURPRISINGLY DIFFICULT TO MAP

Recently a number of non-peptide agonists were discovered in a series of biphenylimidazole compounds in the AT_1 system.[33] Although these compounds clearly are agonists *in vivo*, they act as partial agonists with bell-shaped dose-response curves when tested in cell lines expressing the cloned AT_1 receptor.[34] However, as shown in FIGURE 6, these compounds are in fact dualists, *i.e.*, both agonists and antagonists[35] as they are able at higher concentration to inhibit the angiotensin-induced phosphoinositol turnover.[34] When these dualists are tested in our library of mutated AT_1 receptors they bind nearly normally to all mutants.[32] This is surprising, in view of the fact that multiple substitutions have been reported to seriously affect or even eliminate the binding of peptide agonists and peptide antagonists,[18] whereas other mutations have been shown to impair the structurally similar biphenylimidazole, non-peptide antagonists.[36] In fact, we have not yet identified a single substitution which convincingly impairs the binding of the prototype non-peptide dualist, L612,313, significantly (unpublished observations).

It is the case that either structurally similar non-peptide agonists and antagonists do not bind to overlapping sites in the receptor or, alternatively, most of the mutational mapping with respect to non-peptide ligands has until now been misleading in the AT_1 receptor system—despite the fact that the mutational "hits" do cluster on opposing faces of transmembrane helices and form what appears to be a nice binding site deep between TM-III and VII. In the latter case, non-peptide compounds—agonists as well as antagonists—should then bind elsewhere in the receptor structure in an area not yet explored by receptor mutagenesis. Another possibility could be, that dualists are capable of binding to different sites presented by different receptor conformations—for example, one being active and the other inactive. In that case, it could be envisioned that point mutations may only affect the binding of the dualist to one of the receptor conformations seriously. This would

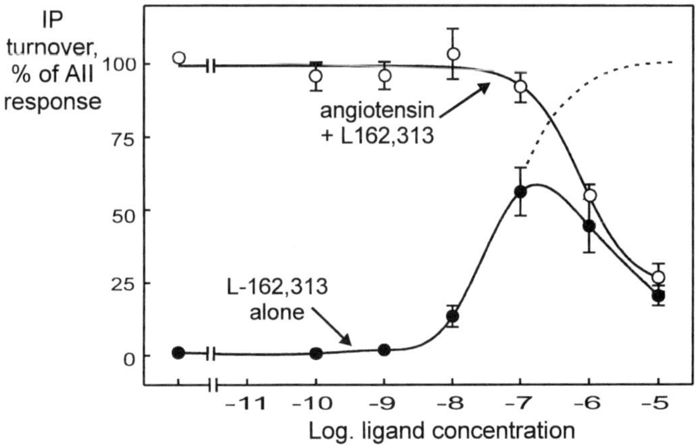

FIGURE 6. Dualistic, combined agonistic and antagonistic property of non-peptide angiotensin AT1 ligands. In COS cells transfected with the AT^1 receptor, the prototype non-peptide "agonist",[33] L162,313, functions as a partial agonist with a bell-shaped dose-response curve.[34] The compound is, in fact also an efficient antagonist of the angiotensin-induced inositolphosphate turnover as shown.[34] As indicated by the dotted curve, it can be envisioned that the compound functions as a dualist,[35] *i.e.*, that it at lower concentrations is able to stimulate the AT_1 receptor (the dotted line) but that this stimulation is turned into the bell-shaped curve by its concomitant inhibitory action, which is evident at slightly higher concentrations of the compound. Interestingly, these "dualist" compounds are very resistant to mutational mapping of binding sites,[34] despite the fact that many presumed interaction sites can be located for structurally homologous non-peptide antagonists[36] or for functionally "similar" peptide agonists.[18–21] In the text it is speculated that this resistance to mutational mapping for dualists may be caused by their apparent ability to exploit several conformations of the same receptor with relatively similar affinity.

not necessarily impair the overall binding of the compound very much unless the substitution also impaired the binding of the ligand to the other receptor conformation.

Insurmountable antagonists also appear to be more resistant to mutational mapping than chemically similar competitive antagonists.[36,37] This phenomenon was unexpectedly discovered during the mutational mapping of presumed binding sites for a large series of biphenylimidazole antagonists in the AT_1 receptor. As a group, the binding affinity of the insurmountable antagonists was affected much less than chemically similar competitive antagonists both by chimeric exchanges and by point mutations in the transmembrane segments.[36–37] Similarly, in the tachykinin receptors it has been surprisingly difficult to identify point mutations which affect certain insurmountable non-peptide ligands, for example the series of homologous but highly selective antagonists developed by Sanofi[38,39] (unpublished observation). Recently, it was shown for one of these compounds SR48,968 that although it was resistant to most point mutations, it was seriously affected by combinations of these.[23]

It could be speculated, that certain ligands—for example dualists and insurmountable antagonists—are able to exploit more than one receptor conformation with reasonably high affinity and therefore can be more resistant to the mutational mapping exercise than, for example, classical competitive antagonists. A particular point substitution may not seriously impair the interactions of such compounds with both receptor conformations.

RECEPTOR MUTATIONS AND SMALL CHEMICAL CHANGES IN THE LIGANDS MAY ALTER THE PHARMACOLOGICAL PROPERTY OF LIGANDS

It is well known that slight chemical alteration of a ligand can change its pharmacological property, for example in the opioid field.[40] In the case of the new series of biphenylimidazole AT_1 ligands discussed above, the addition of a single methyl group can change the compound from being a clear antagonist to being a good agonist (unpublished information). In the CCK receptor system very similar changes in pharmacological properties from agonism to antagonism have been observed in response to a substitution of a single methyl group on the compound (E. E. Sugg, personal communication). In view of the fact that such small chemical changes on the ligands can change their pharmacological phenotype dramatically, it is not surprising that receptor mutations can have similar effects. There are in the literature sporadic cases where such a phenomenon has been reported. For example, in the adrenoceptor field it was described early on that substitution of the notorious Asp^{113} in TM-III with a glutamic acid residue changed the activity of a series of well-known antagonists to being partial agonists.[41] In essence this chemical modification of the receptor is equal to the introduction of a single methyl group, albeit in this case it means that the carboxylic acid group is shifted further out in the receptor structure.

It should be noted, that most receptor mutations usually have been investigated rather superficially, *i.e.*, only addressing the question which the authors initially were interested in. It is reasonable to expect that many more examples of mutationally induced changes in the pharmacological properties of ligands will appear in the literature as this issue will become directly investigated in the future.

CONCLUDING REMARKS

Receptor mutagenesis has provided some surprising pictures of ligand receptor interactions, especially in the peptide field. Some peptides apparently bind totally differently from where monoamines are believed to bind;[1] peptide binding sites do not appear to be as conserved between receptor subtypes as would be expected; competitive antagonists apparently do not have to share binding site with the corresponding agonists;[1] compounds with dual agonistic-antagonistic properties can be surprisingly resistant to mutational mapping of binding sites, *etc.* Either the mutational mapping of binding sites has been misleading in most cases or some of the basic dogmas of ligand-receptor interactions may not be totally correct. However, much of the surprising observations may possibly be explained by an allosteric "two-stage" model for receptor activation and inhibition in which the ligands obtain their properties basically by selecting and stabilizing preformed receptor conformations.[1]

Importantly, small chemical modifications of a ligand can change its pharmacological property—just as mutations of the receptor also can modify the pharmacological properties of ligands. This emphasizes the fact that the pharmacological property of a given ligand is obtained in a close two-way interaction with the receptor. Subtle alterations in this chemical recognition may change the pharmacological property of the ligand. It could be speculated that such alterations could possibly even be caused by interaction of the receptor with different cellular components, *e.g.*, G-proteins, in different tissues. This possibility could open up a new interesting pharmacotherapeutic area.

REFERENCES

1. Schwartz, T. W., U. Gether, H. T. Schambye & S. A. Hjorth. 1995. Molecular mechanism of action of non-peptide ligands for peptide receptors. Curr. Pharmaceut. Design **1:** 355–372.
2. Thirstrup, K., C. E. Elling, S. A. Hjorth & T. W. Schwartz. 1996. Construction of a high affinity zinc-switch in the k-opioid receptor. J. Biol. Chem. **271:** 7875–7878.
3. Trumpp-Kallmeyer, S., J. Hoflack, A. Bruinvels & M. Hibert. 1992. Modeling of G-protein-coupled receptors: Application to dopamine, adrenaline, serotonin, acetylcholine, and mammalian opsin receptors. J. Med. Chem. **35:** 3448–3462.
4. Cronet, P., C. Sander & G. Vriend. 1993. Modeling of transmembrane seven helix bundles. Protein Eng. **6:** 59–64.
5. Baldwin, J. 1993. The probable arrangement of the helices in G protein-coupled receptors. EMBO J. **12:** 1693–1703.
6. Schwartz, T. W. 1994. Locating ligand-binding sites in 7TM receptors by protein engineering. Curr. Opinion Biotech. **5:** 434–444.
7. White, S. H. 1994. Membrane Protein Structure—Experimental Approaches. Oxford University Press. New York.
8. Probst, W. C., L. A. Snyder, D. I. Schuster, J. Brosius & S. C. Sealfon. 1992. Sequence Alignment of the G-Protein Coupled Receptor Superfamily. DNA Cell Biol. **11:** 1–20.
9. Strader, C. D., I. S. Sigal & R. A. F. Dixon. 1989. Structural basis of β-adrenergic receptor function. FASEB J. **3:** 1825–1832.
10. Strader, C. D., T. M. Fong, M. R. Tota & D. Underwood. 1994. Structure and function of G protein-coupled receptors. Ann. Rev. Biochem. **63:** 101–132.
11. Fong, T. M., R. C. Huang & C. D. Stader. 1992. Localization of agonist and antagonist binding domains of the human neurokinin-1 receptor. J. Biol. Chem. **267:** 25664–25667.
12. Gether, U., T. E. Johansen, R. M. Snider, J. A. Lowe III, S. Nakanishi & T. W. Schwartz. 1993. Different binding epitopes on the NK1 receptor for substance P and a non-peptide antagonist. Nature **362:** 345–348.
13. Fong, T. M., M. A. Cascieri, H. Yu, A. Bansal, C. Swain & C. D. Strader. 1993. Amino-aromatic interaction between histidine 197 of the neurokinin-1 receptor and CP 96345. Nature **362:** 350–353.
14. Cascieri, M. A., A. M. MacCleod, D. Underwood, L. Shiao, E. Ber, S. Sadowski, H. Yu, K. J. Merchant, C. J. Swain, C. D. Strader & T. M. Fong. 1994. Characterization of the interaction of N-acyl-L-tryptophan benzyl ester neurokinin antagonists with the human neurokinin-1 receptor. J. Biol. Chem. **269:** 6587–6591.
15. Gether, U., L. Nilsson, J. A. Lowe III & T. W. Schwartz. 1994. Specific residues at the top of transmembrane segment V and VI of the neurokinin-1 receptor involved in binding of the nonpeptide antagonist CP 96,345. J. Biol. Chem. **269:** 23959–23964.
16. Huang, R. C., H. Yu, C. D. Strader & T. M. Fong. 1994. Interaction of substance P with the second and seventh transmembrane domains of the neurokinin-1 receptor. Biochemistry **33:** 3007–3013.
17. Rosenkilde, M. M., M. Cahir, U. Gether, S. A. Hjorth & T. W. Schwartz. 1994. Mutations along transmembrane segment II of the NK-1 receptor affect substance P competition with non-peptide antagonists but not substance P binding. J. Biol. Chem. **269:** 28160–28164.
18. Hjorth, S. A., H. T. Schambye, W. J. Greenlee & T. W. Schwartz. 1994. Identification of peptide binding residues in the extracellular domains of the AT1 receptor. J. Biol. Chem. **269:** 30953–30959.

19. NODA, K., Y. SAAD, A. KONOSHITA, T. P. BOYLE, R. M. GRAHAM, A. HUSAIN & S. S. KARNIK. 1995. Tetrazole and carboxylate groups of angiotensin receptor antagonists bind to the same subsite by different mechanisms. J. Biol. Chem. **270:** 2284–2289.
20. YAMANO, Y., K. OHYAMA, M. KIKYO, T. SANO, Y. NAKAGOMI, Y. INOUE, N. NAKAMURA, I. MORISHIMA, D. GUO, T. HAMAKUBO & T. INAGAMI. 1995. Mutagenesis and the molecular modeling of the rat angiotensin II receptor (AT_1). J. Biol. Chem. **270:** 14024–14030.
21. FENG, Y., K. NODA, Y. SAAD, X. LIU, A. HUSAIN & S. S. KARNIK. 1995. The docking of Arg^2 of angiotensin II with Asp^{281} of AT_1 receptor is essential for full agonism. J. Biol. Chem. **270:** 12846–12850.
22. BHOGAL, N., D. DONNELLY & J. B. C. FINDLAY. 1994. The ligand binding site of the neurokinin 2 receptor—Site-directed mutagenesis and identification of neurokinin A binding residues in the human neurokinin 2 receptor. J. Biol. Chem. **269:** 27269–27274.
23. HUANG, R. C., P. P. VICARIO, C. D. STRADER & T. M. FONG. 1995. Identification of residues involved in ligand binding to the neurokinin-2 receptor. Biochemistry **34:** 10048–10055.
24. ZOFFMANN, S., U. GETHER & T. W. SCHWARTZ. 1993. Conserved His VI-17 of the NK-1 receptor is involved in binding of non-peptide antagonists but not substance P. FEBS Lett. **336:** 506–510.
25. FONG, T. M., H. YU, M. A. CASCIERI, D. UNDERWOOD, C. J. SWAIN & C. D. STRADER. 1994. The role of histidine 265 in antagonist binding to the neurokinin-1 receptor. J. Biol. Chem. **269:** 2728–2732.
26. WALKER, P., M. MUNOZ, R. MARTINEZ & M. C. PEITSCH. 1994. Acidic residues in extracellular loops of the human Y1 neuropeptide Y receptor are essential for ligand binding. J. Biol. Chem. **269:** 2863–2869.
27. SAUTEL, M., R. MARTINEZ, M. MUNOZ, M. C. PEITSCH, A. G. BECK-SICKINGER & P. WALKER. 1995. Role of a hydrophobic pocket of the human Y1 neuropeptide Y receptor in ligand binding. Mol. Cell. Endocrinol. **112:** 215–222.
28. ROSE, P. M., P. FERNANADES, J. S. LYNCH, S. T. FRAZIER, S. M. FISHER, K. KODUKULA, B. KIENZLE & R. SEETHALA. 1995. Cloning and functional expression of a cDNA encoding a human type 2 neuropeptide Y receptor. J. Biol. Chem. **270:** 22661–22664.
29. LUNDELL, I., A. G. BLOMQVIST, M. M. BERGLUND, D. A. SCHOBER, D. JOHNSON, M. A. STATNICK, D. R. GEHLERT & D. LARHAMMER. 1995. Cloning of a human receptor of the NPY receptor family with high affinity for pancreatic polypeptide and peptide YY. J. Biol. Chem. **270:** 29123–29128.
30. BARD, J. A., M. W. WALKER, T. A. BRANCHEK & R. L. WEINSHANK. 1995. Cloning and functional expression of a human Y4 subtype receptor for pancreatic polypeptide, neuropeptide Y, and peptide YY. J. Biol. Chem. **270:** 26762–26765.
31. GERALD, C., M. W. WALKER, P. J.-J. VAYSSE, C. HE, T. A. BRANCHEK & R. L. WEINSHANK. 1995. Expression cloning and pharmacological characterization of a human hippocampal neuropeptide Y/peptide YY Y2 receptor subtype. J. Biol. Chem. **270:** 26758–26761.
32. CLACKSON, T. & J. A. WELLS. 1995. A hot spot of binding energy in a hormone-receptor interface. Science 383–386.
33. KIVLIGHN, S. D., W. R. HUCKLE, G. J. ZINGARO, R. A. RIVERO, V. J. LOTTI, R. S. L. CHANG, T. W. SCHORN, N. KEVIN, R. G. JOHNSON, JR., W. J. GREENLEE & P. K. S. SIEGL. 1995. Discovery of L-162,313: A nonpeptide that mimics the biological actions of angiotensin II. Am. J. Physiol. **268:** R820–R823.
34. PERLMAN, S., H. T. SCHAMBYE, R. A. RIVERO, W. J. GREENLEE, S. A. HJORTH & T. W. SCHWARTZ. 1995. Non-peptide angiotensin agonist—Functional and molecular interaction with the AT_1 receptor. J. Biol. Chem. **270:** 1493–1496.
35. ARIËNS, E. J. 1954. Affinity and intrinsic activity in the theory of competitive inhibition—Part 1: Problems and theory. Arch. Int. Pharmacodyn. **XCIX:** 32–49.
36. SCHAMBYE, H. T., S. A. HJORTH, D. J. BERGSMA, G. SATHE & T. W. SCHWARTZ. 1994. Differentiation between binding sites for angiotensin II and non-peptide antagonists on the AT1 receptors. Proc. Natl. Acad. Sci. USA **91:** 7046–7050.
37. SCHAMBYE, H. T., B. V. WIJK, S. A. HJORTH, W. WIENEN, M. ENTZEROTH, D. J.

BERGSMA & T. W. SCHWARTZ. 1994. Mutations in transmembrane segment VII of the AT1 receptor differentiate between closely related insurmountable and competitive angiotensin antagonists. Br. J. Pharmacol. **113:** 331–333.

38. EMONDS-ALT, X., P. VILAIN, P. GOULAOUIC, V. PROIETTO, D. V. BROEK, C. ADVENIER, E. NALINE, G. NELIAT, G. LE FUR & J.-C. BRELIÉRE. 1992. A potent and selective non-peptide antagonist of the neurokinin A (NK2) receptor. Life Sci. **50:** 101–106.
39. EMONDS-ALT, X., C. ADVENIER, T. CROCI, L. MANARA, G. NELIAT, M. PONCELET, V. PROIETTO, V. SANTUCCI, P. SOUBRIÉ, D. VAN BROECK, P. VILAIN, G. LE FUR & J.-C. BRELIÉRE. 1993. SR 48968, a neurokinin A (NK_2) receptor antagonist. Regulatory Peptides **46:** 31–36.
40. PORTOGHESE, P. S. 1993. Selective nonpeptide opioid antagonists. Handb. Exp. Pharmacol. **104:** 276–296.
41. STRADER, C. D., M. R. CANDELORE, W. S. HILL, R. A. F. DIXON & I. S. SIGAL. 1989. A single amino acid substitution in the β-adrenergic receptor promotes partial agonist activity from antagonists. J. Biol. Chem. **264:** 16470–16477.

A Genome-based Receptor Nomenclature

PAUL R. HARTIG

DuPont Merck Pharmaceutical Co.
Wilmington, Delaware 19880-0400

GENOME COMPARISONS

At the present time, over 4,000 genes out of an estimated 70,000–100,000 genes in the human genome have been identified.[1] The human genome program should lead to complete sequence information on all 3 billion base pairs of our genome by the year 2005.[1,2] Once completed, this map of all human genes and their corresponding proteins is likely to become the preferred basis set for biomedical research. By reviewing the current picture of protein and receptor properties presented by comparisons of mammalian genomes, we can hope to construct nomenclature rules that will simplify our use of the numerous receptor subtypes, species homologues, and splice variants we have identified. In addition, evaluations of receptor properties in different genomes should help prepare us for genome-based research programs of the future.

Two receptor superfamilies have had the greatest impact on current neurobiological and CNS (Central Nervous System) pharmacological research: the G protein–coupled or 7TM (7 transmembrane segment) receptor superfamily and the superfamily of ligand-gated multisubunit ion channel receptors. Over 400 different clones of 7TM receptors have been described from rodent or human tissues, which represent more than 150 receptor subtypes plus many orthologous genes (species homologues which are homologous in structure and function but have been cloned from a different species). Almost every 7TM receptor described in the rat genome has a counterpart in the human genome (see below for exceptions), as expected from the fact both genomes appear to contain similar numbers of genes.[3] These 7TM receptors are all single subunit proteins with a single ligand binding site, and therefore each receptor subtype is associated with a single gene (usually at one copy per genome). By comparing the characteristics of these clones between mammalian genomes, we can determine whether a gene-based nomenclature system would be practical.

Of the several hundred receptor genes cloned to date, there are only a few examples where a species-specific gene has been found to exist in one mammalian genome but not in others. At least two cases are known where the rat genome contains a functional receptor gene which is absent from or inactivated in the human genome. In one case, there are two very closely related type 1 angiotensin II receptor (AT1) genes in the rat genome, designated AT1a and AT1b, which are located on different chromosomes.[4] The human genome only contains a single AT1 receptor gene.[4] Thus, the AT1 receptor gene appears to have duplicated and diverged in the rate genome sometime after the evolutionary separation between rat and man (some 80 million years ago). In another example, the serotonin 5-HT$_{5B}$ receptor gene appears to have been rendered inactive in the human genome by the introduction of several nonsense mutations (stop codons within the protein coding sequence) which are not found in the rat gene.[5,6] Therefore, the 5-HT$_{5B}$ receptor gene appears to

TABLE 1. Estimates of Genome Sequence Identities

Genome Comparison	All Nucleotides	Protein Coding Regions
Individual humans	99.8%	
Human vs. primate	99%	
Human vs. rat		79%
Mouse vs. rat		93.9%
(see text for references)		

have acquired a "fatal" mutation sometime after the rat-human divergence. In terms of unique human genes, I am not aware of any examples of 7TM receptors which are known to be unique to the human genome, despite our 80 million years of presumed evolutionary superiority over rats.

Expanding our comparison beyond receptor genes leads to additional examples of differences between the rat and human genomes. For example, there is a second insulin gene present as a tandem repeat (a duplicated gene adjacent to the original gene) in the rat genome[7] and humans have evolved a new hemoglobin gene, delta globin, not found in rodents.[8] It is surprising, however, that few differences have been found so far between the set of genes present in rodent vs. human genomes. Furthermore, many known differences (*e.g.*, AT1, 5-HT_{5B}) are simple modifications to genes, which appear to have arisen from the types of processes believed to underly the evolution of genomes. Data from the globin and T cell receptor gene families[9] show that this evolutionary process includes occasional gene duplications, separation of the tandomly duplicated genes (translocation), evolutionary divergence of the duplicated gene due to random mutation, and fixation in the genome of those genes which acquire favorable mutations.[9,10]

Support for this view that 7TM receptor genes are usually highly conserved between mammalian species also comes from broader comparisons of genomes, as shown in TABLE 1. It has been estimated that individual human genomes differ only at an average of one nucleotide position out of each 600 base pairs,[11] producing an average 99.8% genome identity between each of us. It also has been estimated that we share approximately 99% nucleotide sequence identity with our closest living evolutionary relatives, the primates (Wills,[10] p. 219). In contrast, our human genomes differ from rodent genomes at most nucleotide positions, as expected from the estimate that only 3% of the human genome encodes proteins and that much of the remaining sequence serves no known function.[a] During the 80 million years of evolution that separate us from rodents, some 70–80% of such nonfunctional DNA would be expected to change owing to random mutations,[7,12] thus accounting for the low conservation of most noncoding regions between rat and human genomes. Genomes apparently evolve as mosaics, however, with some noncoding regions highly conserved within certain gene families while other gene families show no such conservation.[13]

Within the protein coding regions of homologous rodent and human genes, a high degree of amino acid conservation is found. Based on a comparison of over 300 cloned genes, approximately 80% of the amino acid positions are identical between homologous rodent and human proteins.[14] For the subset of cloned 7TM

[a] Recent evidence has indicated that some of this so-called "junk DNA" may have vital functional roles for genes or for genomes.[1]

receptors, a sample of 11 homologous human and rat monoamine receptor subtypes yielded an average 90% overall amino acid sequence identity and 97% identity in the critical transmembrane domains, which contain the ligand binding site.[15] As a comparison, a diverse group of 363 homologous mouse and rat protein coding sequences were found to encode the same amino acid at 94% of all codons examined.[16] Thus, the individual amino acids of homologous receptor genes are highly conserved between rodent and human genomes, especially in critical ligand binding and functional domains, which greatly simplifies the task of determining which clones represent species homologues of already-characterized genes.

If receptor genes, and indeed most genes, appear to be well conserved among mammalian genomes, how do we account for the many obvious differences between rats and humans, such as the 1,000-fold higher number of cortical synapses found in the human brain?[17] It may be that a few key unique genes (which might be partly found among certain hypervariable loci in our genomes, such as the HLA antigens[18]) along with changes in regulatory regions of genes controlling gene expression (*e.g.*, promoters and enhancers) could be responsible for most of these critical evolutionary differences.[10] Additional differences may come from the 10–20% evolutionary amino acid substitutions between homologous human and rodent proteins. Most of these changes are found at less critical, less constrained positions in the protein (such as the amino terminus of 7TM receptors), where substitutions can be tolerated. Important changes in structure, distribution or function of receptor genes might be introduced by these known genome differences, but our research to date indicates that such differences are rare and do not obscure obvious homology relationships. Thus, we can tentatively conclude that higher mammals, including humans, possess nearly the same basis set of receptor genes. Our receptor nomenclature and our multi-species mammalian receptor research can, therefore, be genome based with the molecular definition of a receptor subtype tied to its corresponding gene.

ONE GENE/ONE SUBTYPE

For the 7TM receptors, the concept that each receptor subtype is associated with a single gene in each mammalian genome offers few difficulties. The ligand-gated ion channel receptors, however, form a heterogeneous set of multisubunit receptors assembled from the products of multiple genes. We can adopt a similar nomenclature strategy for multisubunit receptors by treating them as assemblies of subunits, with each subunit named as a single gene product. For example, at least eight different α subunits and four different β subunits have been identified in the nicotinic cholinergic receptor family.[19] A common molecular subtype of neuronal nicotinic cholinergic receptor is the α_4-β_2 complex, which is commonly known by this name. In this manner, gene cloning can be used to determine the basis set of all possible receptor subunits, and experimentation will determine which of the many possible combinations of subunits are functionally important in living systems. This principle of using cloned receptor and subunit genes to identify receptor subtypes is already in common use and has been previously suggested by others as the appropriate basis for receptor nomenclature.[20,21]

Several issues have arisen from the use of genes to define receptor subtypes. Some receptors contain introns in their protein coding regions, leading to several examples of differential splicing reactions producing multiple receptor products from a single gene. The dopamine D_2 receptor,[22] and the serotonin 5-HT_4 receptor[23] are both differentially spliced to produce two different receptor forms. Furthermore, genes can contain multiple transcription initiation sites or multiple polyadenylation sites, leading to different mRNA and protein products from a single gene. These cases are the exception rather than the rule for currently characterized 7TM receptor genes since it appears that most of these receptor genes do not generate multiple transcripts.[24] These complications can be handled, when necessary, by naming alternate transcripts and their proteins as variants of a single gene's receptor subtype. No single nomenclature is yet in use for this purpose, although simple cases where only two splice variants exist have often been named as short and long variants of the subtype (*e.g.*, dopamine D_{2S} and D_{2L}).

Another problem can arise when a gene is highly polymorphic within and between populations, such that different species may contain different sets of alleles and different alleles may be isolated from individuals of the same species. As was the case with species homologues of receptor genes, these allelic variations generally do not obscure the fundamental identity of the clone since they often occur as small changes to a basic form. As an example, the dopamine D_4 receptor exhibits at least 25 different alleles (haplotypes) in the human population.[25] These polymorphs arose from simple expansions and modifications of a 48 base pair nucleotide element that is imperfectly repeated in tandem arrays within the third intracellular loop coding region of this receptor gene.[25] These expanded repeat elements within the D_4 receptor gene are found in primate but not in rodent genomes. Each primate species appears to have derived its own set of alleles subsequent to its divergence from other species (a founder effect).[26] Owing to the obvious relationship of these allelic variants to a primordial founder gene, these different primate D_4 receptor alleles can be treated as variants of a basic receptor gene, and an appropriate nomenclature devised. Population studies of these different allelic forms are likely to provide important breakthroughs in classifying individuals for disease susceptibility (as shown for the ApoE alleles and Alzheimer's disease[27]), and for classifying patients into subgroups that may show different disease progressions or different responses to various therapeutic options.

PHARMACOLOGICAL SUBTYPES AND GENETIC SUBTYPES

Another issue for a genome-based nomenclature concerns the different definitions of receptor subtypes as viewed from a pharmacological perspective versus a genetic perspective. We could use the term "genetic subtype" to indicate a single molecular form of a receptor and its associated gene, as discussed above. The term "pharmacological subtype" would then be an operational definition for a receptor, based on a particular pattern of responses (electrophysiology, physiology, receptor binding or biochemical responses) to an array of pharmacological compounds. Years of research on the pharmacology of receptors has led to the current set of "pharmacological subtypes" (such as D_2, 5-HT_3, or nicotinic cholinergic) that form the backbone of our present nomenclature. From the perspective of modern molecular neuroscience, it is a great credit to earlier and current generations of classical pharmacologists that their indirect measures of individual receptor responses so often led to correct deductions about the nature of single molecular subtypes, as recently proven by receptor cloning. Thus, pharmacological subtypes should be carefully and smoothly integrated into any molecular nomenclature scheme, since *in vivo* data on a receptor's response properties and pharmacological profile in native tissues is essential to understanding its role in the brain. This has been occurring over the past few years, as indicated by the widespread view that receptor nomenclature should incorporate all aspects of known receptor properties including structural (gene and protein sequence information), transductional (second messenger coupling), and operational (response pattern to a set of pharmacological ligands).[28]

The integration of pharmacological and genetic views of receptor subtypes has not always been straightforward. In the subfamily of serotonin 5-HT_{1D} and 5-HT_{1B} receptors and species homologues, confusing differences between pharmacological and genetic subtypes have only recently been fully resolved. Within this receptor subfamily, different mammalian species show either one (in human) or two (in rat) pharmacological subtypes for the two genetic subtypes known as 5-HT_{1D} and 5-HT_{1B}.[30] This species difference has been shown by several groups to reside in a single amino acid difference in the seventh transmembrane domain of these receptors (reviewed in ref. 29). A similar dramatic change in receptor pharmacology due to a species difference at one amino acid position of the fifth transmembrane domain has been demonstrated for the serotonin 5-HT_{2A} receptor (reviewed in ref. 30). These examples demonstrate that dramatic differences in pharmacological properties can be introduced by single amino acid changes during the course of evolution. Conversely, multiple amino acid differences between genetic subtypes may not lead to any significant differences in their binding and response properties to an array of compounds if all changes are remote from the ligand binding site, as is the case for the human serotonin 5-HT_{1D} and 5-HT_{1B} subtypes (formerly 5-$HT_{1D\alpha}$ and 5-$HT_{1D\beta}$).[29] Full understanding of genetic subtypes, pharmacological subtypes and species homologues of genes is needed to resolve these complexities into the integrated molecular nomenclature schemes we are currently building.

CONCLUSIONS

Based on the issues raised in this discussion, a molecular nomenclature scheme which is aligned with our current understanding of mammalian genomes can be constructed. Organizing principles for such a nomenclature scheme are listed in TABLE 2 and are reviewed below.

Receptor subtypes should be based on cloned receptor genes (one gene, one subtype) with full consideration given to the structural, operational and transductional properties of receptor proteins in choosing names which group them into receptor families and subfamilies. This is not difficult for 7TM receptors since the functional receptor contains only a single subunit from a single gene product, and one ligand binding site per protein monomer. This principle can be extended to multisubunit receptors by naming functional receptor complexes as combinations of individual types from each contributing subunit. Nomenclature conventions

TABLE 2. Principles of a Genome-based Nomenclature

- One gene defines one receptor subtype (or one receptor subunit of a multisubunit receptor)
- Multisubunit receptors are assemblies of individually named subunits
- Multiple transcripts, alleles and protein products of one receptor gene are variants of a gene-based subtype (or subunit)
- Species homologues receive the same receptor subtype name
- Receptor subtype names group receptors into families based on common structural, transductional and operational properties
- Characteristics of the human receptor subtype take priority in selecting names

should be developed which name splice variants from a single gene, and allelic polymorphs as simple variants of a basic gene-based subtype.

Species homologues of a receptor subtype (orthologous genes) should receive the same name. When significant differences exist in the properties of receptors from different mammalian species, authors should be careful to specify which species they are describing (*e.g.*, use hu 5-HT_{1B} to differentiate the human from the rat 5-HT_{1B} subtype). In choosing a name for receptor subtypes which exhibit dramatic species differences, the properties of the human subtypes should determine the nomenclature (human species primacy[29]).

Adoption of a genome-based nomenclature will anchor our classification schemes on the ultimate reference source: the list of all proteins expressed by the human genome. This anchor point should provide a stable molecular reference for all future research and one that is least subject to lab-to-lab variations since any disagreements or errors should be readily correctable. Rapid progress in the sequencing of the human genome, and in directed cloning from multiple mammalian genomes has already provided a basic overview of the nature of our genetic inheritance. By adopting a gene-oriented nomenclature at this time, we can simplify and clarify our current research and be confident that the standards we establish should endure well into the future.

REFERENCES

1. Nowak, R. 1994. Science **263:** 608–610.
2. Carey, J., J. O. Hamilton, J. Flynn & G. Smith. 1995. Business Week, May 8: 72–78.
3. Miklos, G. L. G. & G. R. Rubin. 1996. Cell **86:** 521–529.
4. Szpirer, C., M. Riviere, J. Szpirer, G. Levan, D. F. Guo, N. Iwai & T. Inagami. 1993. J. Hypertension **11(9):** 919–925.
5. Rees, S., I. den Daas, S. Foord, S. Goodson, D. Bull, G. Kilpatrick & M. Lee. 1994. FEBS Lett. **355:** 242–246.
6. Grailhe, R., S. Ramboz, U. Boschert & R. Hen. 1995. Soc. Neurosci. Ab. **21:** 1856.
7. Alberts, B., D. Bray, J. Lewis, M. Raff, K. Roberts & J. D. Watson. 1994. Molecular Biology of the Cell, 3rd edition. Garland Publishing Co. New York. pp. 387–390.
8. Dickerson, R. E. & I. Geis. 1983. Hemoglobin: Structure, Function, Evolution and Pathology. Benjamin-Cummings. Menlo Park, CA.
9. Hood, L., L. Rowen & B. Koop. 1995. Ann. N. Y. Acad. Sci. **758:** 390–412.
10. Wills, C. 1993. The Runaway Brain. Basic Books (division of Harper Collins Publishers). New York.
11. Nickerson, D. 1996. Human Genome News **7(6):** 9.
12. Hartig, P., H.-T. Kao, M. Macchi, N. Adham, J. Zgombick, R. Weinshank & T. Branchek. 1990. Neuropsychopharmacology **3(5/6):** 335–347.
13. Koop, B. F. 1995. Trends Genetics **11(9):** 367–371.
14. Ellison, J. W., X. Li, U. Francke & L. J. Shapiro. 1996. Mammalian Genome **7:** 25–30.
15. Hartig, P. R., T. A. Branchek & R. L. Weinshank. 1992. Trends Pharmacol. Sci. **13:** 152–159.
16. Wolfe, K. H. & P. M. Sharp. 1993. J. Mol. Evolution **37:** 441–456.
17. Colonnier, M. 1981. Electron Microscopy of the Cortex. *In* The Organization of the Cerebral Cortex. F. O. Schmitt, F. G. Worden, G. Adelman & S. G. Dennis, Eds. The MIT Press. Cambridge, MA. p. 127.
18. Hunkapiller, T. & L. Hood. 1989. Adv. Immunol. **44:** 1–63.
19. Sargent, P. B. 1993. Ann. Rev. Neurosci. **16:** 403–443.
20. Kenakin, T. P., R. A. Bond & T. I. Bonner. 1992. Pharmacol. Rev. **44(3):** 351–362.
21. Watson, S. P. & D. Girdlestone. 1995. Trends Pharmacol. Sci. **16:** 15–17.
22. Monsma, F. J., L. D. McVittie, C. R. Gerfen, L. C. Mahan & D. R. Sibley. 1989. Nature **342:** 926–929.

23. GERALD, C., N. ADHAM, H.-T. KAO, M. A. OLSEN, T. M. LAZ, L. E. SCHECHTER, J. A. BARD, P. J. J. VAYSSE, P. R. HARTIG, T. A. BRANCHEK & R. L. WEINSHANK. 1995. EMBO J. **14(12):** 2806–2815.
24. VAN RHEE, A. M. & K. JACOBSON. 1996. Drug Devel. Res. **37:** 1–38.
25. LICHTER, J. B., C. L. BARR, J. L. KENNEDY, H. H. M. VAN TOL, K. K. KIDD & K. J. LIVAK. 1993. Human Mol. Genetics **2(6):** 767–773.
26. LIVAK, K. J., J. ROGERS & J. B. LICHTER. 1995. Proc. Natl. Acad. Sci. USA **92:** 427–431.
27. STRITTMATTER, W. J. & A. D. ROSES. 1995. Proc. Natl. Acad. Sci. USA **92:** 4725–4727.
28. HUMPHREY, P. P. A., P. HARTIG & D. HOYER. 1993. Trends Pharmacol. Sci. **14:** 233–236.
29. HARTIG, P. R., D. HOYER, P. P. A. HUMPHREY & G. R. MARTIN. 1996. Trends Pharmacol. Sci. **17:** 103–105.
30. HARTIG, P. R. 1997. Molecular biology and transductional characteristics of 5-HT receptors. *In* Serotonergic Neurons and 5-HT Receptors in the CNS. H. G. Baumgarten & M. Gothert, Eds. Handbook of Exp. Pharmacol. In press.

Do Recent Operational Studies Indicate That a Single State Model Is No Longer Applicable to G Protein–coupled Receptors?

RICHARD A. BOND

Department of Pharmacological and Pharmaceutical Sciences
College of Pharmacy, University of Houston
4800 Calhoun
Houston, Texas 77204-5515

INTRODUCTION

For the superfamily of guanine nucleotide binding regulatory protein (G protein)–coupled receptors, traditional receptor theory has postulated that agonist binding to the receptor induces a conformational change of the receptor resulting in the agonist-receptor (AR) complex's possessing increased affinity for the G protein. This increase in affinty has been proposed to define a ligand's intrinsic efficacy.[1-3] Unoccupied receptors are believed to be in a single, quiescent state until they are activated by agonist binding. According to this theory, antagonists are believed to bind to the receptor and to prevent agonist binding, but not to alter the (lack of) affinity of the complex for the G protein.

However, recent evidence obtained in functional studies suggests that some unliganded G protein–coupled receptors exhibit constitutive activity, that is, high affinity for the G protein and the ability to signal a cellular response in the absence of hormone (agonist) (see ref. 4 for review). This article concerns the **functional** evidence (involving second messenger or effector mechanisms) obtained using various systems such as transfected cells, untransfected native cells, and transgenic mice. This information suggests the existence of a spontaneously active unliganded receptor. Previous studies using radioligand-binding assays and reconstituted or transfected cell systems that monitored GTPase activity or ligand modification of guanosine-5′-O-(3-thio)triphosphate (GTPγS) binding had already implied the existence of a receptor state that was coupled to the G protein in the absence of ligand.[5-8] However, these studies did not directly show any involvement of an effector mechanism(s). This article will also limit itself to evidence obtained with systems expressing the wild type receptor, as there is indirect evidence for constitutive activity by receptor mutations in the third cytoplasmic loop.[9-11] Besides constitutive activity of receptors, reports have also shown certain compounds (herein termed inverse agonists) to produce decreases in spontaneous baseline activity in various assays that include effector mechanisms (*i.e.*, adenylyl cyclase activity, phosphoinositide (PI) hydrolysis, *etc.*[12-15]).

BIOCHEMICAL EVIDENCE IN TRANSFECTED CELL SYSTEMS AND NATIVE CELL SYSTEMS

In Sf9 cells infected with the human β_2-adrenoceptor using the baculovirus expression system, baseline adenylyl cyclase activity and receptor density were shown to increase proportionately with time of infection, and several compounds previously classified as β-adrenoceptor antagonists were shown to be inverse agonists and to inhibit baseline adenylyl cyclase activity in the apparent absence of agonist stimulation.[12] In an attempt to rule out competition with contaminating endogenous hormone as the cause of the inhibition, these authors presented three lines of evidence. 1) The increase in the baseline and inverse agonist activities of some ligands could be observed even in the absence of serum from the medium. 2) The magnitude of the inverse agonist effect, E_{inv}, was not correlated with the binding affinity of the ligands for the β_2-adrenoceptor, a correlation that would have been predicted if competition with endogenous hormone were the mechanism for inhibition of baseline activity. 3) β-adrenoceptor antagonists that had little effect on baseline activity (herein termed neutral antagonists) were able to block the inhibitory effect of inverse agonists. Again, if competition with endogenous catecholamines were the primary cause of inhibitory effect on baseline activity, then a combination of two "antagonists" should have augmented the response by producing greater competition with the hormone.[12]

Similar results were obtained for the 5-hydroxytryptamine $(5\text{-HT})_{2C}$ receptor expressed in NIH 3T3 fibroblasts, using phosphoinositide (PI) hydrolysis as the functional endpoint. Cells expressing approximately 5 pmol/mg receptor protein showed an increase in baseline inositol monophosphate (IP) formation relative to non-transfected cells, and some ligands previously classified as 5-HT antagonists functioned as inverse agonists and decreased baseline IP formation.[12,15] In these studies, the authors also attempted to rule out competition with endogenous 5-HT as the mechanism for inhibition of baseline activity. They measured the amount of 5-HT in the culture medium by high performance liquid chromatography (HPLC) and concluded that the amount of 5-HT present (0.1 nM) could not solely account for the increase in baseline.[12] Furthermore, they used a neutral antagonist, 2-bromolysergic acid diethylamide (BOL), to block the effects of the inverse agonists, mianserin and mesulergine.[12] Finally, the authors attempted to correlate the inverse agonist activity of the compounds upon IP formation with their ability to produce down-regulation of 5-HT_{2C} receptors in choroid plexus epithelial cells and showed that mianserin, but not BOL produced a decrease in receptor density.[6] However, a more extensive comparison of this type (using nine 5-HT ligands) suggests that a ligand's down-regulation of the 5-HT_{2C} receptor in the choroid plexus and its inverse agonist effect on IP formation may not involve the same mechanism.[15]

Aside from the β_2-adrenoceptor and the 5-HT_{2C} receptor, some evidence for constitutive activity of receptors expressed in cell lines has also been shown for D_1 and D_5 dopamine receptors.[13] In regard to the D_1 and D_5 dopamine receptors, the authors found that despite displaying a very similar pharmacology, as determined by the affinities of various compounds for both receptor subtypes, the receptors were markedly different in their ability to increase baseline signalling in the absence of hormone. The D_5 receptor produced greater increases in basal adenylyl cyclase activity at comparable levels of expression,[15] which suggests that one reason receptor subtypes exist may be to set the "signal to noise ratio" in different tissues.

Constitutively active receptors and compounds functioning as inverse agonists have also been demonstrated in native, untransfected cells such as the β-adrenoceptor of turkey erythrocytes, using adenylyl cyclase activity as the endpoint,[16] and for β-adrenoceptors and muscarinic receptors of cardiac myocytes from various species in which electrophysiologic measurements of Ca^{2+} and K^{+} currents was assessed.[17,18]

EVIDENCE FROM TRANSGENIC MICE OVEREXPRESSING THE HUMAN β_2-ADRENOCEPTOR

Three lines of transgenic mice have been developed with cardiac myocyte-specific overexpression of the human β_2-adrenoceptor.[19] These include lines TG35, TG33, and TG4 which have, respectively, 50, 100, and 200 times the β-adrenoceptor density of control mice.[19]

Biochemical Evidence

Basal adenylyl cyclase activity in cardiac membranes was significantly greater in the TG4 line relative to control mice. The basal measurement in the TG4 line was comparable to the maximal agonist-stimulated (isoproterenol) response in control mice.[19] Evidence of receptor and G protein–coupling ($G_{\alpha s}$ subunit) in the absence of apparent agonist stimulation[20] also suggests a constitutively active conformation of the β_2-adrenoceptor. Following immunoprecipitation of $G_{\alpha s}$ from cardiac membranes of TG4 and control mice, radioligand-binding with the β-adrenoceptor ligand, ^{125}Iodocyanopindolol (^{125}I-CYP) was performed on the immunoprecipitate. The density of β-adrenoceptors in the immunoprecipitate of the TG4 line was approximately 50-fold higher than in the control line.[20]

Isolated Tissue Evidence

Isometric contractility was measured in the isolated, paced, left atria of control and TG4 mice. In a manner similar to the results obtained for adenylyl cyclase activity, baseline contractility in the atria of TG4 mice was significantly elevated compared to controls, and the baseline of the TG4 line was comparable to controls following maximal β-adrenoceptor stimulation with isoproterenol (10 nM).[19] However, unlike the adenylyl cyclase data in which further increases in enzyme activity were observed after isoproterenol administration, the TG4 atria did not respond to isoproterenol. Thus, atrial contractility was maximal in the TG4 line and could not be further enhanced by β-adrenoceptor activation.[19]

In the three lines of transgenic mice (TG35, TG33, and TG4), but not in the control line, the β_2-adrenoceptor ligand, ICI-118,551, produced concentration-dependent inhibition of left atrial tension, and the maximal inhibition correlated with the receptor density, suggested a receptor-mediated event.[21] To again rule out competition with endogenous hormone as the mechanism of inhibition, three lines of evidence were investigated. First, if the inhibition produced by the antagonist were due to displacement of endogenous catecholamines, then the inhibition produced by various antagonists should be equal if the amount of such displacement were equal. However, at variance with expectations of the competition of endogenous hormone explanation, four different β_2-adrenoceptor antagonists all produced varying amounts of inhibition when used at concentrations that were 300 times their respective K_B values for the β_2-adrenoceptor.[21] Second, animals were treated with an intraperitoneal (i.p.) dose of reserpine (0.3 mg/kg 24 h prior to sacrifice), which produced a greater than 98% depletion of cardiac catecholamines as measured by HPLC coupled to electrochemical detection. This treatment with reserpine failed to alter the concentration-response curve (CRC) to ICI-118,551, again indicating that displacement of endogenous catecholamines was not the mechanism of inhibition by ICI-118,551.[21] Third, a neutral antagonist, alprenolol, was used to block the inhibitory effects of ICI-118,551. Alprenolol (100 nM) produced an approximately 30-fold shift to the right of the ICI-118,551 CRC. This shift is appropriate for alprenolol's affinity for the β_2-adrenoceptor and is indicative of a competitive interaction between the two compounds for a single receptor.[21] These data also rule out the possibility that the inhibitory effect of ICI-118,551 is due to a non-specific effect of the drug. If a non-specific effect were the cause, the combination of two blockers should have exacerbated the effect or, at least, left the larger effect intact. Thus, the ICI-118,551 effect was exerted via the β-adrenoceptor but not by competition with endogenous agonist.

In Vivo *Data*

Cardiac catheterization and hemodynamic measurements were performed in anesthetized TG4 and control mice. The maximum first derivative of left ventricular pressure (LV dp/dt_{max}) was used as an index of *in vivo* cardiac contractility. The TG4 mice had significantly higher LV dP/dt_{max} values compared to controls. Again, the inverse agonist ICI-118,551 produced inhibition of this parameter and this inhibitory effect could be blocked by prior administration of the neutral antagonist, alprenolol.[21]

Apparent Violation of the Occupancy-Response Theory

The occupancy-response theory is the cornerstone of classical receptor theory. Introduced in 1933 by A. J. Clark, it mathematically modeled graduated responses by suggesting that the magnitude of a receptor-mediated response was proportional to the number of receptors occupied.[22] Thus, increasing the number of receptors is predicted to increase the potency of ligands and decreasing the amount of receptors (by alkylation for example) is predicted to decrease potency.[23,24] However, in the isolated atria of TG4, mice β-adrenoceptor alkylation with bromoacetyl alprenolol menthane (BAAM) increased the potency of the inverse agonist, ICI-118,551. These results leave the options of either rejecting the occupancy-response theory or of having more than one state of the receptor.

SUMMARY

Recent evidence suggests that unliganded G protein–coupled receptors exist in at least two states, an inactive conformation and an active conformation possessing affinity for the G protein even in the absence of agonist. The data accumulated so far for wild-type receptors imply that this is true for receptors for several hormones and receptor subtypes, and theoretically for all G protein–coupled receptors. The data now consist of studies implicating not only spontaneous receptor-G protein coupling, but also effector mechanisms and, in the case of transgenic mice overexpressing the human β_2-adrenoceptor, physiologic responses at the level of the isolated tissue and *in vivo*. Furthermore, there appear to be ligands (inverse agonists) that can decrease the level of the constitutively active conformation of the receptor, and neutral antagonists can not only block classical agonist responses, but also inhibit the response of inverse agonists.

REFERENCES

1. KENAKIN, T. P. 1988. Are receptors promiscuous? Intrinsic efficacy as a transduction phenomenon. Life Sci. **43:** 1095–1101.
2. LEFF, P. & D. HARPER. 1989. Do pharmacological methods for the quantification of agonists work when the ternary complex mechanism operates? J. Theor. Biol. **140:** 381–397.
3. MACKAY, D. 1988. Continuous variation of agonist affinity constants. Trends Pharmacol. Sci. **9:** 156–157.
4. MILLIGAN, G., R. A. BOND & M. LEE. 1995. Inverse agonism: Pharmacological curiosity or potential therapeutic strategy? Trends Pharmacol. Sci. **16:** 10–13.
5. COSTA, T. & A. HERZ. 1989. Antagonists with negative intrinsic activity at d opioid receptors coupled to GTP-binding proteins. Proc. Natl. Acad. Sci. USA **86:** 7321–7325.
6. COSTA, T., J. LANG, C. GLESS & A. HERZ. 1990. Spontaneous association between opioid receptors and GTP-binding regulatory proteins in native membranes: Specific regulation by antagonists and sodium ions. Mol. Pharmacol. **37:** 383–394.
7. DELEAN, A., J. M. STADEL & R. J. LEFKOWITZ. 1980. A ternary complex model explains the agonist-specific binding properties of the adenylate cyclase-coupled β-adrenergic receptor. J. Biol. Chem. **255:** 7108–7117.
8. WANG-NI, T., E. DUZIC, S. M. LANIER & R. C. DETH. 1994. Determinants of α_2-adrenergic receptor activation of G proteins: Evidence for a precoupled receptor/G protein state. Mol. Pharmacol. **45:** 524–531.
9. SAMAMA, P., S. COTECCHIA, T. COSTA & R. J. LEFKOWITZ. 1993. A mutation-induced activated state of the β_2-adrenergic receptor. J. Biol. Chem. **268:** 4625–4636.
10. SAMAMA, P., G. PEI, T. COSTA, S. COTECCHIA & R. J. LEFKOWITZ. 1994. Negative antagonists promote an inactive conformation of the β_2-adrenergic receptor. Mol. Pharmacol. **45:** 390–394.
11. LEFKOWITZ, R. J., S. COTECCHIA, P. SAMAMA & T. COSTA. 1993. Constitutive activity of receptors coupled to guanine nucleotide regulatory proteins. Trends Pharmacol. Sci. **14:** 303–307.
12. CHIDIAC, P., T. E. HEBERG, M. VALIQUETTE, M. DENNIS & M. BOUVIER. 1994. Inverse agonist activity of β-adrenergic antagonists. Mol. Pharmacol. **45:** 490–499.
13. TIBERL, M. & M. G. CARON. 1994. High agonist-independent activity is a distinguishing feature of the dopamine D1B receptor subtype. J. Biol. Chem. **269:** 27925–27931.
14. BARKER, E. L., R. S. WESTPHAL, D. SCHMIDT & E. SANDERS-BUSH. 1994. Constitutively active 5-hydroxytryptamine$_{2C}$ receptors reveal novel inverse agonist activity of receptor ligands. J. Biol. Chem. **269:** 11687–11690.
15. LABRECQUE, J., A. FARGIN, M. BOUVIER, P. CHIDIAC & M. DENNIS. 1995. Serotonergic antagonists differentially inhibit spontaneous activity and decrease ligand binding capacity of the rat 5-hydroxytryptamine type 2C receptor in Sf9 cells. Mol. Pharmacol. **48:** 150–159.
16. GOTZE, K. & K. H. JAKOBS. 1994. Unoccupied β-adrenoceptor-induced adenylyl cyclase stimulation in turkey erythrocyte membranes. Eur. J. Pharmacol. **268:** 151–158.
17. MEWES, T., S. DUTZ, U. RAVENS & K. H. JAKOBS. 1993. Activation of calcium currents in cardiac myocyte by empty β-adrenergic receptors. Circulation **88:** 2916–2922.
18. HANF, R., Y. LI, G. SZABO & R. FISCHMEISTER. 1993. Agonist-independent effects of muscarinic antagonists on Ca2+ and K+ currents in frog and rat cardiac cells. J. Physiol. (Lond.) **461:** 743–765.
19. MILANO, C. A., L. F. ALLEN, P. DOLBER, H. ROCKMAN, K. R. CHIEN, T. D. JOHNSON, R. A. BOND & R. J. LEFKOWITZ. 1994. Enhanced myocardial function in transgenic

mice with cardiac overexpression of the human β_2-adrenergic receptor. Science **264:** 582–586.
20. GURDAL, H. & M. JOHNSON. 1995. Personal communication.
21. BOND, R. A., T. D. JOHNSON, C. A. MILANO, P. LEFF, H. A. ROCKMAN, T. MCMINN, S. APPARSUNDURAM, M. F. HYEK, T. P. KENAKIN, L. F. ALLEN & R. J. LEFKOWITZ. 1995. Physiologic effects of inverse agonists in transgenic mice with myocardial overexpression of the beta$_2$-adrenoceptor. Nature **374:** 272–276.
22. CLARK, A. J. 1933. The Mode of Action of Drugs on Cells. London.
23. KENAKIN, T. P. 1989. Challenges for receptor theory as a tool for drug and drug receptor interactions. Trends Pharmacol. Sci. **10:** 18–22.
24. MACEWAN, D. J., G. D. KIM & G. MILLIGAN. 1995. Analysis of the role of receptor number in defining the intrinsic activity and potency of partial agonists in neuroblastoma × glioma hybrid NG108-15 cells transfected to express differing levels of the human β_2-adrenoceptor. Mol. Pharmacol. **48:** 316–325.

Agonist Efficacy and Allosteric Models of Receptor Action[a]

H. ONGUN ONARAN[b] AND TOMMASO COSTA[c,d]

[b]*Department of Pharmacology*
Ankara University Faculty of Medicine
Sihhye. Ankara, Turkey

[c]*Department of Pharmacology*
Istituto Superiore di Sanità
Rome, Italy

INTRODUCTION

Theoretical modeling of receptor action can be defined as the problem of finding the connecting rule between the strength of a biological signal to the receptor and output response. Finding a rule means to understand a general underlying principle, which evidently is a more ambitious task than merely explain a particular set of experimental observations.

Many classical models of receptor action were built robust enough to be applicable to any receptor for which a biological response can be measured.[1,2] In fact they depend neither on intramolecular events nor on the biochemical identity of the signal transduction steps that are taking place. The receptor with its intramolecular world are left inside a "black box"[2] which is not the subject of direct experimental manipulation.

However, progress in gene cloning and protein engineering directs new questions to theories of receptor action. Molecular biology entices us to open that "black box" in which the receptor macromolecule has been standing thus far, and this seems to shift the problem of receptor modeling from that of finding the linking relation between ligand's molecular structure and biological response, to that of seeking the mechanistic dependence between ligand-recognition and functional domains within a single macromolecule of receptor.

Yet, experimental and conceptual problems remain unsolved. Experimentally, there is a painful discrepancy between the exactness of the changes site-directed mutagenesis can produce in the receptor molecule, and the lack of resolution by which the result of those changes can be evaluated. In fact, like most integral membrane proteins, receptors need to be examined in the native environment of the plasma membrane following transfection into cells.

[a] This work is supported in part by the following grants: Turkish Scientific and Technical Research Council, SBAG-1203, to H.O.O., and CEE Biotechnology project BIO2CT-930083 (Euroceptor) to T.C.

[d] Address correspondence to: Dr. Tommaso Costa, Laboratory of Pharmacology, Istituto Superiore di Sanità, Viale Regina Elena, 299, 00161 Rome, Italy; Tel: 0039-6-49902386; Fax: 0039-6-49387104; E-mail: (tomcosta @net.iss.it).

Conceptually, the theoretical formalism apt to a quantitative description of the intramolecular environment of a protein—such as the partition functions of statistical mechanics, or the computations of molecular dynamics, for example—is far too complex for the definition of simple operational parameters, such as agonist affinity and efficacy. Thus, the more detailed the attempt to describe microscopically the physical properties of the receptor macromolecule, the steepest the fall into a status of total standstill from the experimental point of view.

AGONIST EFFICACY AS A LIGAND-INDUCED ALLOSTERIC EFFECT

Most quantitative models of receptor action revolve around two fundamental quantities, affinity and efficacy, both of which depend on the molecular interaction between ligand and receptor.[1,2] Affinity is often thought of as the way to express the strength of the bonds holding together ligand and receptor into a bimolecular complex. But, operationally, it can only be defined by the mass action law notion of equilibrium association constant. Efficacy, instead, refers to the functional consequences of the binding process.[1–4] At the molecular level, denotes the magnitude of the change the receptor undergoes as a result of the agonist binding process. Unlike affinity, agonist's efficacy was not defined as chemical constant in the classical model of receptor action. It appears, there, as a numerical multiplier of the bound receptor concentration, needed to define the magnitude of the stimulus induced by the agonist.[3,4] This is concordant with the freedom from molecular details the model was built upon. Exhausting experimental evidence gained from the study of structural analogues of ligands, indicates that affinity and efficacy can be manipulated independently. Which means that if we only consider the structure of the ligand (leaving that of the receptor inside the "black box"), we find affinity and efficacy encoded through different sets of topochemical elements into the molecule of the hormone, although both apparently depend on an identical macroscopic interface of interaction. It is not known, however, whether such lack of correlation would still be maintained when examining the structure of the receptor. However, as we set as target the discrimination of affinity from efficacy among the structural epitopes of the receptor, at once we discover the need to define both quantities with a comparable chemical scale.

At the molecular level the notion of agonist efficacy is identical to that of ligand-induced allosteric effects in biology.[5] In fact, the idea that agonist binding to a recognition domain results in a change of a spatially distinct functional zone of the receptor, is exactly alike that of an enzymatic modulator which can alter catalytic properties by binding to a site other than that for the substrate. Therefore, the receptor models proposed to describe the chemical nature of agonist efficacy are direct translations of those relating ligand binding to intramolecular changes in protein chemistry. Two are the principal underlying concepts: the allosteric transition of Wyman,[6,7] and the free energy coupling of Weber.[8] Although these two concepts start from different views of the macromolecule, both are defined according to the macroscopic principles of thermodynamics. Therefore, they do not address the intramolecular "inside" of the protein directly, but only tell which relations it imposes on the external macroscopic world of the reacting partners. Yet, as we draw mechanistic inferences from receptor models derived from those concepts, we reach divergent conclusions on the operational significance of affinity and efficacy. We shall illustrate this point examining receptor models grouped according to the two concepts from which they originate.

THE CONCEPT OF ALLOSTERIC TRANSITIONS AND THE TWO-STATE ALLOSTERIC RECEPTOR MODEL

The idea was originally proposed to explain how ligands can affect the open or closed state of a ion channel,[9] but the first thermodynamic description was provided by Monod, Wyman and Changeaux.[6] Many authors have then extensively illustrated its pharmacological implications.[9–13] The core idea is that in order to explain how a ligand binding process can result in a functional change of a macromolecule—the receptor in our case—we can think of the protein as existing in equilibrium between two conformationally and also functionally different tautomeric states. This can be altered by ligand binding according to mass action law. The reaction scheme drawn in FIGURE 1 explicitly shows the linkage between the second-order ligand binding process and the first-order receptor isomerization event. The two receptor states R and R* have different functional properties. If a channel, for instance, R and R* may indicate two different conducting states; if an enzyme, two different catalytic rates. The equilibrium between R and R* is given by the stability constant J, while the equilibrium binding constant K indicates the affinity of the ligand H for the "less active" form R. The third constant, β, is necessary to describe thermodynamically the system. It indicates the energetic linkage that the existence of two interconverting

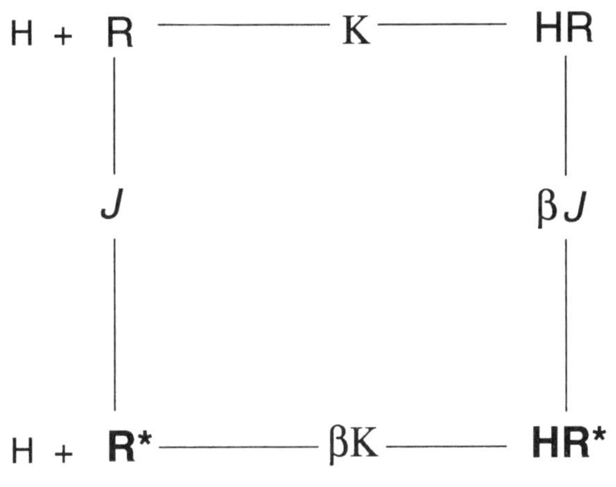

$$J = \frac{[R^*]}{[R]}; \quad K = \frac{[HR]}{[H][R]}; \quad \beta J = \frac{[HR^*]}{[HR]}; \quad \beta K = \frac{[HR^*]}{[H][R^*]}$$

$$\beta = \frac{[HR^*][R]}{[HR][R^*]}$$

FIGURE 1. The two-state allosteric receptor model. The scheme describes the equilibria between a receptor existing in two tautomeric forms R and **R*** reacting with a ligand H. The connecting lines between species indicate reversible reaction paths. The associated constants are either first-order stability constants (J and βJ) or second-order association constants (K and βK), as indicated by the corresponding mass equations.

conformational states imposes on the macromolecule's reactivity with an external ligand. Whatever effect the ligand binding process has on the interconversion between the two receptor conformers, it must be energetically identical to that the isomerization has on ligand binding. Although dimensionless, β is a true constant according to mass-law principles. In fact, from the mass relations shown in FIGURE 1, it is evident that β is the equilibrium constant for the reaction:

$$[HR] + [R^*] \rightarrow [HR^*] + [R]$$

Hence, it tells how the ligand-bound form is distributed at equilibrium between the two receptor conformers. From the ligand side, we see β as a measure of how well the ligand "selects" the different allosteric states (K vs. βK). From the receptor side we think of β as the extent to which the functionally relevant state is "induced" by the ligand (J vs. βJ).

The identity of β as definition of ligand's efficacy is then clear. If $\beta > 1$, the ligand promotes (or selects) the functional (R*) receptor form, i.e. it's an agonist. It does the opposite if $\beta < 1$, nothing, if $\beta = 1$.

The interesting question is whether affinity and efficacy can be measured independently. The apparent affinity observed in a binding experiment is:

$$K_{app} = \frac{[HR] + [HR^*]}{[H][R] + [H][R^*]} = \frac{[HR] + \beta J[HR]}{[H]([R] + J[R])} = \frac{[HR]}{[H][R]} \cdot \frac{(1 + \beta J)}{(1 + J)} = K\,\frac{(1 + \beta J)}{(1 + J)} \quad \textbf{(1)}$$

thus, it always depends on both β and J, and cannot be separated from the first-order conversion of the receptor into the active form. In other words, this model links inextricably the quantities binding affinity and efficacy, unless we postulate as pharmacologically meaningful to define "true" or "reference" the affinity K for the "inactive" state R of the receptor. In which case, there is no chance to measure it directly. If, however, the fraction of receptor in R* state could be experimentally measured, (say, the fraction of open and closed forms of a channel), by comparing that in the absence and presence of saturating ligand, J and β can be measured, and the "reference" K computed from apparent affinity. In fact, the proportion F* of receptor in R* form is:

$$F^* = \frac{[R^*] + [HR^*]}{[R] + [HR]} = J\,\frac{(1 + \beta K[H])}{(1 + K[H])} \quad \textbf{(2)}$$

hence, $F^* = J$ when $[H] = 0$, and $F^* = \beta J$ when $[H] \rightarrow \infty$.

FIGURE 2 shows the asymmetric nature (implicit in eq. 2) of the relation between the ligand-dependent isomerization of the receptor (*i.e.*, the ratio F*/J) and the normalized concentration of ligand (*i.e.*, K[H]), plotted at different values of β. In case of a "strong" agonist ($\beta = 100$), it is sufficient to increase [H] 10-fold over its reference dissociation constant for the "closed" state to obtain a full shift of receptors into the active form. But that constant ought to be exceeded three orders of magnitude to drive receptors to an equal extent into the inactive form, using a comparably "strong" antagonist ($\beta = 0.01$). Obviously, asymmetry simply results here from our insistence in considering relevant the parameter K, which is an incomplete descriptor of the binding process outlined by this system.

WEBER'S FREE ENERGY COUPLING AND THE TERNARY COMPLEX MODEL

Many receptors lack the ability to generate intrinsic biochemical signals. Following agonist binding, in fact, G protein linked receptors, bind and activate a transducer

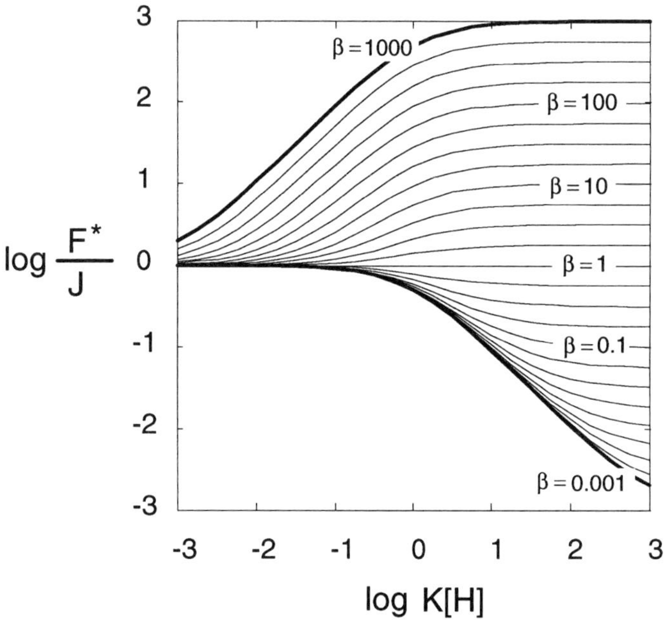

FIGURE 2. Relation between apparent fraction of ligand-induced receptor in R* form and normalized concentration of the ligand. *Ordinate*: Decimal logarithm of the ratio between the molar fraction of receptor in R* form ([R*] + [HR*]/[R] + HR]) and the stability constant J (i.e. [R*]/[R]). This ratio represents the net shift in receptor isomerization due to the ligand. *Abscissae*: Decimal logarithm of the product between ligand concentration and its affinity for the R state, *i.e.*, the normalized concentration of the ligand. Each curve corresponds to a different value of β, which was varied by 0.25 log steps within the indicated range.

regulatory protein to transfer the signal onto second-messenger generating effector molecules. In this case, the relevant cooperative effect underlying agonist efficacy occurs between two ligands (the agonist and the G protein) binding to distinct sites of the receptor, rather than between agonist binding and an intrinsic biochemical task executed by the receptor itself. The model best describing such a situation was named ternary complex model to emphasize the importance of the trimolecular complex, agonist, receptor and G protein, in the process of signal transfer. Initially applied to receptors to explain the anomalous binding isotherms of agonists and the effects of GTP,[14] the pharmacological implications of the model have been subsequently studied by many authors.[2,15–19] However, the original idea and thermodynamic treatment is from Weber[8,20] and constitutes a special case of the more general linkage relations of Wyman.[21] The composition of the chemical species at equilibrium (FIG. 3), highlights three relevant equilibrium constants. Two of these, M and K, the association constants of, respectively, G and H for the receptor, are called by Weber "unconditional", because they describe reactions that do not depend on the condition of a second ligand being bound to the receptor, as, instead, αM and αK do. The constant α states the first-principle derived notion that the overall free energy change for the transition from all-free species to the ternary bound one is independent of the path of the reaction. Thus, K × αM must be equal to M × αK, or, equivalently, the difference in free energy between each conditional and unconditional reaction path for the two ligands must be identical. This invariant quota of energy change of the reaction was named free energy-coupling,[8,20] since it measures the degree of coupling that the macromolecule itself imposes on the two macroscopic binding events. Put in common jargon, it gauges the "allosteric" effect between two cooperative binding sites.

Also in this case, α describes an equilibrium reaction:

$$[HR] + [RG] \rightarrow [HRG] + [R]$$

stating to which extent we find hormone and G protein at equilibrium either associated with two different molecules of receptor, or both bound to an identical one. $\alpha > 1$ means that ligand H (and reciprocally ligand G) "promotes" or "stabilizes" the ternary species, the reverse when $\alpha < 1$, random binding when $\alpha = 1$. Thus, α is in fact a chemical definition of hormonal (or drug) efficacy since "agonism" results from promotion of receptor binding to the G protein transducer. The important question, again, is under which experimental protocols can affinity and efficacy be measured. Unlike the two-state model, the ternary complex model tells that there are two affinities to be concerned with, that for H, which is generally accessible to experimental investigation, and that for G, which usually it is not so. But, given the reciprocal nature of the free energy change related to α, the model

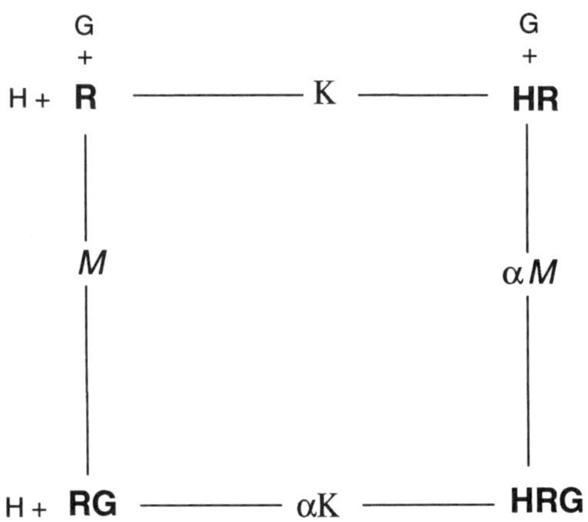

$$K = \frac{[HR]}{[H][R]}; \quad M = \frac{[RG]}{[G][R]}; \quad \alpha K = \frac{[HRG]}{[H][RG]}; \quad \alpha M = \frac{[HRG]}{[G][HR]};$$

$$\alpha = \frac{[HRG][R]}{[HR][RG]}$$

FIGURE 3. The ternary complex model. Reacting species are ligand H, G protein, G, and the receptor R. Connecting lines indicate reversible association reaction at equilibrium and the corresponding constants are given by the mass equations.

indicates that K and α can be measured even when M is unknown, if the concentration of G can be sequentially changed between two extremes: close to zero and close to saturation of its receptor binding site. In fact, the observed binding affinity of the ligand H is given by:

$$K_{app} = \frac{[HR] + [HRG]}{[H][R] + [H][RG]} = \frac{[HR] + \alpha M[HR][G]}{[H][R](1 + M[G])}$$

$$= \frac{[HR]}{[H][R]} \cdot \frac{(1 + \alpha M[G])}{(1 + M[G])} = K \frac{(1 + \alpha M[G])}{(1 + M[G])} \qquad \textbf{(3)}$$

and, $K_{app} = K$, when $[G] \approx 0$, while $K_{app} = \alpha K$ when $[G] \rightarrow \infty$. Therefore the shift in apparent ligand affinity from ligand binding isotherms determined at both saturating and infinitesimal concentrations of G protein (*i.e.*, $\alpha K/K$), directly measures α.

Once again, we must point to the relation in equation (3) between G protein-induced affinity shift (K_{app}/K) and normalized concentration of G protein (M[G]), which is formally identical to that between ligand-dependent receptor activation and concentration of ligand in equation 2. (To view that, go to FIGURE 2, and just replace α for β, in the captions, K_{app}/K for F*/J in ordinate, and M[G] for K[H] in abscissa.) Here we find that for a "strong" agonist it takes little excess of G over its dissociation constant to reach a maximal shift, but for an equally strong antagonist the required concentration of G protein may well be beyond the realms of experimental manipulation. Which has two implications. First, it is very difficult using binding experiments to detect ligands having the property $\alpha < 1$ (we may call them antagonists with "negative efficacy", for they exert a negative cooperative effect on G binding[19]). Second, if we have faith in the idea that the addition of GTP to the system (or pertussis toxin modification of the G protein, where it applies), is functionally equivalent to drive either M or [G] towards zero, then the "GTP shifts" of agonist binding isotherms experimentally measured are a good approximation of the magnitude of α. Such shifts (ranging anywhere between 20–100-fold), correspond to free energy changes of 3–5 *RT* units and are thus only a fraction of the free energy changes of 15–20 and even larger magnitudes associated with binding affinities. As Weber has shown,[20] there is a logistic relation between the magnitude of free energy coupling and the proportion of protein in ternary-bound form, so that it takes only −4.6 *RT* units of free energy change to promote more than 95% double occupancy of the receptor by both ligands, but the free energy change required for full coupling is infinite. It is therefore interesting that the probable magnitude of agonist efficacy as deduced by considerations of recorded GTP shift, is consistent with the upper limits predicted by the thermodynamic principles of the model.

DIFFERENCES BETWEEN THE TWO-STATE AND THE TERNARY COMPLEX MODEL

Although we have drawn the two models in a way that is both chemically and mathematically similar, there are obvious differences.

1. The ternary complex model does not explicitly recognize distinct "allosteric states" of the receptor, it only accounts for the existence of a "cooperative effect". Free energy transfer between separated areas of the protein macromolecule is the

only requirement to account for the existence of free energy coupling. As Weber puts it, more than a conformational change, it is the tendency of a protein to resist such change when bound by multiple ligands that may best explain the existence of cooperativity. The two-state allosteric model, instead, separates the receptor into two structurally different conformers, to which different functions are attributed, and thus inevitably links any binding process to any functional state associated with the two receptor forms.

2. As a consequence, the two models differ in the operational dissection of affinity and efficacy. In the two-state model these are inseparable entities: the proportion of allosteric forms changes as the receptor is saturated with ligand. Even if we elect as reference affinity that for only one of those forms, by conceptually "freezing" their transition, such affinity remains a virtual quantity, which can be only calculated, but never measured. In contrast, the ternary complex model identifies agonist affinity as an experimental parameter that can be easily measured using a solubilized receptor separated from the G protein. The efficacy β of the two-state model only depends on hormone and receptor. The α of the ternary complex depends on at least three molecules, hormone receptor and G protein. Therefore, given the degree of promiscuity for the interactions between seven transmembrane receptors and G proteins, we can only define efficacy for a specified pair of receptor and G protein subtypes, and may expect to find it different when either partner is exchanged.

In essence, the two ideas differ for the choice of where indetermination is brought to the concepts of affinity and efficacy. The allosteric transition dissolves affinity but leaves efficacy intact. Free energy coupling, rescues affinity, but defines efficacy as a relative quantity, which can only be defined according to the particular repertoire of reacting partners for the receptor under study. However, these differences only reflect the diverse angles of view by which we try to express the phenomenological principle that the microsystem ⟨protein⟩ influences the relations among the components of the macrosystem in which it is located. When we suppose an allosteric transition, our thinking is dominated by the experimental observation that the protein apparently undergoes discontinuous shuttling among different functional states. Consequently we impose arbitrary boundaries within the conformational space of the protein by generating "aggregate" forms to which we assign different activities, but also, inevitably, different chemical potentials. When, instead, we invoke free energy coupling, the cooperative effects among binding ligands apparently transmitted by the protein molecule itself, is the dominant phenomenon we wish to explain. The net energetic changes involved in such effects are so small compared to those of the binding processes from which they originate, that we cannot avoid to think of them as resulting from continuous fluctuations of the balance of weak non-covalent interactions among the internal residues of the protein itself. Thus, instead of imposing artificial breaking points within the conformational space of the macromolecule, we accept it in the entire complexity of its stochastic continuum, as Weber taught us to do with the notion of free energy coupling.

The limitations of each of the two concepts are also clear: the allosteric transition idea forces us to fractionate the conformational space of the receptor into an endless number of states as we add more biological functions and more ligands, and this inevitably results in the deceptive vision of the receptor protein as a mechanical object "clicking" into different shapes depending on "key" switches hidden among the many residues of its sequence. The free energy coupling idea leaves us speechless on try to account how can the change of functional status be related to any individual binding event.

We shall show next that these limitations do not complement each other when these two concepts are included into the same model.

TWO-STATE ALLOSTERIC TERNARY COMPLEX MODEL

The ternary complex model does not establish any explicit relation between ligand binding and structural change of the receptor protein. Whether the single ligand binding process is true second-order, or also includes first-order contributions of the receptor, is neither excluded nor specified. As discussed before for the allosteric model, such distinction is irrelevant from the experimental point of view. Thus, free energy coupling only addresses that contribution of receptor internal equilibria which can be experimentally observed: i.e., the reciprocal effect of each ligand on the binding of the other. In other words, since there is no intrinsic receptor functional status to be concerned with, any effect the binding of one ligand may have on receptor structure, is there to be seen only when another ligand enters the scene.

A problem arises, however, if we imagine to produce a mutation of the receptor structure which interferes directly or indirectly with the molecular mechanisms involved in free energy coupling. The resulting mutant receptor would show changes in all the parameters of the model, but the model itself could not explain the origin of the observed changes. Experimental evidence of this kind was obtained by Kjelsberg *et al.*[22] who showed that a point mutation in the α_1-adrenergic receptor induced constitutive receptor activity. Ligand-independent receptor activation was induced—albeit to a different extent—by replacement of virtually any aminoacid other than the natural one, suggesting that that point of the molecule is critical in preventing the receptor from interacting spontaneously with the G protein. Further studies extended to similarly engineered members of the adrenergic family,[23] revealed that the behavior of such mutants cannot be explained by an increase of the unconditional affinity constant M of the receptor for the G protein only. In fact, the mutation also increased K in a manner correlated to agonist efficacy. Such a correlation cannot be predicted by the model, and suggests that the mutation facilitates that agonist-dependent change of receptor structure which favors the binding of the G protein. To make explicit an agonist-dependent effect on receptor conformation, independently of the presence of G in the ternary complex model, we may merge the concept of allosteric transition with that of free energy coupling (FIG. 4). The resulting two-state ternary complex model[23] describes an allosteric transition between two receptor conformers in the presence of two cooperative binding events. This splits the ternary species into two kinds, with free energy coupling relations that are linked by the isomerization process. Even if we "force" the model to a less complex formulation (as in fact it was originally done[23]), by assuming—for example—that only one of the two conformers offers a binding site to the second ligand G (*i.e.*, disregard the grey lines and associated constants in FIG. 4), the degree of operational indetermination that the allosteric transition injects into the ternary complex model remains. In fact agonist efficacy is described by two distinct constants α and β, and agonist affinity is now linked to the receptor structural change as seen for the two-state model.

Therefore, two-state ternary complex model is not a "better" model than the ternary complex model itself for the description of G protein-linked receptor action. Instead, it is a more verbose version, which we may be forced to adopt only if an irreversible change (*e.g.*, by protein engineering) of the receptor itself, or, alternatively, the inclusion in our experimental determination of a molecular effect intrinsically present within the receptor, asks to do so. Otherwise, the receptor behavior predicted by the two models is identical. In fact the parameters of the ternary complex model can always be expressed in terms of its two-state

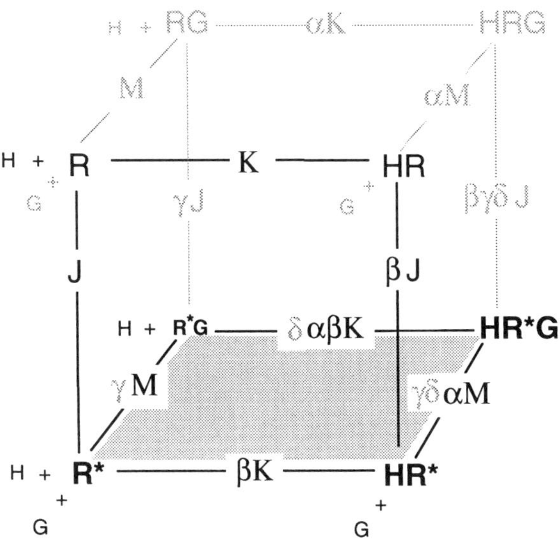

$$\gamma M = \frac{[R^*G]}{[R^*][G]};\quad \gamma J = \frac{[R^*G]}{[RG]};\quad \alpha\gamma\delta M = \frac{[HR^*G]}{[HR^*][G]};\quad \beta\gamma\delta M = \frac{[HR^*G]}{[HRG]};\quad \alpha\beta\delta K = \frac{[HR^*G]}{[H][R^*G]};$$

$$\delta = \frac{[HR^*G][R^*]}{[HR^*][GR^*]} \bullet \frac{[HR][RG]}{[HRG][R]}$$

FIGURE 4. Two-state ternary complex model. The equilibrium constants K, M, J, α and β are defined as in FIGURES 1 and 3. The others are given in the mass equations above. Note that δ is a "second order" free energy coupling constant stating the reciprocity of the three chemical events occurring in the receptor (two ligand binding processes and the isomerization). It indicates the overall effect that the interconversion has on the simultaneous occupation by the two ligands, or, equivalently, how the binding of both ligands affects the interconversion. As shown by the mass equation above, δ is a ratio of the free energy couplings of two ternary complex cycles: that for R* and that for R (bottom and top side, respectively, of the cube above). The dark paths and associated species indicate how the model simplifies if we assume that there is no binding of the ligand G to the R form of the receptor.

extension (FIG. 5):

$$M_{tcm} = M\left(\frac{J}{1+J}\right); K_{tcm} = K\left(\frac{1+\beta J}{1+J}\right); \alpha_{tcm} = \alpha\beta\left(\frac{1+J}{1+\beta J}\right) \tag{4}$$

These equations clearly illustrate how the presence of a conformational transition within the receptor (J), impose a correlation that ties together the independent parameters of the ternary complex model.

Two-state ternary complex model　　　　Ternary Complex Model

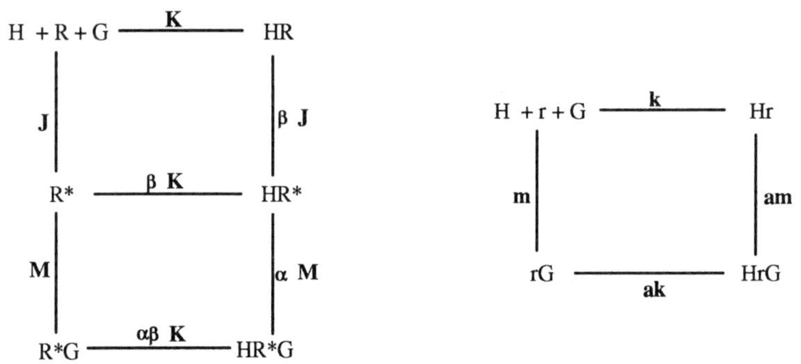

$[r]=[R]+[R^*]$ and $[Hr]=[HR]+[HR^*]$　　　　$(r_t=R_t)$

with the boundary conditions of reversible interconvertibility:

$$[R]=\frac{1}{J}[R^*] \qquad [HR]=\frac{1}{\beta J}[HR^*].$$

- $$k=\frac{[Hr]}{[H][r]}=\frac{[(HR)+(HR^*)]}{[H][(R)+(R^*)]}=\frac{[HR^*]}{[H][R^*]}\cdot\frac{1+\beta J}{\beta(1+J)}=K\left(\frac{1+\beta J}{1+J}\right),$$
- $$m=\frac{[rG]}{[r][G]}=\frac{[R^*G]}{[(R)+(R^*)][G]}=\frac{[R^*G]}{[R^*][G]}\cdot\frac{J}{1+J}=M\left(\frac{J}{1+J}\right),$$
- $$ak=\frac{[HrG]}{[H][rG]}=\frac{[HR^*G]}{[H][R^*G]}=\alpha\beta K,$$
- $$am=\frac{[HrG]}{[G][Hr]}=\frac{[HR^*G]}{[G][(HR)+(HR^*)]}=\frac{[HR^*G]}{[G][HR^*]}\cdot\frac{\beta J}{1+\beta J}=\alpha M\left(\frac{\beta J}{1+\beta J}\right),$$
- $$a=\frac{ak}{k}=\frac{am}{m}=\alpha\beta\left(\frac{1+J}{1+\beta J}\right)$$

FIGURE 5. Relation between parameters of the ternary complex model and its two-state extension. It is assumed that the receptor r contains two states R and R*. The equations calculate how the implicit presence of two states affects the apparently independent parameters of the ternary complex model when the receptor is observed as a single species r. Parameters of the ternary complex are given by small letters, those for the two-state extension in capitals or greek letters. The equations allow calculation of how each parameter of the ternary complex model can be expressed in term of the parameters of the corresponding two-state modification. Note that the two state extension impose a correlation on all the parameters of the ternary complex model.

A UNITARY PROBABILISTIC VIEW OF ALLOSTERIC TRANSITION AND FREE ENERGY COUPLING

Do we always need different theoretical frameworks to conceptualize efficacy defined by free energy coupling or by the idea of allosteric transition? To answer this question, we shall first attempt to dissect the concept of receptor state into its ultimate microscopic components. Let's consider a protein existing in a very large number of interconvertible microscopic states, and define as conformational space of that protein the universe of all *possible* microstates that can be theoretically achieved without imposing an irreversible change. Each microstate is associated with a probability of occurrence in equilibrium, which obviously is a function of environmental variables and the internal configuration of the protein. For a single molecule, this probability shows the fraction of time that the protein spends in a particular microscopic state. But, for an ensemble of molecules, the ergodicity assumption of statistical mechanics[24] allows us to see this probability as the fraction of molecules in a particular state, or simply as the activity (or concentration) of that state. Therefore we shall define $[P_i]$ as the average number of molecules existing in the ith microstate, and the probability p_i that the protein is in the ith state as the fraction of molecules in that state (or, equivalently, the fraction of time the molecule spends in that state as time becomes large). It follows that:

$$p_i = \frac{[P_i]}{[P_0] + [P_1] + \cdot\cdot\cdot + [P_n]}, \text{ and obviously: } \sum_{i=0}^{n} p_i = 1$$

The probability distribution 'p' is conditional in the sense that it only concerns the unbounded form of the protein.

If we pass from microscopic to macroscopic observation, the macroscopic state of the unligated protein can be seen as the collection of the average activities of all its microstates. Thus, for a conformational space of $n + 1$ elements:

$$[\mathbf{P}] = \sum_{i=0}^{n} [P_i], \text{ where } [P_i] \text{ can be given as } [\mathbf{P}]\, p_i$$

and the total concentration of the protein can then be described as a probability distribution on the domain of its conformational space. Suppose the protein can reversibly bind the ligand X. If there is a conditional distribution p related to the unligated protein, there should also be another conditional distribution p^x for the ligand-bound form of the same protein, defined on the same domain, for both share the same conformational space. Obviously, p and p^x need not to be equal. The total concentration of ligand-bound protein ($[\mathbf{XP}]$) is partitioned as $[XP_i] = [\mathbf{XP}]\, p_i^x$, where p_i^x is the conditional probability of the same i^{th} state, given the condition that the protein is ligated. Hence, the total concentration of bound protein is:

$$[\mathbf{XP}] = \sum_{i=0}^{n} [XP_i] \quad \text{with} \quad \sum_{i=0}^{n} p_i^x = 1.$$

We are now ready to interpret the macroscopic binding reaction of the protein with the ligand X as a relation between two conditional probability functions. The system

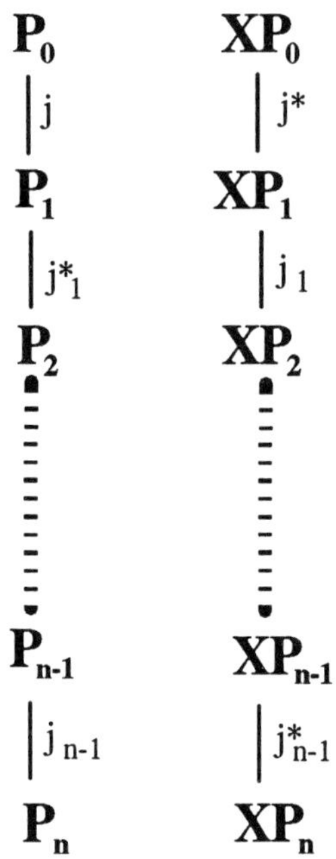

FIGURE 6. Ligand binding changes the state distribution of P. Schematic view of arbitrarily ordered states transitions among n microstates of a protein, with corresponding transition constants j. **P** and **XP** indicate unligated and ligated protein, respectively.

can be schematically drawn (FIG. 6) as two series of arbitrarily ordered state transitions with corresponding transition constants. By selecting from the conformational space an arbitrary state as reference, we can define the conditional transition constant l with respect to that particular state as the ratio of the concentration of the ith state to the reference one:

$$l_i = \frac{[P_i]}{[P_0]},\ l_i^x = \frac{[XP_i]}{[XP_0]};\ \text{or, in terms of probabilities: } l_i = \frac{p_i}{p_0},\ l_i^x = \frac{p_i^x}{p_0^x}$$

In the macroscopic observation at equilibrium, the conditional transition constants add up to generate conditional partition functions defined as:

$$\Omega_P = 1 + l_1 + l_2 + \cdot\ \cdot\ \cdot + l_n; \quad \text{and} \quad \Omega_{XP} = 1 + l_1^x + l_2^x + \cdot\ \cdot\ \cdot + l_n^x$$

Further, if we define: $a_i = \dfrac{l_i^x}{l_i}$ for $i = 1 \ldots n$, as a measure of how much the conditional probability of each state is affected by the ligand binding process, then:

$$\Omega_{XP} = 1 + a_1 l_1 + a_2 l_2 + \cdot\ \cdot\ \cdot + a_n l_n$$

For the equilibrium transition: **P** $\rightarrow$ **XP**, the ratio between the bound and free macroscopic states can be described as:

$$\frac{[\mathbf{XP}]}{[\mathbf{P}]} = \frac{[XP_0]\Omega_{\mathrm{XP}}}{[P_0]\Omega_{\mathrm{P}}},$$

which indicates how the average microstate composition of the empty protein is perturbed in the presence of the reversibly bound ligand X.

Let us now introduce "biological function" by identifying within the macroscopic observables an active (P*) and inactive (P) state for this protein, which interconvert reversibly. For example opening and closing of an ion channel. These functional macroscopic states represent, then, two clusters of microstates drawn within the entire conformational space of the protein. We may arbitrarily set microstates numbers 0 to k as P and k + 1 to n, as P*. The "macroscopic" interacting species can then be expressed in terms of the conditional partition functions of the corresponding subsets of microstates they include as follows:

$$[\mathbf{P}] = [P_0]\ \Omega_P\ , \quad \Omega_P = 1 + l_1 + l_2 + \cdots + l_k$$

$$[\mathbf{P^*}] = [P_0]\ \Omega_{P^*}\ , \quad \Omega_{P^*} = l_{k+1} + l_{k+2} + l_{k+3} + \cdots + l_n$$

$$[\mathbf{XP}] = [XP_0]\ \Omega_{XP}\ , \quad \Omega_{XP} = 1 + a_1 l_1 + a_2 l_2 + \cdots + a_k l_k$$

$$[\mathbf{XP^*}] = [XP_0]\ \Omega_{XP^*}, \quad \Omega_{XP^*} = a_{k+1} l_{k+1} + a_{k+2} l_{k+2} + a_{k+3} l_{k+3} + \cdots + a_n l_n$$

$$[\mathbf{XP}_{total}] = [\mathbf{XP}] + [\mathbf{XP^*}] \qquad [\mathbf{XP}_{total}] = [XP_0](\Omega_{XP} + \Omega_{XP^*})$$

$$[\mathbf{P}_{total}] = [\mathbf{P}] + [\mathbf{P^*}] \qquad [\mathbf{P}_{total}] = [P_0]\ (\Omega_P + \Omega_{P^*}).$$

Where the macroscopic forms [**P**] and [**P***], or [**XP**] and [**XP***] share the same reference state because they still originate from subdivisions of the entire conformational space of P.

Then, the combined effects of bound ligand and functional change on the average microstate composition of the receptor protein is now given by the ratios:

$$\frac{[XP_0]\Omega_{XP^*}}{[P_0]\Omega_{P^*}} \Bigg/ \frac{[XP_0]\Omega_{XP}}{[P_0]\Omega_P} = \frac{\Omega_P \Omega_{XP^*}}{\Omega_{XP}\Omega_{P^*}} = \frac{[\mathbf{XP^*}][\mathbf{P}]}{[\mathbf{XP}][\mathbf{P^*}]} = \beta_{\text{(macroscopic)}} \tag{5}$$

Equation 5 is the probabilistic version of the allosteric constant β in FIGURE 1. It tells us that if the average distribution in microstates in the inactive (or active) form of the protein is different between bound and unbound conditions, then we will observe an "allosteric effect" of the ligand. The nature of this result does not depend on how the conformational space is divided among subsets, because such a subdivision is a matter of macroscopic measurability. If the only appreciable quantity is ligand bound, then efficacy disappears, since in the macroscopic observables we only consider the single arrow: **P—XP** (FIG. 7). But it can be decomposed in an endless way within the conformational space of the protein as we add macroscopic functional forms (FIG. 7).

We shall now apply the same analysis to derive the cooperative effects between ligands described by the ternary complex model. In the presence of two ligands, say X and Y, binding to two distinct sites on P, we have four macroscopic states (**P**, **XP**, **PY** and **XPY**) for which we can write corresponding conditional partition functions:

$$\Omega_P = 1 + l_1 + l_2 + \cdots + l_n$$

$$\Omega_{XP} = 1 + a_1 l_1 + a_2 l_2 + \cdots + a_n l_n$$

$$\Omega_{PY} = 1 + b_1 l_1 + b_2 l_2 + \cdots + b_n l_n$$

$$\Omega_{XPY} = 1 + a_1 b_1 l_1 + a_2 b_2 l_2 + \cdots + a_n b_n l_n$$

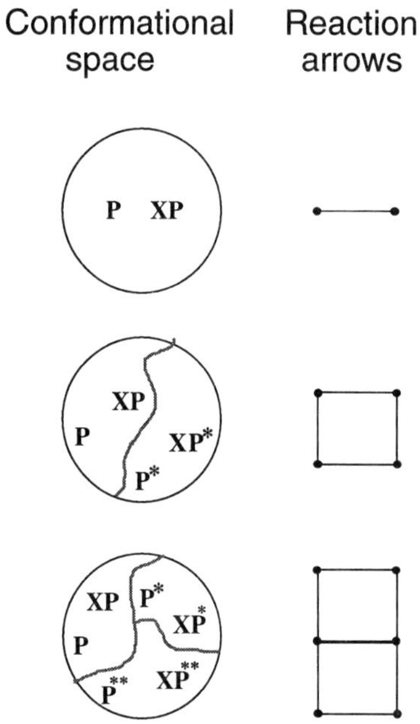

FIGURE 7. Subdivision of the conformational space and corresponding reaction species. The conformational space of the protein is schematically illustrated as the area of a circle which is shared by both unbound (**P**) and bound protein (**XP**). The experimental identification of different functional forms selects collections of microstates within the conformational space of the protein. The scheme on the left is the "reactivity" of the macroscopic forms generated by the attribution of biological functional diversity to collections of microstates. The figure illustrates how the number of macroscopic forms of the protein P selected within its conformational space determines the number and the complexity of the reacting paths at macroscopic equilibrium.

where l_i is defined as before, and 'a_i' and 'b_i' represent the effects that each ligand independently imparts on the probability of occurrence of microstate i. The implicit assumption in the definition of Ω_{XPY} is that the effects of ligands X and Y on (p_i) are independent of each other. That is, there is no direct long-range repulsion or attraction between the bound ligands, nor is there any other interaction mediated *via* the protein between the bound ligands, unless the protein changes its *"actual microscopic state"*. This assumption, in fact, sets the scale of microscopicity of the analysis. This also implies that:

$$[XP_0]/[P_0] = [XP_0Y]/[YP_0], \quad \text{or} \quad [YP_0]/[P_0] = [XP_0Y]/[XP_0]$$

which means that for any given microstate the proportion of average number of molecules unbound and bound by one of the ligands does not depend on the state of ligation of the other ligand, as long as the protein is in that particular microstate.

Then, the domain of all the four conditional partition functions spans the entire conformational space of P, and can once again be expressed in terms of the probabilities of the reference state. Thus, the macroscopic transitions **PY** → **XPY** and **P** → **PX** are described by the terms:

$$\frac{[XP_0Y]}{[YP_0]}\frac{\Omega_{\mathrm{XPY}}}{\Omega_{\mathrm{YP}}} \bigg/ \frac{[XP_0]}{[P_0]}\frac{\Omega_{\mathrm{XP}}}{\Omega_{\mathrm{P}}} = \frac{\Omega_{\mathrm{XPY}}\Omega_{\mathrm{P}}}{\Omega_{\mathrm{XP}}\Omega_{\mathrm{YP}}} = \frac{[\mathbf{P}][\mathbf{XPY}]}{[\mathbf{XP}][\mathbf{PY}]} = \alpha_{(\mathrm{macroscopic})}\,.$$

As a numerical example, consider a conformational space consisting of 1000 elements. The conditional distributions for **P**, **XP**, and **YP** were chosen arbitrarily assuming that the most abundant states differ among the unbound, X-bound, and Y-bound forms. The distribution in **XPY**, however, was deduced from the others using the relation: $[XP_iY] = a_ib_il_i/\Omega_{XPY}$. The conditional distribution of the double-ligated species **XPY** looks like neither that of **XP** nor of **YP** (FIG. 8). Simulations of this kind showed that this is generally the case when the two ligands exert negative cooperative effects at the macroscopic level ($\alpha < 1$, *i.e.*, negative antagonism). The same analysis, however, predicts that the heterotropic interaction becomes positive ($\alpha > 1$, *i.e.* agonism) as the distributions for **XP** and **YP** tend to overlap (data not shown). The lower the spontaneous probability in **P** of those states having relatively high probabilities in **XP** or **YP** forms, the larger the resulting α.

It should now be clear how the probabilistic approach proposed here is the exact inversion of the macroscopic principles behind allostery. Instead of putting the intramolecular transitions of the receptor in the background, and defining their existence as connecting rules that link intermolecular interactions, we do the opposite. The lens of our camera is now pasted on the single receptor molecule and its entire repertoire of microscopic transitions defined explicitly, while the intermolecular interactions are left in the dark. This analysis therefore provides a simple tool to formulate allosteric behavior at the microscopic level by using probabilistic arguments instead of thermodynamics. According to this model, both allosteric transition and free energy coupling can be regarded as the case of matching

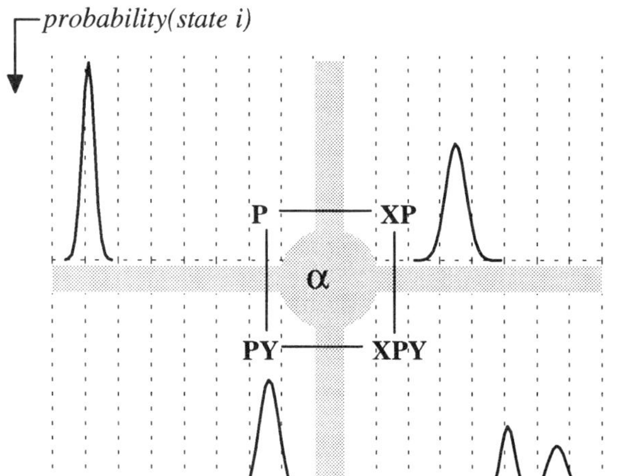

FIGURE 8. Distributions of microstates and the corresponding ternary complex cycle of the macroscopic forms. Computer simulation of a ternary complex model with a protein having a conformational space of 1000 microstates. **P**, **XP**, **PY** and **XPY** indicate the empty protein and the corresponding ligated forms, respectively. The conditional distributions of microstates for each macroscopic form are shown in the ordinate, and indicate the probability that the protein is in the ith microstate, where i ranges from 0 to 1000. The distributions for **P**, **PX** and **PY** were calculated using three independent Gaussian density functions, each having different variance and mean. The distribution for **XPY** is computed from the other three as explained in the text. Arbitrary shapes of the independent distributions, and the presentation of the conformational space as an arbitrarily ordered sequence in one dimension are only for schematization purposes. The resulting coupling factor α from these distributions is less than unity and predict a negative heterotropic interaction between the two ligands, *i.e.*, negative antagonism by the ligand X.

or mismatching between conditional probabilities of the collections of microstates associated with each macroscopic form of the receptor. In case of allosteric change in a channel receptor, the match (or lack of it) is between the most probable microstates in the ligand-bound form of the receptor and those included within the "open" macroscopic form. In case of G protein-coupled receptors, the target configuration is set by the collection of microstates which are most probable in the G protein-bound form of the receptor. The efficacy of a ligand will thus depend on the extent to which the probability of microstates in the receptor bound to that ligand will match the G protein-bound target configuration.

CONCLUSION

We may draw three brief conclusions.

First, the pharmacological distinction between affinity and efficacy as separate and independent entities inevitably vanishes as we step even minimally into the intramolecular environment of the receptor. This apparently reflects diverse views by which the relation between protein structure and function has been formalized in chemistry, but, ultimately, our ignorance of the dynamics of the intramolecular environment.

For this reason the "classical" definition of agonist efficacy as non-chemical entity is still the most useful for both receptor classification and medicinal chemistry. If, however, our experiments manipulate the structure of the receptor itself, we are forced to seek a chemical definition of efficacy. In which case there is no best model. The choice primarily reflects which aspect of receptor behavior is dominant in the experimental investigation.

Finally, in the last part of this paper, we propose a probabilistic view of the conformational space of the receptor which provides a pragmatic solution to the dualism in the concepts of allosteric transition and free energy coupling. Although the conformation space that we define here is mathematical rather than physical, the process by which the population of microstates is related to the macroscopic world of experimental observation is similar to that inherent in molecular dynamics simulations. There, snapshots of the molecular motion trajectory of the free receptor are compared with those of the bound receptor, to find concerted changes between molecular recognition and average properties of macroscopic states. Thus, this probabilistic view of allosterism may be a first step towards reaching a simple theoretical formulation of agonist efficacy based on dynamical properties of the macromolecular structure of receptors.

REFERENCES

1. Kenakin, T. P. 1992. Pharmacological analysis of drug-receptor interaction. Raven Press. New York.
2. Black, J. W. & P. Leff. 1983. Proc. R. Soc. London Ser. B **220:** 141–162.
3. Stephenson, R. P. 1956. Arch. Int. Pharmacodyn. Ther. **99:** 32–49.
4. Furchgott, R. F. 1966. *In* Advances in Drug Research, Vol. 3. N. J. Harper & A. B. Simmonds, Eds.: 21–55. Academic Press.
5. Monod, J., J. P. Changeux & F. Jacob. 1963. J. Mol. Biol. **6:** 306–329.
6. Monod, J., J. Wyman & J. P. Changeux. 1965. J. Mol. Biol. **12:** 88–118.
7. Wyman, J. 1987. J. Am. Chem. Soc. **89:** 2202–2218.
8. Weber, G. 1972. Biochemistry **11:** 864–878.
9. Del Castillo, J. & B. Katz. 1957. Proc. R. Soc. London Biol. Sci. **146:** 369–381.

10. KARLIN, A. 1967. J. Theor. Biol. **16:** 306–320.
11. THRON, C. D. 1973. Mol. Pharmacol. **9:** 1–9.
12. COLQUHOUN, D. 1973. *In* Drug Receptors. H. P. Rang, Ed.: 149–182. MacMillan.
13. LEFF, P. 1995. TIPS **16:** 89–97.
14. DE LEAN, A., J. M. STADEL & R. J. LEFKOWITZ. 1980. J. Biol. Chem. **255:** 7108–7117.
15. WREGGET, K. A. & A. DE LEAN. 1984. Mol. Pharmacol. **26:** 214–227.
16. EHLERT, F. J. 1985. Mol. Pharmacol. **28:** 410–421.
17. MACKAY, D. 1990. J. Theor. Biol. **142:** 415–427.
18. KENAKIN, T. P. & P. H. MORGAN. 1989. Mol. Pharmacol. **35:** 214–222.
19. COSTA, T., Y. OGINO, P. MUNSON, H. O. ONARAN & D. RODBARD. 1992. Mol. Pharmacol. **41:** 549–560.
20. WEBER, G. 1973. Adv. Protein Chemistry **29:** 1–83.
21. WYMAN, J. & D. W. ALLEN. 1951. J. Polymer Sci. **7:** 499–518.
22. KJELSBERG, M. A., S. COTECCHIA, J. OSTROWSKI, M. CARON & R. J. LEFKOWITZ. 1992. J. Biol. Chem. **267:** 1430–1433.
23. SAMAMA, P., S. COTECCHIA, T. COSTA & R. J. LEFKOWITZ. 1993. J. Biol. Chem. **268:** 4625–4636.
24. HILL, T. l. 1956. Statistical Mechanics. McGraw-Hill. New York.

Protean Agonists

Keys to Receptor Active States?

TERRY KENAKIN

Department of Receptor Biochemistry
GlaxoWellcome
5 Moore Drive
Research Triangle Park, North Carolina 27709

While it is a reasonable assumption that agonism is related to a conformational change in seven transmembrane receptors, it presently is not clear whether agonism results from the enrichment of a single "active" (with respect to G-protein activation) receptor state or the enrichment of some members of a library of active states. In the latter case, the prospect of agonist-selective receptor active states becomes a possibility with the resulting ramifications regarding the design of more selective agonists than available at present. The logical prediction of agonist-selective receptor active states would be that they could be more efficacious or less efficacious with respect to G-protein coupling. The former scenario would be difficult to define experimentally, but the latter case suggests an experimentally accessible avenue for hypothesis testing. Specifically, the effects of the agonist-selective active state should be recognizably different from those of the natural, spontaneous, receptor active state. Under these conditions, the effects of the agonist producing a less efficacious active state should be different, depending upon the setpoint of the receptor system. Such agonists will be referred to in this paper as "protean" agonists.

PROTEAN AGONISTS

Like the god of Greek mythology, Proteus, who could change shape at will depending on how he was viewed, a protean agonist would produce an agonist response in a quiescent receptor system (consisting of receptors the majority of which are in the inactive state) and be an inverse agonist in a constitutively active receptor system (one in which the receptors have been induced into the active state with respect to G-protein coupling). The key element in observing this type of profile is the control of the receptor system, *i.e.*, the same receptor is involved but the relative proportion of activation is varied. The creation of such systems now is possible with transfection of receptor cDNA into surrogate cells and the resulting control of receptor/G-protein stoichiometry. Another key element in these systems is the nature of 7TM receptors. Made to transmit chemical information from the extracellular space to the cytoplasm, 7TM receptors are reactive proteins with respect to G-proteins and can produce measurable cellular responses in the absence of agonist. Thus, enrichment of the natural active receptor state can be achieved biochemically (*i.e.*, removal of Na^+ [1-4]), or by overexpression of either receptors[5,6] or G-proteins.[7] Thus, receptors can be thought to demonstrate "behavior" when they are in different systems. This behavior can be divided into "intrinsic" behavior (the ability of receptors to exist in different conformational states) and "interactive" behavior (the ability of these active states to interact with G-proteins and other components of the cell).

AGONIST-SPECIFIC RECEPTOR STIMULUS TRAFFICKING

The possibility to be explored here is the study of receptor intrinsic behavior by exploitation of the interactive behavior. The definition of agonist-specific receptor active states opens the possibility of defining agonist selective receptor trafficking of stimulus.[8,9] Specifically, this is defined as the stimulation of one G-protein pathway by agonist activation of the receptor, and the stimulation of another G-protein pathway by another agonist activating the same receptor (see FIG. 1). Such effects are relevant only to receptors known to promiscuously interact with more than one G-protein in a given system. Many such systems have been described, although it presently is not clear to what extent this phenomenon occurs naturally or is the product of receptor overexpression in surrogate cell lines. The distinction between such systems is not relevant to this discussion since either would furnish an adequate system to detect agonist-selective receptor active states.

The key to the detection of agonist-selective trafficking of stimulus is the vantage point for the viewing of agonist response. Historically, the integrated cellular response to an agonist was being used to gauge receptor stimulus and, under these conditions, all agonists were considered to produce a uniform activation of the receptor differing only in degree, *i.e.*, the only difference between agonists for a given receptor was one of strength of intrinsic efficacy. However, when agonist response is dissected into events preceding the production of syncytial cellular response, then interesting dissimulations can be seen in some systems. For example, in LLC PK1 cells transfected with a single splice variant of the PACAP receptor,[10] it can be seen that the PACAP fragments PACAP-27 and PACAP-38 activate both adenylate cyclase and IP_3 pathways, but with a reverse order of potency. Such a reversal of agonist potency for two pathways in the same cell under the control of a single receptor is difficult to account for in terms of a single receptor activation state, and raises the possibility that these two fragments produce different receptor active states which then go on to interact with G-proteins in different ways. Other such stimulus trafficking behavior has been reported for the *Drosophila* octopamine-tyramine receptor transfected into CHO cells.[11]

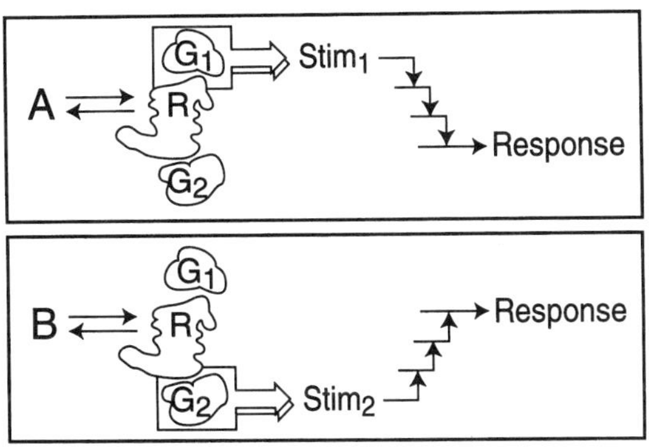

FIGURE 1. Agonist trafficking of receptor stimulus. Agonist A produces a receptor active state that preferentially activates G_1 even though the receptor can activate both G-proteins. Agonist B produces another active state to preferentially activate G_2.

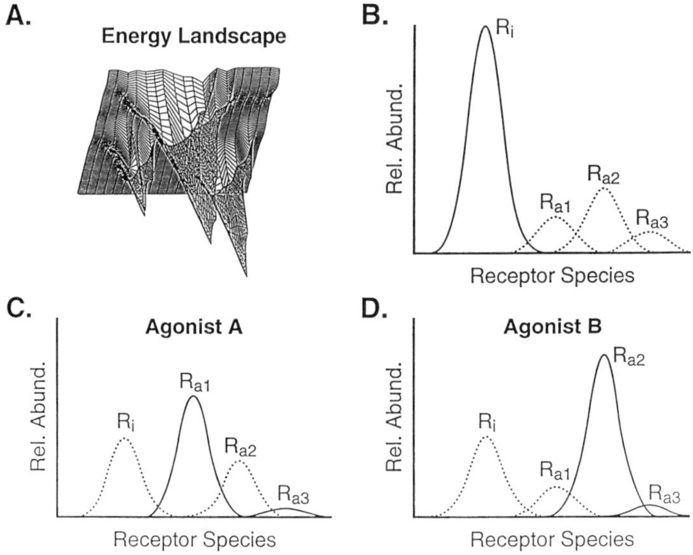

FIGURE 2. Selective receptor active states as wells in an energy landscape. **A**: Depiction of a surface showing energy wells for various conformational states of a receptor. **B**: A snapshot of the relative proportion of the receptor population in various states. Shown are most of the receptors in the inactive (R_i) state with the potential to form three active states. **C**: A hypothetical agonist A produces active states R_{a2} and R_{a3} at the expense of R_i. **D**: Another agonist B enriches the R_{a1} and R_{a3} states at the expense of R_i.

A proposed molecular mechanism of agonist directed stimulus trafficking would be the selective stabilization of an array of receptor active states by different agonists. Thus, receptor proteins would be viewed as traversing an "energy landscape" with the possibility of existing in a vast number of tertiary conformations[12,13] (FIG. 2A). It should be noted that these receptor states are intrinsic to the receptor protein and do not depend upon the presence of G-proteins. Indeed, receptor allosteric states can be distinguished in solubilized receptor preparations in the absence of G-protein (see Wregget and Wells, this volume). At any instant in time, the relative proportions of receptor states could be approximated as a collection of Boltzman distributions. Thus, FIGURE 2B shows a quiescent receptor system with all of the receptors existing in an inactive ($[R_i]$) state with the potential to form three active states ($[R_{a1}]$, $[R_{a2}]$, $[R_{a3}]$). Agonist A enters this array and selectively stabilizes the $[R_{a1}]$ and $[R_{a3}]$ form (FIG. 2C) while Agonist B stabilizes the $[R_{a2}]$ and $[R_{a3}]$ forms (FIG. 2D). Depending upon the differential equilibrium constants of each active receptor form for different G-proteins, a differential activation of G-protein pathways would be elicited by each agonist.

There are practical reasons for defining agonist-selective stimulus trafficking in terms of the discovery of more selective agonists. Specifically, it is not clear whether pleiotropic responses to agonists in cells leads to therapeutic and side effects. However, the design of more selective agonists (*i.e.*, less pleiotropic G-protein activation) may open the possibility of testing the hypothesis that more selective

G-protein activation will lead to a reduction in side effects and a more useful therapeutic activity.[14]

PROTEAN AGONISTS AS DIFFERENTIATORS OF RECEPTOR ACTIVE STATES

Given the technical capability of creating receptor systems with different steady-state amounts of natural active receptor, the hypothesis to be tested is that an agonist that produces a unique active state of lesser intrinsic efficacy (with respect to G-protein coupling) can be detected under conditions of different constitutive receptor activity. Thus, the scenario for a single receptor active state (*i.e.*, the agonist simply stabilizes the natural active state of the receptor) would be as follows. In a quiescent system (most of the receptors in the inactive form), the agonist would produce an increase in receptor/G-protein coupling (FIG. 3A). In a constitutively active system, a steady-state amount of active state receptor is created either biochemically or by stoichiometric manipulation. However, as long as the amount of activated receptor was submaximal with respect to the response measuring characteristics of the system, the agonist would still increase response by virtue of the

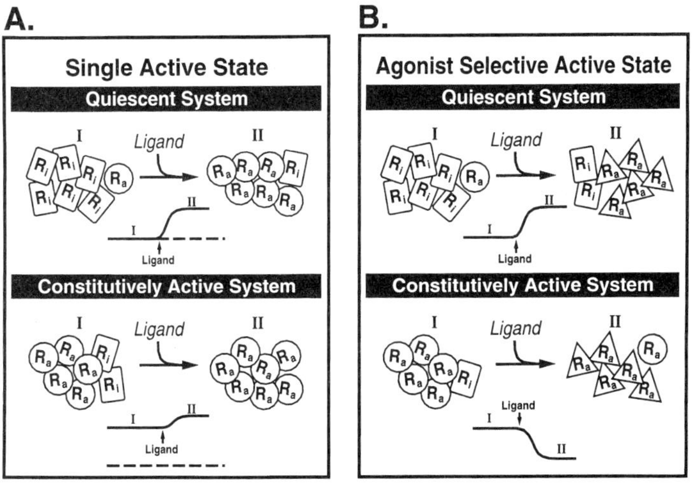

FIGURE 3. Schematic diagram of receptor states for two systems. **A**: There are only two receptor states, a naturally active state (R_a *circle*) and an inactive state (R_i *rectangle*). When most of the receptors are in the inactive state (quiescent system), an agonist enriches the naturally active state and an increased agonist response results. In a constitutively active system (most of the receptors in the naturally active state), this same agonist will produce a further increase in response if there is enough scale to do so. **B**: The agonist produces a pseudo-active state that promotes response but with less efficacy than the naturally active state (R_a *triangle*). In a quiescent system, the increase in R_a *triangle* produces an increased agonist response. However, in a constitutively active system, the agonist converts the highly active natural active state receptor to a less efficacious active species thereby reversing constitutive agonism.

production of an increase in the amount of single receptor activated state (FIG. 3A, lower).

In contrast, if the agonist produced a unique activation state of the receptor that was of lower intrinsic efficacy for G-protein coupling, then the agonist would still produce a measurable increased response in the quiescent system. This would occur because the lower efficacy active state would still promote coupling to G-protein to a greater extent than the inactive state of the receptor (FIG. 3B). The difference would be seen in a constitutively activated receptor system. Under these conditions, a basal response would be observed because of a steady-state amount of the natural activated receptor species. The intrusion of the agonist would change this condition to one of a steady-state amount of a less efficacious activated receptor species. Under these circumstances, and under the right conditions, an inverse agonism would be observed when the agonist was added to the system (FIG. 3B, lower). This reversal of agonism is what will be referred to as protean behavior for an agonist, and will be considered to be presumptive evidence of agonist-selective active receptor states.

THE CUBIC TERNARY COMPLEX MODEL

The predicted behavior of a protean agonist was simulated with the cubic ternary complex model. This model differs from the extended ternary complex model[5] in that the inactive receptor is allowed to interact with the G-protein under certain conditions.[15,16] For the purposes of this simulation, the constants were chosen such that the predictions did not depend on this condition and thus, the results would be formally identical with either model.

The cubic ternary complex model allows the receptor to exist in an inactive $[R_i]$ and active state $[R_a]$ the various proportions of which can be controlled by selective binding of a ligand [A] (see face I in FIG. 4). In addition, the two receptor states can interact with G-protein (face II, FIG. 4) and ligand can affect this interaction (face III, FIG. 4). When these interactions are put together, a cubic symmetry results with a number of constants controlling the various proportions of the species and propensity of reactions to occur under conditions of ligand binding. In general, ligands are thought to have two properties in such systems, affinity and efficacy. The affinity determines the equilibrium dissociation constant of the ligand-receptor species (K_a) while efficacy is defined as the ability of a ligand to change the various proportions of different receptor species (namely $[R_i]$, $[R_a]$, and $[R_aG]$) with binding. A drug with zero efficacy views the $[R_i]$, $[R_a]$ and $[R_aG]$ forms of the receptor as identical and does not alter the equilibrium proportions of each as it binds. In contrast, a drug with efficacy changes the proportions of each species according to the multipliers of the system equilibrium constants (α, γ, and δ). It should be noted that ligands could have positive efficacy (enhancement of receptor/G-protein coupling) or negative efficacy (destabilization of the $[R_aG]$ form with ensuing reversal of constitutive basal receptor response). Under these conditions, the observed affinity of an efficacious ligand will be affected by the setpoint of the system (*i.e.*, equilibrium value of receptor precoupling).

Constants for this model can be divided into those that describe the ligand (and is relationship to the receptor active state) and those that describe the system (TABLE 1). Under these circumstances, conditions were explored whereby experimental manipulation of the receptor system would reveal selective ligand constants and thus, probe receptor activation states. Specifically, this model was used to

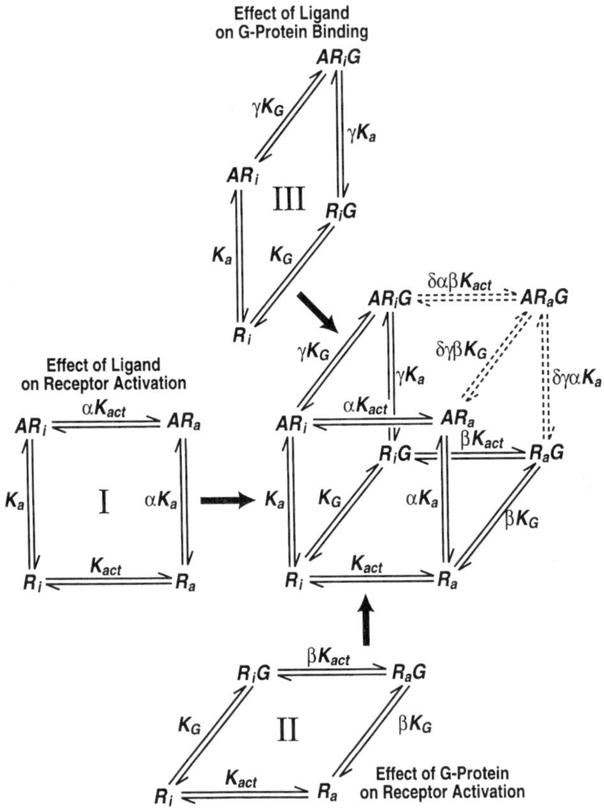

FIGURE 4. The cubic ternary complex model. This model simulates the effect of ligands on an inactive R and active R state of the receptor (Face I), the interaction of these receptor states with G-protein (Face II), and the effect of ligands on the receptor interaction with G-proteins (Face III). (From Weiss *et al.*[15,16])

predict observed effects of ligands that produced an active receptor state that was less efficacious than the natural one (*i.e.*, protean ligands). In terms of the cubic ternary complex model, this was affected by setting the constant for receptor activation (α) greater than unity (to promote production of $[R_a]$) but setting the subsequent coupling constant for that activated receptor for G-proteins to be less than

TABLE 1. Constants for the Cubic Ternary Complex Model

	Systems Constants		Ligand Constants
K_G	R_i/G-Protein	K_A	Affinity
β	Activation Effect on K_G	α	Ligand Activation of R_i
K_{act}	Receptor Activation	γ	Ligand Activation of G
[R]	Receptor Density	δ	Synergy α/γ

K_{act}, [R], and [G] can be manipulated experimentally. α, γ, δ are active state dependent.

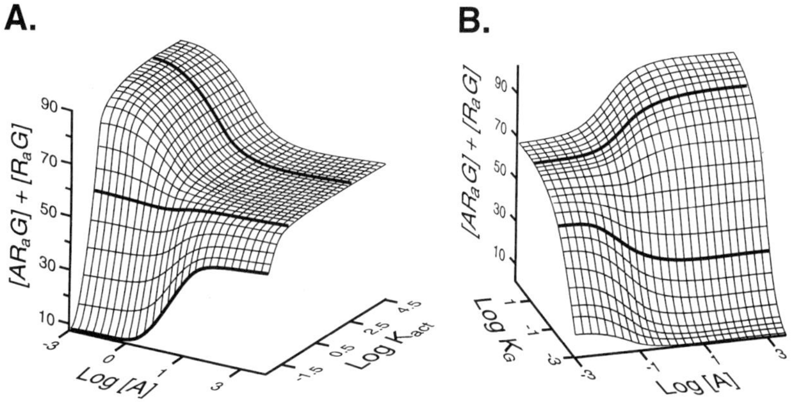

FIGURE 5. The effects of changes in system parameters on observed effect to a protean agonist. **A**: Varying K_{act} (propensity of production of the natural active state of the receptor) produces systems where the protean ligand produces agonist response (low K_{act}), no response, and an inverse agonism (high K_{act}). **B**: Varying the equilibrium constant between the receptor and G-protein can produce a system where the protean ligand produces a positive agonist response (when K_G favors interaction of the receptor and G-protein). In contrast, if there is low constant of interaction between the receptor and the G-protein (as perhaps produced by addition of GTPγS), then an inverse agonism would result.

unity ($\gamma < 1$). Under these circumstances, the active receptor species produced by the agonist would be less likely to couple to G-proteins than the natural $[R_a]$ state.

The response producing elements ($[R_aG]$ and $[AR_aG]$) were calculated for such an agonist with the model under different values of systems constants. FIGURE 5A shows the effect of differing values for K_{act}, the equilibrium constant controlling the equilibrium between natural inactive and active receptors. It can be seen from this figure that if the system is quiescent (*i.e.*, low K_{act} indicating that most of the receptors are in the $[R_i]$ state) then the protean agonist will produce positive agonism. In contrast, when K_{act} is large (a large spontaneous equilibrium amount of natural $[R_a]$), then the protean agonist will be an inverse agonist. Similarly, the effects of varying K_G, the equilibrium constant between the receptor and G-protein, was explored. As shown in FIGURE 5B, when K_G is substantial, the protean agonist produces positive agonist responses. In contrast, when K_G is very small, inverse agonism is observed. This may be a useful experimental manipulation since a very low K_G may be approximated experimentally by addition of GTPγS to cancel receptor/G-protein coupling. Therefore, a reversal in agonist effect with GTPγS could indicate protean activity, which in turn, could indicate an agonist-specific active receptor state.

Further simulations were done to estimate the observed responses to various agonists in a number of cellular receptor systems. Thus, for a given array of ligand constants, the systems constants [R], [G] and K_{act} were randomly assigned by computer and the response producing elements ($[R_aG]$ and $[AR_aG]$) calculated. This was repeated 2000 times to simulate the response to the ligand in 2000 different cellular expression systems for the receptor. FIGURE 6A shows the results of these simulations for positive agonist. The ordinate axis refers to the change in response producing elements in the presence of a receptor saturating concentration of the ligand. As can be seen from this figure, the response either increased or was unchanged. This latter effect occurs in systems where the constitutive activity is already so high that an agonist could not produce further increases. Of note is the fact that there are no values obtained below basal. The data shown in the top half of FIGURE 6A is correlated with the corresponding value of K_{act} in the bottom half of FIGURE 6A. As can be seen from this figure, response is increased in quiescent systems (low K_{act}) and the magnitude of the maximal response decreases with increasing constitutive activity (increasing K_{act}).

A contrasting profile was obtained for a protean agonist. FIGURE 6B shows the response producing elements for a saturating concentration of protean agonist in 2000 different theoretical cell types. As can be seen from this figure, response can be increased, unchanged or decreased depending upon the system. Thus, in constitutively active systems, inverse agonism is observed while in quiescent systems, increased response results. The bottom half of this figure shows the corresponding correlation of response with K_{act}. It can be seen that the resulting response to a protean agonist is much more variable than that for a positive agonist (or for a pure inverse agonist, data not shown).

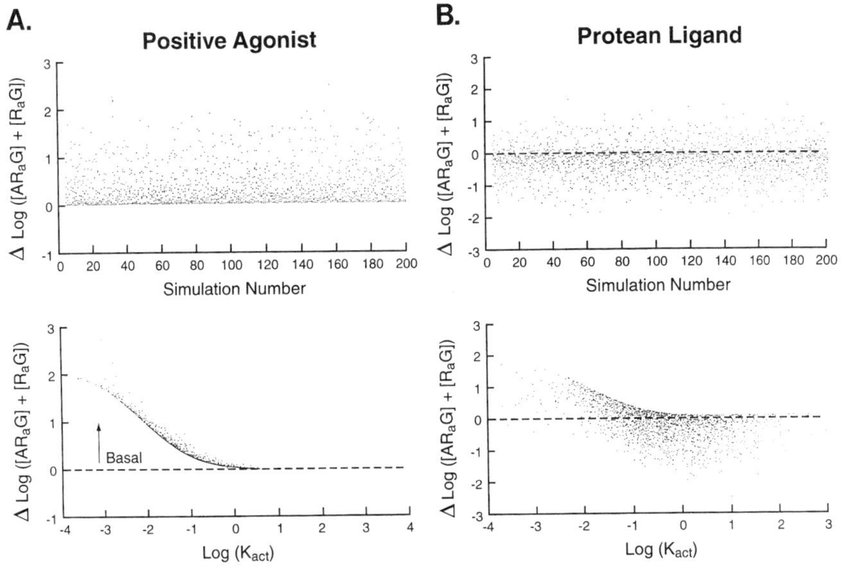

FIGURE 6. Multiple simulations of the maximal change in response producing elements (Δ [R_aG] + [AR_aG]) for 2000 variable combinations of [R], K_{act} and [G] (simulating 2000 different cellular expression systems). **A**: Simulations for a positive agonist ($\alpha = 10, \gamma = 10$). In all cases, no response is seen (in systems already maximally stimulated by constitutive receptor activity) or an increased response is seen. On lower graph, the response is correlated with K_{act}. It can be seen from this figure that quiescent systems (low K_{act}) show an increased basal response. **B**: Simulations for a protean agonist ($\alpha = 10, \gamma = 0.1$). In contrast, this drug can produce a positive response, no response or negative response in different systems. The lower graph shows the greater variability of response to a protean ligand in systems of different K_{act}.

RAMIFICATIONS FOR RECEPTOR CLASSIFICATION

These ideas have ramifications for the use of cellular expression systems for the classification of drugs and receptors. It is clear that the effects of the host cell on receptors can be overriding (especially for agonists) in that different ligand profiles can be obtained for the same receptor expressed in different cell lines. In essence, the cellular host becomes a detection system for receptor-activated states, and the observed profile of a ligand may depend upon the components of that detection system.

If a ligand promotes the creation of a minority (in the absence of ligand) activated state that is less efficacious at coupling to G-proteins, then the resulting observed effects of that ligand will be variable and system dependent. This can be dissimulating for the classification of the receptor being studied and the classification of the ligand. On the other hand, such behavior, if it can be observed in a controlled system, can be a hallmark of an agonist-specific receptor active state which, in turn, could lead to insights into activation mechanisms and the design of more selective agonists.

REFERENCES

1. COSTA, T & A. HERZ. 1989. Antagonists with negative intrinsic activity at d-opioid receptors coupled to GTP-binding proteins. Proc. Natl. Acad. Sci. USA **86:** 7321–7325.
2. COSTA, T., Y. OGINO, P. J. MUNSON, H. O. ONARAN & D. RODBARD. 1992. Drug efficacy at guanine nucleotide-binding regulatory protein linked receptors. Thermodynamic interpretation of negative antagonism and of receptor activity in the absence of ligand. Mol. Pharmacol. **41:** 549–560.
3. TIAN, W.-N., E. DUZIC, S. M. LANIER & R. C. DETH. 1994. Determinants of α_2-adrenergic receptor activation of G-proteins: Evidence for a precoupled receptor/G-protein state. Mol. Pharmacol. **45:** 524–531.
4. SHI, A.-G. & R. C. DETH. 1994. Precoupling of alpha 2B-adrenergic receptors and G-proteins in transfected PC-12 cell membranes: Influence of pertussis toxin and a lysine-directed cross-linker. J. Pharmacol. Exp. Ther. **271:** 1520–1527.
5. SAMAMA, P., S. COTECCHIA, T. COSTA & R. J. LEFKOWITZ. 1993. A mutation induced activated state of the β_2-adrenergic receptor: Extending the ternary complex model. J. Biol. Chem. **268:** 4625–4636.
6. CHIDIAC, P., T. E. HEBERT, M. VALIQUETTE, M. DENNIS & M. BOUVIER. 1994. Inverse agonist activity of β_2-adrenergic antagonists. Mol. Pharmacol. **45:** 490–499.
7. SENOGLES, S. E., A. M. SPIEGEL, E. PARDRELL, R. IYENGAR & M. CARON. 1990. Specificity of receptor G-protein interactions. J. Biol. Chem. **265:** 4507–4514.
8. KENAKIN, T. P. 1995. Agonist-receptor efficacy. I. Mechanisms of efficacy and receptor promiscuity. Trends Pharmacol. Sci. **16:** 188–192.
9. KENAKIN, T. P. 1995. Agonist-receptor efficacy. II. Agonist trafficking of receptor signals. Trends Pharmacol. Sci. **16:** 232–238.
10. SPENGLER, D., C. WAEBER, C. PANTOLINI, F. HOLSBOER, J. BOCKAERT, P. H. SEEBURG & L. JOURNOT. 1993. Differential signal transduction by five splice variants of the PACAP receptor. Nature **365:** 170–175.
11. ROBB, S., T. R. CHEEK, F. L. HANAN, L. M. HALL, J. M. MIDGLEY & P. D. EVEANS. 1994. Agonist-specific coupling of a cloned *Drosophila* octopamine/tyramine receptor to multiple second-messenger systems. EMBO J. **13:** 1325–1330.
12. FRAUENFELDER, H., F. PARAK & R. D. YOUNG. 1988. Conformational substates of proteins. Annu. Rev. Biophys. Chem. **17:** 451–467.
13. FRAUENFELDER, H., S. G. SLIGAR & P. G. WOLYNES. 1991. The energy landscapes and motions of proteins. Science **254:** 1598–1603.
14. GETTYS, T. W., T. A. FIELDS & J. R. RAYMOND. 1994. Selective activation of inhibitory

G-protein α-subunits by partial agonists of the human 5-HT_{1A} receptor. Biochemistry **33:** 4283–4290.

15. Weiss, J. M., P. H. Morgan, M. W. Lutz & T. P. Kenakin. 1995. The cubic ternary complex receptor-occupancy model. I. Model description. J. Theor. Biol. **178:** 151–167.
16. Weiss, J. M., P. H. Morgan, M. W. Lutz & T. P. Kenakin. 1995. The cubic ternary complex receptor-occupancy model. II. Understanding apparent affinity. J. Theor. Biol. **178:** 169–182.

Is Promiscuity of G Protein Interaction an Issue in the Classification of Receptors?

GRAEME MILLIGAN[a]

Molecular Pharmacology Group
Division of Biochemistry and Molecular Biology
Institute of Biomedical and Life Sciences
University of Glasgow
Glasgow G12 8QQ Scotland, United Kingdom

Cellular integration of information must result following the binding of an agonist to a guanine nucleotide binding (G protein)-coupled receptor so that the response to the perception of extracellular information, usually signaled by an alteration in the concentration of the agonist, can be correctly performed. Transfer of such information across the plasma membrane is produced by a minimum of three proteinaceous elements, the receptor, a heterotrimeric G protein, and an effector element, which is capable of altering the intracellular environment either by changing the concentration of second messenger molecules or by altering the ionic permeability of the plasma membrane. Initial concepts on the function of such systems tended to envisage a linear transduction pathway in which one receptor species activated a single type of G protein resulting in modification of a single effector. To some degree such concepts were based on the notion that such simplicity must be inbuilt to prevent the response to a single hormone being extremely complex. With the identification of considerable numbers of each of individual receptors for a single ligand, G proteins, and effector species isoforms it appeared that the long appreciated cellular variations in response to individual agonists could still be explained by a single receptor activating a single G protein and thence a single effector but that the molecular diversity of these components was simply larger than initially envisaged. However, direct experiments have shown, both in reconstituted and more native systems, that many agonist-occupied receptors have the capacity to interact with multiple G protein species. Within this article I will examine whether the coupling of a defined receptor to different G proteins can alter the observed pharmacology at the receptor ligand binding site and thus whether so-called "promiscuity" of G protein interactions can, or indeed should, be alluded to in the classification of G protein-coupled receptors.

G PROTEIN DIVERSITY AND EXPRESSION

Examination of a wide range of cells and tissues has demonstrated the co-expression of many heterotrimeric G proteins. To date some 20 distinct G protein

[a] *Address correspondence to*: Dr. Graeme Milligan, Davidson Building, University of Glasgow, Glasgow G12 8QQ Scotland, UK; Fax (44) 141 330 4620.

α subunits have been identified in mammalian cells. The bulk of these are derived from individual genes, but in the case of both $G_s\alpha$, the G protein most usually associated with receptor-mediated stimulation of adenylyl cyclase, and $G_o\alpha$, which has been implicated in the regulation of voltage-operated Ca^{2+} channels, further variation is produced via differential splicing mechanisms. Although some G proteins, such as gustducin which appears only to be expressed within the taste buds and the transducins which are found virtually exclusively in photoreceptor-containing cells, show a very restricted pattern of distribution, many G protein α subunits are very widely, if not universally, expressed in mammalian cells. These include, $G_s\alpha$, $G_{i2}\alpha$, $G_{i3}\alpha$, $G_q\alpha$, $G_{11}\alpha$, $G_{12}\alpha$ and $G_{13}\alpha$. The overall sequence identity (40–50%) of these polypeptides is high, as befits members of a family of proteins which bind and hydrolyze GTP to define their state of activation, and in the cases of particular family members (G_{i2} vs. G_{i3} and G_q vs. G_{11}) this figure rises to greater than 90%.

Although many observations of the regulation of levels of multiple second messengers have been reported following activation of a single receptor species, it is not a precondition that such observations must imply activation of multiple G protein species.[1] With the appreciation that the activity of isoforms of adenylyl cyclase may be modified by G protein $\beta\gamma$ subunits (see below) and via phosphorylation by protein kinase C and protein kinase A and the knowledge that both the α subunit and the $\beta\gamma$ complex derived from a single heterotrimeric G protein may each regulate different effectors, then a single receptor activating a single G protein can be envisaged to have consequences for the cellular levels of multiple second messengers.[1] Despite these caveats, many studies have described the activation of multiple G protein α subunits upon agonist occupancy of a single receptor subtype.

PROMISCUITY OF RECEPTOR-G PROTEIN INTERACTIONS: STUDIES WITH TRANSFECTED AND NATIVE CELL SYSTEMS

Given the high degree of similarity between closely related G proteins, it is not surprising that many examples of receptor co-activation of these G proteins have been observed. Despite these qualitative observations (see below for examples) few studies have made a priority to examine the absolute levels of expression of multiple co-expressed G proteins and their relative affinities of interaction with a single receptor. Following stable co-transfection of the α_{2A}-adrenoceptor and the G_{o1} splice variant of $G_o\alpha$ in Rat 1 cells (which express $G_{i2}\alpha$ endogenously), Grassie and Milligan[2] demonstrated that occupancy of the receptor with the agonist UK14304 resulted in activation of both $G_{i2}\alpha$ and $G_{o1}\alpha$ and that concentration-effect curves for production of the ternary complex and for activation of these two G proteins were coincident. Similar conclusions have been reached for activation of $G_q\alpha$ and $G_{11}\alpha$ by a range of receptors, including isoforms of α_1-adrenoceptors[3] and the gonadotrophin-releasing hormone receptor,[4] which function via increasing hydrolysis of inositol-containing phospholipids and the generation of the second messengers inositol 1,4,5 trisphosphate and *sn* 1-2 diacylglycerol. No data have been reported which address whether different agonists derived from different chemical structures would show selectivity between activation of these different G proteins or display differences in efficacy when measuring activation of the individual G proteins. It may be that for these particular examples the G proteins are sufficiently similar that the conformational changes of a receptor required to cause their activation upon binding of agonists are very similar. This, however, is an intriguing question given studies from Eason *et al.*[5] who examined this question following expression of the α_2C2, α_2C4 and α_2C10 adrenoceptors in CHO cells. They noted that although all of the receptors produced marked inhibition of forskolin-amplified adenylyl cyclase activity, subsequent to treatment of the cells with pertussis toxin to attenuate receptor-mediated interactions with the G_i-like G proteins, distinct stimulation of adenylyl cyclase activity could be recorded for all of the α_2-adrenoceptor subtypes.[5] $G_s\alpha$ and $G_i\alpha$ are relatively dissimilar G proteins and as the majority of adenylyl cyclase-linked receptors do not seem to concurrently activate both of them then different agonist ligands which bind in distinct ways to the receptor may favor activation of one over another. Moreover, marked differences in the efficacy for stimulation of adenylyl cyclase was observed for ligands at the α_2C10-adrenoceptor. All the ligands examined functioned as full agonists when inhibition of forskolin-amplified adenylyl cyclase activity was measured. However, although adrenaline, noradrenaline and UK14304 were all capable of producing the same maximal activation of adenylyl cyclase, oxymetazoline, BHT-920 and BHT-933 acted only as weak partial agonists.[5] A range of differences were also observed for the other α_2-adrenoceptor subtypes with oxymetazoline acting as an efficacious agonist for activation of $G_s\alpha$ at the α_2C2-adrenoceptor but displaying no agonist properties at the α_2C4 adrenoceptor.[5] Such results clearly indicate that promiscuity of G protein interaction by a receptor can result in different detailed pharmacologies depending upon the assay used to identify and examine the receptor. As such, different conclusions as to the identity of a receptor might be reached in the absence of further knowledge. It would, however, be equivalent to the opening of Pandora's box to suggest that receptor classification be altered and enlarged to include variation from such differences in G protein coupling and it is a moot point as to whether receptor classifications based on agonist pharmacologies and efficacies can actually provide compelling reasons to subdivide receptor families beyond broad classifications.

A considerable benefit in assessing questions of G protein promiscuity and receptor classification in α_2-adrenoceptor expressing systems is the availability of a large and detailed pharmacology for the receptors. This is also true for many other receptors which have been studied in detail, such as the β-adrenoceptors. Such an extensive pharmacology should be of particular use in examining potential differences in the pharmacology of the observed interactions of the β_2-adrenoceptor in rat ventricular myocytes with both $G_s\alpha$ and $G_i\alpha$.[6] This report[6] is particularly interesting, as pharmacological approaches were used to indicate that the interaction of the β-adrenoceptor population of the heart with G_i was restricted to the β_2 and not the β_1-adrenoceptor population.[6] It is also likely that the β_2-adrenoceptor has the ability to couple to $G_{13}\alpha$[7] to cause activation of a Na^+-H^+ exchanger. This hypothesis is further strengthened by the observation that a mutant version of the β_2-adrenoceptor, which cannot couple to $G_s\alpha$, can regulate the Na^+-H^+ exchanger in a guanine nucleotide-dependent manner.[8] As $G_s\alpha$ and $G_{13}\alpha$ are amongst the least highly related mammalian heterotrimeric G-protein α subunits in terms of sequence identity, it must be considered likely that pharmacological differences will be observed for agonist activation of these transducers if such selective agonist channeling towards ternary complexes with different G proteins is a reality. However, without stringent controls, any attempt to utilize agonist efficacy as a means to classify receptors is fraught with potential artifacts as, for example, efficacy of β-adrenoceptor partial agonists increases with levels of receptor expression.[9] Thus, in terms of classification and G protein promiscuity differences in detailed agonist pharmacology are likely to exist dependent upon both the absolute levels of the receptor and the relative levels of expression of two relevant G proteins in a cell or tissue.[10–11]

Although the β-adrenoceptor of turkey erythrocytes (which most closely resembles the mammalian β_1-adrenoceptor) has the capacity to activate both adenylyl cyclase, via $G_s\alpha$, and phospholipase C, via $G_{11}\alpha$, and isoprenaline displays a 30-fold lower EC_{50} to activate adenylyl cyclase compared to phospholipase C there was no indication from antagonist studies of differences in pharmacology in these two G protein-coupled responses to indicate that they were being produced by molecularly distinct receptors.[12] Such a difference in isoprenaline potency for activation of the two G proteins could easily be interpreted as an effect produced at the level of G protein expression, but quantitative estimations of the levels of expression of $G_s\alpha$ and $G_{11}\alpha$ demonstrated them to be present at similar levels. Kenakin[11] has noted that the degree of coupling observed to different G proteins will depend both on the levels of receptor expression and on the relative affinity constants for the interaction of the receptor with each G protein, but it equally may represent nothing more than differences in receptor reserve for the two effector cascades as little is currently known about the levels of expression of adenylyl cyclases and phospholipases in different cells. Furthermore, the effector isoform profile in individual cell types can define the second messenger output profile from a single molecularly defined receptor. To date the most convincing examples of this are derived from the adenylyl cyclase field. At least 9 isoforms of adenylyl cyclase are known to exist.[13] While all of these are activated by $G_s\alpha$, further levels of regulation are known to occur. For example, the type II and type IV isoforms of adenylyl cyclase are known to be conditionally stimulated by G protein $\beta\gamma$ complexes.[13] Thus, in cells expressing these isoforms to a large fraction of the total cellular adenylyl cyclase population, receptors which are traditionally viewed as inhibitors of adenylyl cyclase through activation of G_i-like G proteins can actually cause stimulation of cyclic AMP generation.[14–17] At a purely descriptive level a receptor with a pharmacology of an α_2-adrenoceptor or an opioid receptor which caused activation, rather than inhibition, of cAMP production would be likely to be classified as a novel subtype even though from classification based on molecular information this would not be appropriate.

Other receptors which are usually classified as being linked to the stimulation of adenylyl cyclase, such as the glucagon receptor,[18] have also been reported to elicit certain responses which are attenuated by treatment with pertussis toxin, again providing data in support of the concept that receptors have the capacity to interact with multiple G proteins that do not belong to the same G protein subfamilies. A further set of studies have indicated that receptors including the luteinizing hormone receptor,[19] the calcitonin receptor,[20] and the thyrotropin receptor[21] have the capacity to interact with and activate both G_s (to stimulate adenylyl cyclase) and members of the G_q/G_{11} family (to cause an increase in the rate of generation of inositol phosphates). However, as rather limited pharmacologies exist for these receptors, no data are yet available to address whether different agonist profiles of activation of these different G proteins and effector systems might be observed.

Despite these comments, for a number of receptors which can be shown to couple to multiple G proteins there are clear examples where the agonist potency profile varies markedly dependent upon the second messenger output measured. In the case of an octopamine/tyramine receptor from *Drosophila melanogaster*, when this was expressed in CHO cells, both inhibition of adenylyl cyclase, produced via a pertussis toxin-sensitive transduction mechanism, and elevation of intracellular Ca^{2+}, produced by a mechanism resistant to the effects of pertussis toxin, were produced by both octopamine and tyramine.[22] However, in the studies on inhibition of adenylyl cyclase activity tyramine was markedly more potent than octopamine whereas when measuring elevation of intracellular Ca^{2+} the two agonists had very similar potencies.[22] Such results indicate that these two agonists are likely to bind to the receptor in sufficiently different ways to promote selective channeling of the receptor towards cascades mediated by different G proteins. In the case of the originally identified type I receptor for pituitary adenylyl cyclase activating polypeptide (PACAP) both stimulation of adenylyl cyclase and of inositol phospholipid breakdown was observed but although the peptides PACAP-27 and PACAP-38 were reported to be equipotent in causing stimulation of adenylyl cyclase, only PACAP 38 was capable of causing high affinity activation of inositol phosphate generation,[23] although it was substantially less potent in causing activation of inositol phosphate generation than stimulation of adenylyl cyclase.

An intriguing extension of these studies is that a number of different splice variants of the PACAP type I receptor have been recorded.[23] These various forms contain differences in the C-terminal region of the third intracellular loop resulting from the presence or absence of coding information derived from two cassettes which have been termed "hip" and "hop."[23] The variable region in these isoforms is in an area which has been indicated to play a key role in the G protein coupling capacity of many G protein-linked receptors (see below). The presence of the "hip" cassette resulted in lower agonist potencies for stimulation of adenylyl cyclase by both PACAP-27 and PACAP-38, without alterations in their rank order of potency. However, the presence of the "hip" cassette resulted in a large decrease in potency of PACAP 38 to cause inositol phosphate generation, while the "hop" cassette had little effect. Such splice variation may thus contribute to the sensitivity of response of the PACAP I receptors to agonist and clearly such alterations in the pharmacology would be expected to result in differences in classification. In this example the anticipated alterations in classification can be justified at a molecular level, but might easily have been missed if analysis was restricted to Northern blotting or polymerase chain reaction (unless the primer pairs selected were on either side of the splice variation).

An even more complex situation in terms of G protein promiscuity and splice variation has been recorded for the EP3 prostanoid receptor.[24] Four splice variant subtypes, EP3A–EP3D, were recorded in which the variation was restricted to the C-terminal tail of the receptor. Following expression in CHO cells an agonist was able to regulate stimulation or inhibition of adenylyl cyclase and stimulation of inositol phospholipid hydrolysis. The details of which second messenger systems and which G protein classes were utilized was dependent upon the identity of the specific splice variant analyzed. Even more interestingly, the EP3C splice variant has been reported to interact with G_s and G_o following reconstitution of receptor expressing CHO cell membranes with these G proteins. However, while the receptor stimulated the GTPase activity of G_s it inhibited that of G_o.[25] Although unusual, this is not the only report of agonist-mediated inhibition of the GTPase activity of a G protein. Ueda *et al.*[26] have noted an ability of the κ opioid receptor to produce a similar effect. Although ligand binding to the EP3 receptor splice variants has been reported to be identical,[27] there is insufficient pharamacology currently available at this receptor to fully justify this statement or to make attempts to explore agonist variation with G protein interaction a viable approach at this time.

ELEMENTS OF RECEPTOR STRUCTURE INVOLVED IN G PROTEIN REGULATION

Given the physical location of G proteins at the intracellular face of the plasma membrane it is not surprising that the intracellular loops and C terminal tail of G protein-coupled receptors have provided the focus for analysis of elements involved in receptor-G protein coupling.[28] Studies involving deletions and site-directed mutagenesis of the adrenergic receptors led to a model in which elements located in the third intracellular loop close to both the fifth and sixth transmembrane elements play key roles in the specificity of G protein interaction.[28] However, this region is clearly not the only important element in these interactions. Data from a variety of approaches, including mutagenesis and the use of peptides corresponding to regions of the sequence of receptors, have indicated that segments of each of the second intracellular loop and the C-terminal tail are also involved. Criticisms have been leveled at some of the studies involving mutagenic approaches as the alterations involved might interfere with elements of structure beyond those involved in receptor-G protein contacts, but it is difficult to imagine that such a body of information in relation to this question could have been generated by other means. In certain other receptors, regions other than the third intracellular loop appear to play key roles in these interactions. For example, in the case of the metabotropic glutamate receptors, strong evidence indicates major roles for both the second intracellular loop and the C terminal tail in the regulation of G protein contacts[29] and as the splice variation within the EP3 receptor is restricted entirely to the C terminal tail this must be a key site of interaction for this receptor. The use of peptides derived from the end of the fifth transmembrane element and the early part of the third intracellular loop of the α_{2A}-adrenoceptor have also recently provided evidence to indicate not only that receptors may show promiscuity of coupling to G proteins[30] but also that certain receptors may have the capacity to inhibit the GTPase of certain G proteins. Peptides corresponding to amino acids 212–227 and 214–227 of the α_{2A}-adrenoceptor were able to inhibit the basal GTPase activity of both recombinant $G_o\alpha$ and $G_s\alpha$.[30] As peptides derived from very similar regions of this receptor can cause stimulation of G protein GTPase activity,[31] such results appear to indicate that rather subtle changes in receptor structure and conformation may be sufficient to change the way in which the cellular G protein population may be regulated by receptor occupation by agonists.

SUMMARY

It is clear that the details of binding of different classes of agonist ligands to the same receptor may be distinct. This could result in subtle differences in the profile of G protein activation by individual ligands at the same receptor due to agonist-induced selection of conformational states of the receptor which favor interaction and ternary-complex formation with different G proteins. This can result in differences in the details of receptor pharmacology when measured at the level of effector output. Such differences are also likely to be dependent upon both the levels of expression of the receptor and the relevant G proteins and the G protein and effector enzyme isoform expression profile of a particular tissue. It would be foolish, however, to use such differences, in isolation, as a means to expand the classification of G protein-coupled receptors particularly when more molecular approaches to receptor classification are readily available.

REFERENCES

1. MILLIGAN, G. 1993. Trends Pharmacol. Sci. **14:** 239–244.
2. GRASSIE, M. A. & G. MILLIGAN. 1995. Biochem. J. **306:** 525–530.
3. WISE, A., T. W. LEE, D. J. MACEWAN & G. MILLIGAN. 1995. J. Biol. Chem. **270:** 17196–17203.
4. SHAH, B. H. & G. MILLIGAN. 1994. Mol. Pharmacol. **46:** 1–7.
5. EASON, M. G., M. T. JACINTO & S. B. LIGGETT. 1994. Mol. Pharmacol. **45:** 696–702.
6. XIAO, R-P., X. JI & E. G. LAKATTA. 1995. Mol. Pharmacol. **47:** 322–329.
7. VOYNO-YASENETSKAYA, T., B. R. CONKLIN, R. L. GILBERT, R. HOOLEY, H. R. BOURNE & D. L. BARBER. 1994. J. Biol. Chem. **269:** 4721–4724.
8. BARBER, D. L. & M. B. GANZ. 1992. J. Biol. Chem. **267:** 20607–20612.
9. MACEWAN, D. J., G. D. KIM & G. MILLIGAN. 1995. Mol. Pharmacol. **48:** 316–325.
10. KENAKIN, T. 1995. Trends Pharmacol. Sci. **16:** 188–192.
11. KENAKIN, T. 1995. Trends Pharmacol. Sci. **16:** 232–238.
12. JAMES, S. R., C. VAZIRI, T. R. WALKER, G. MILLIGAN & C. P. DOWNES. 1994. Biochem. J. **304:** 359–364.
13. TAUSSIG, R. & A. G. GILMAN. 1995. J. Biol. Chem. **270:** 1–4.
14. LUSTIG, K. D., B. R. CONKLIN, P. HERZMARK, R. TAUSSIG & H. R. BOURNE. 1993. J. Biol. Chem. **268:** 13900–13905.
15. FEDERMAN, A. D., B. R. CONKLIN, K. A. SCHRADER, R. R. REED & H. R. BOURNE. 1992. Nature **356:** 63–65.
16. TSU, R. C., R. A. ALLEN & Y. H. WONG. 1995. Mol. Pharmacol. **47:** 835–841.
17. TSU, R. C., J. S. C. CHAN & Y. H. WONG. 1995. J. Neurochem. **64:** 2700–2707.
18. JELENIK, L. K., S. LOK, G. B. ROSENBERG, R. A. SMITH, F. J. GRANT, S. BIGGS, P. A. BENSCH, J. L. KUIJPER, P. O. SHEPPARD, C. A. SPRECHER, P. J. H. O'HARA, D. FOSTER, K. M. WALKER, L. H. J. CHEN, P. A. MCKERNAN & W. KINDSVOGEL. 1993. Science **259:** 1614–1616.
19. GUDERMANN, T., M. BIRNBAUMER & L. BIRNBAUMER. 1992. J. Biol. Chem. **267:** 4479–4488.
20. CHABRE, O., B. R. CONKLIN, H. Y. LIN, H. F. LODISH, E. WILSON, H. E. IVES, L. CATANZARITI, B. A. HEMMINGS & H. R. BOURNE. 1992. Mol. Endocrinol. **6:** 551–556.
21. VANSANDE, J., E. RASPE, J. PERRET, C. LEJEUNE, C. MANHAUT, G. VASSART & J. E. DUMONT. 1990. Mol. Cell. Endocrinol. **74:** R1–R6.
22. ROBB, S., T. R. CHEEK, F. L. HANNAN, L. M. HALL, J. M. MIDGLEY & P. D. EVANS. 1994. EMBO J. **13:** 1325–1330.
23. SPENGLER, D., C. WAEBER, C. PANTALONI, F. HOLSBOER, J. BOCKAERT, P. H. SEEBURG & L. JOURNOT. 1993. Nature **365:** 170–175.
24. NAMBA, T., Y. SUGIMOTO, M. NEGISHI, A. IRIE, F. USHIKUBI, A. KAKIZUKA, S. ITO, A. ICHIKAWA & S. NARUMIYA. 1993. Nature **365:** 166–170.
25. NEGISHI, M., T. NAMBA, Y. SUGIMOTO, A. IRIE, T. KATADA, S. NARUMIYA & A. ICHIKAWA. 1993. J. Biol. Chem. **268:** 26067–26070.
26. UEDA, H., S. UNO, J. HARADA, I. KOBAYASHI, T. KATADA, M. UI & M. SATOH. 1990. FEBS Lett. **266:** 178–182.
27. SUGIMOTO, Y., M. NEGISHI, Y. HAYASHI, T. NAMBA, A. HONDA, A. WATABE, M. HIRATA, S. NARUMIYA & A. ICHIKAWA. 1993. J. Biol. Chem. **268:** 2712–2718.
28. SAVERESE, T. M. & C. M. FRASER. 1992. Biochem. J. **283:** 1–19.
29. PIN, J.-P., C. JOLY, S. F. HEINEMANN & J. BOCKAERT. 1994. EMBO J. **13:** 342–348.
30. WAGNER, T., C. OPPI & G. P. TOCCHINI VALENTITI. 1995. FEBS Lett. **365:** 13–17.
31. OKAMOTO, T. & NISHIMOTO. 1992. J. Biol. Chem. **267:** 8342–8346.

Kinetic Models of Ion Channels

BARRY S. PALLOTTA[a]

Department of Pharmacology
The University of North Carolina at Chapel Hill
Chapel Hill, North Carolina 27599

INTRODUCTION

Agonist binding to an ion channel receptor is followed by a series of conformational changes that ultimately result in the passage of ionic current through the conducting conformation.[1] The current that flows through a single ion channel is readily detected with patch-clamp techniques, and the kinetics of channel opening and closing ("gating") in the presence of agonist reveal many of the steps between binding and channel activation. Here, I'll briefly describe how single-channel currents are used to construct kinetic models of receptor activation and desensitization. Unfortunately, a single unifying kinetic model does not exist to describe the behavior of all ligand-gated channels, and given that similar experimental methods and analyses were applied, it is likely that real differences exist between channel types rather than experimenter bias towards one model or another. In surveying the models that have been proposed for calcium-activated potassium (BK) channels,[2] $GABA_A$,[3] and NMDA receptors[4,5] there is both good news and bad news, and they are the same: many kinetic states have already been identified, and we expect even greater complexity in channel behavior to be revealed by future studies.

WHY BUILD MODELS?

Before discussing some of the methods used for constructing kinetic models, it is important to understand why such an activity is useful. One reason is that a model concisely summarizes the experimental measurements and provides a common framework for interpreting future experiments. In other words, a model is an unambiguous summary of the data that other experimenters can test. This framework also helps to interpret the actions of various types of drugs that might influence receptor function. Another justification is that models can predict channel behavior under conditions that might be difficult to attain experimentally. This might include, for example, the high concentrations of neurotransmitter reached, however briefly, in the normal synapse. Finally, a kinetic model can suggest a physical picture of channel operation and structure, the most well-known example being the Hodgkin-Huxley four-subunit model derived from the timecourse of potassium channel currents (for reviews see ref. 6). Although kinetic states are not identical to confor-

[a] Address for correspondence: CB #7365 FLOB, UNC-CH, Chapel Hill, NC 27599; Tel.: 919-962-1059; Fax: 919-966-5640; E-mail: (pallotta@med.unc.edu).

mational states, they are closely related, and at the very least, the patch clamp unambiguously distinguishes "open" from "closed" conformations.

BUILDING KINETIC MODELS

The raw materials for the kinetic models described here are the single channel measurements of channel open and closed times. These can be obtained at a fixed agonist concentration, during which time the channel opens and closes for as long as agonist is present. Single-channel currents from most ligand-gated ion channels all bear a superficial resemblance to each other. Some openings seem to be grouped together into "bursts", and the bursts themselves grouped further into "clusters". Periods of this type of activity are typically followed by periods of inactivity during which time the channel occupies liganded closed states. Liganded closed states of relatively long duration correspond to "desensitized" states, and these are often punctuated by occasional, brief duration openings. As long as agonist is present, the channel cycles between conducting and nonconducting kinetic states. To begin model construction, the durations of each opening (and closed time) are measured and incorporated into interval duration histograms. These histograms immediately reveal the complexities of channel behavior. First, we see that the durations of the openings are exponentially distributed. This is apparent in the single-channel recordings themselves: each opening of the channel has a different duration. Second, histograms of channel open times are often fitted with the sum of two or more exponential components, while closed time histograms can require five or more exponential components.

To understand the information contained in these histograms, consider an imaginary channel that upon binding agonist, opens. But for how long? Channels don't possess stopwatches (as far as we know), but they do possess a form of random number generator inherent in the thermal fluctuations that can propel a channel from one conformation to another.[7] Imagine now that this particular channel possesses a six-sided die, and that one side is marked "shut". The channel then throws this die until the side marked "shut" comes up. If the die is thrown once every millisecond, then 4 throws would result in a 4 ms open lifetime. The next time the channel opens it again throws the die, but most likely, the duration of the next and each successive lifetime will each be different from the preceding one. In fact, a duration histogram of a large number of such open lifetimes shows that these lifetimes are exponentially distributed with a mean lifetime of 6 ms (FIG. 1). (Actually, the distribution is a *geometric* because throws of the die is a discrete variable, unlike time, which is a continuous variable). Thus, we interpret the exponential distributions of channel open and closed times as arising from the random, or stochastic, behavior of the channel. We interpret open duration histograms that have three exponential components as evidence of three different kinetic open states, each with a characteristic average open time. Similarly, multi-exponential closed interval histograms provide evidence for a number of closed kinetic states, each with a characteristic lifetime. The lifetimes of all these states are obtained by fitting sums of exponentials to the histograms.[7]

Interval histograms provide a method for counting the numbers of open and closed states. As shown in FIGURE 2, NMDA receptors have at least three open and five shut states. Similar numbers of states have been found for nAChR,[8] BK channels,[2] and $GABA_A$ receptors.[3] However, the histograms contain no information about how these states are interconnected. The reason is that the measured durations of each type are simply added to their respective histograms regardless of when they occurred in the experimental record, and with no regard to the surrounding open or shut intervals. For this reason, a variety of *conditional* approaches have been developed to measure the relationships between intervals. One method, autocorrelation analysis, detects whether successively occurring open times (or closed times) are correlated with each other. The existence of significant correlations indicates that there are two or more pathways between the aggregate open states and the (aggregate) closed states.[9] In one form of adjacent-states analysis[10] the average durations of the open times are plotted vs. the average duration of the immediately adjacent shut time. For several different channels, this analysis has shown that short duration open states tend to associate with long duration shut states. This correlation arises because the relatively long duration open states tend to follow each other in "bursts", while the shortest duration open states tend to occur in isolation during long periods of relative inactivity. Another approach has been to combine the above approaches into a single step by constructing (two-dimensional) distributions that contain both the interval durations themselves, as well as the correlations between them.[11]

Information about agonist-binding steps can come from the concentration dependencies of the interval distributions and correlations, or from experiments in which the agonist concentration is suddenly "jumped" from one concentration to another. Macroscopic (whole-cell) currents have proven especially effective in the latter types of experiments because the summed behavior of many hundreds (or thousands) of single channels is measured during a single agonist application. Estimating the numbers of kinetic states from macroscopic measurements is more difficult than from single-channel data, and models derived from such measurements tend to underestimate the number of states, particularly the number of open states. On the other hand, obtaining sufficient single-channel data at a number of agonist concentrations is a technically difficult task. The reason for this is that large numbers of intervals (at each concentration) are needed for reliable histograms. When desensitization is frequent, as it is for most, if not all ligand-gated channels, obtaining the appropriate numbers of intervals can take longer than the average patch "lifetime". Fortunately, there has been sufficient experimenter persistence to yield models for several important ion channels.

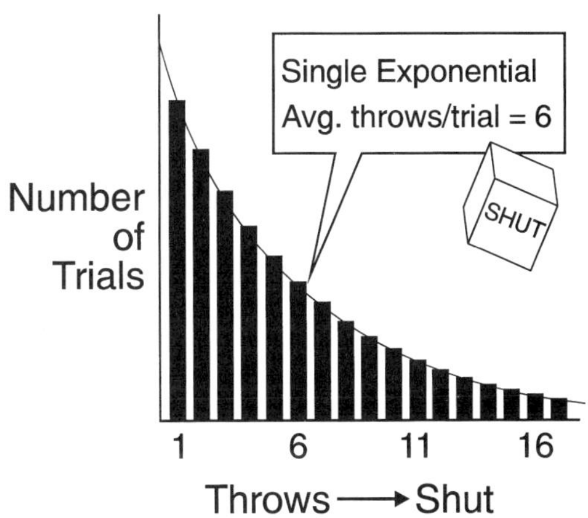

FIGURE 1. Exponential distributions reflect the inherent randomness of channel gating. The histogram shows the distribution of "lifetimes" that would be obtained from a six-sided die with one face marked "shut". In this histogram, each bar represents the number of separate trials that required 1, 2, 3 . . . *etc.* throws of the die until "shut"ing. The solid line shows the exponential fit (but see text) to the histogram obtained from a large number of simulated lifetimes. Since the probability of "shut"ing is 1/6, the average lifetime will be 6 throws.

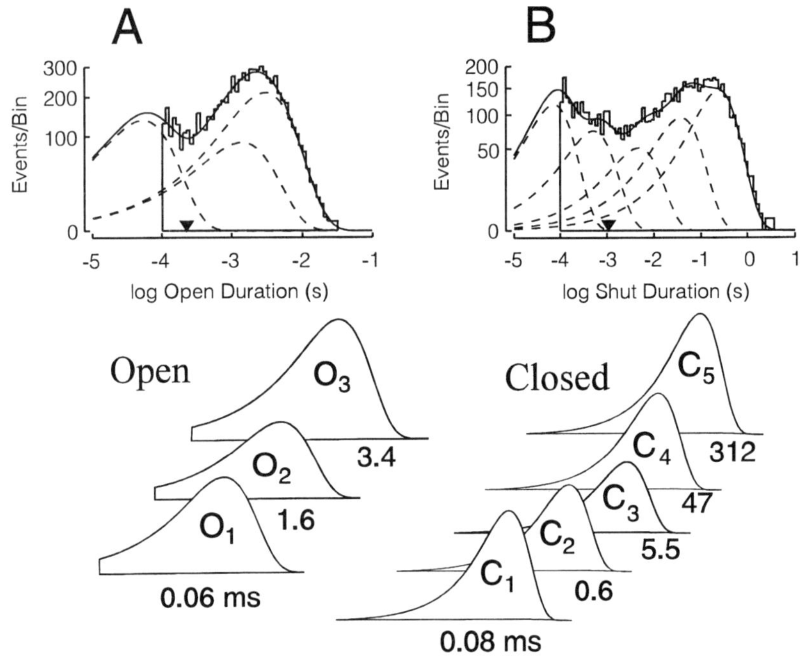

FIGURE 2. Distributions of open **(A)** and closed **(B)** times from an NMDA receptor in rat brain cortical neurons (data from Kleckner & Pallotta[4]). Unlike the linear-binned histogram shown in FIGURE 1, these histograms have been plotted with a square root vertical axis and bin sizes that increase logarithmically along the horizontal axis.[24] When plotted in this manner, the exponential components (*dotted lines*) appear as peaks that are centered at the component's average lifetime. **A**: The distribution of open times is composed of the sum (*solid line*) of three exponential components (*dotted lines*), shown separately below the histogram. The average lifetimes of the three kinetic open states are shown below each component. **B**: The distribution of shut times is composed of the sum (*solid line*) of five exponential components (*dotted lines*). The average lifetimes of the five kinetic closed states are shown below each component.

THE NMDA RECEPTOR

The NMDA receptor requires two different agonist molecules to open,[12] and for this reason alone, one would expect a kinetic model for its behavior to be especially complex. In addition, macroscopic current studies show that when NMDA and glycine are used to activate the channel, two molecules of each are required.[13] A model that incorporates each binding step contains nine closed states but only one open state (FIG. 3, *upper panel*). On the other hand, Lester and Jahr[14] proposed a model with one open and only four closed states that describes the time course of macroscopic currents quite well. The model works because macroscopic currents are relatively insensitive to the short duration open and closed times that are readily detected in single-channel recordings. Single-channel studies, however, showed the existence of at least three open states.[15–17]

To construct a model of channel bursting behavior, Kleckner and Pallotta[4] recorded long segments of single-channel activity at a fixed concentration of NMDA and glycine. Duration histograms were then constructed for the first opening of every burst, the second opening of every burst, and so on (excluding bursts of only one opening). We found that the first opening of each burst had a three-component distribution: 20% of the first openings were brief (0.08 ms), 60% were of intermediate duration (2.4 ms), and the remainder were long duration (4.7 ms). The second opening of each burst had the same distribution, as did the third, fourth, and fifth openings. When the bursts themselves were sorted according to the numbers of openings they contained, the same result was found. That is, each opening of a burst was drawn from the same kinetic "pool" of short, intermediate, or long-duration openings. A similar result was found when we examined the shut intervals during bursts: each shut interval was drawn from the same kinetic pool of either short (0.06 ms) or long intervals (0.64 ms), regardless of when the interval occurred within the burst.

To account for these results, we proposed a "gateway" model to describe bursts of two or more openings.[4] In this model (FIG. 3, *lower panel*), bursts begin when the channel enters the gateway closed state (C_L), from which it can enter any of the open

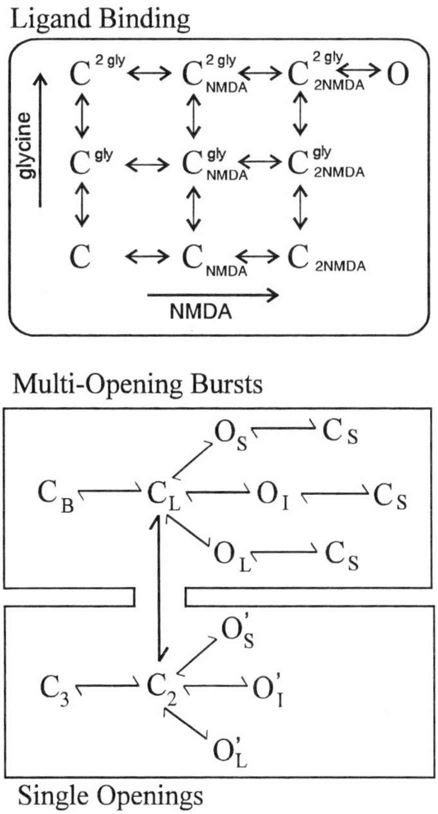

FIGURE 3. Kinetic models for NMDA receptor channel gating. A complete model for this channel will have to incorporate the kinetics of ligand binding,[13] multi-opening bursts,[4] and isolated openings.[5] Symbols: C, closed state; O open state.

states. With this sort of arrangement, the "selection" of open state is always the same regardless of which openings have already occurred, or the length of the burst. Each open state is also connected to a brief closed state (C_S), each one having the same (short) lifetime. Since these states have the same lifetime, they contribute only a single component to the distribution of shut times during bursts. The gateway state itself contributes the second component to that distribution.

While the above model accounts for the distributions of openings and closings during bursts of two or more openings, it does not account for single opening bursts (which themselves are described by a three-component distribution) or for desensitization. To account for these behaviors, we modified our model by adding two closed states and three more open states[5] (FIG. 3, *lower panel*). The two closed states are relatively long-lived and allow the model to "desensitize". The three new open states that are placed near these closed states give rise to isolated openings, since they are more likely to occur when the channel recovers from the desensitized states than when the channel is occupying the "bursting" states.

All told, the macroscopic and single-channel models contain a total of sixteen closed and seven open states. Some of these states might be redundant in that one or more of the closed states from the single-channel model might correspond to one or more states from the binding model. In any case, a combined model is certain to be extremely complicated. Yet, the NMDA receptor shows even more behaviors such as glycine-sensitive and -insensitive desensitization,[18–20] and subconductance levels.[21,22] Models that incorporate these additional phenomena have yet to be proposed.

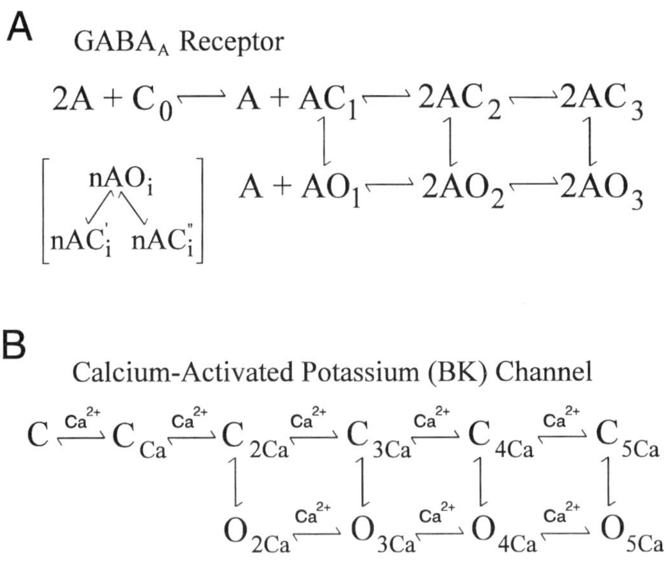

FIGURE 4. Models proposed for the GABA$_A$ receptor[3] and calcium-activated potassium (BK) channel.[2] **A**: In this model, each liganded open state (AO_i) is also connected to two additional liganded closed states (shown in brackets). Symbols are *A*, agonist; *C*, closed state; *O*, open state. **B**: In this model for the BK channel, calcium ions can bind to all but two states. This is one of several models that satisfactorily account for many aspects of BK channel gating.

COMPLEXITY OF MODELS

The complexities of the illustrated models reflect many aspects of channel structure and function. They probably do not reflect simply the choice of model by the experimenter since these models were derived using similar methods. In fact, we found that a variety of models of the type proposed for the nicotinic acetylcholine receptor, BK channel, and $GABA_A$ receptor (FIG. 4) were inconsistent with the kinetics of NMDA receptor bursts.[4] In other words, the receptors for which detailed models have been proposed do indeed exhibit different kinetics. What gives rise to the observed kinetic states? Most likely, each kinetic state reflects one or more closely related conformational states whose aggregate lifetime is long enough for us to detect with existing patch clamps. These conformational changes might occur primarily within individual subunits, but could reflect interactions between subunits as well. As the time resolution of the patch clamp is gradually improved, we'll probably uncover even more kinetic states. The challenge remains, however, to relate specific structural changes to the observed kinetics.

It is apparent now that no single kinetic model is adequate to describe the kinetics of ion channels. This should not be surprising, given that each channel has probably evolved to perform one or more specific functions, and that to perform these functions the molecule must harvest sufficient binding energy from its ligands.[23] While the most common kinetic behaviors among channels are the bursts of openings and desensitization, it is in the details of these processes that the different ion channels diverge.

REFERENCES

1. MILLER, C. 1987. How ion channel proteins work. *In* Neuromodulation: The Biochemical Control of Neuronal Excitability. L. K. Kaczmarek & I. B. Levitan, Eds.: 39–63. Oxford University Press. New York.
2. MCMANUS, O. B. & K. L. MAGLEBY. 1991. Accounting for the Ca^{2+}-dependent kinetics of single large-conductance Ca^{2+}-activated K^+ channels in rat skeletal muscle. J. Physiol. **443:** 739–777.
3. TWYMAN, R. E., C. J. ROGERS & R. L. MACDONALD. 1990. Intraburst kinetic properties of the $GABA_A$ receptor main conductance state of mouse spinal cord neurones in culture. J. Physiol. **423:** 193–220.
4. KLECKNER, N. W. & B. S. PALLOTTA. 1995. Burst kinetics of single NMDA receptor currents in cell-attached patches from cultured rat brain cortical neurons. J. Physiol. **486.2:** 411–426.
5. DONNELLY, J. L. & B. S. PALLOTTA. 1995. Single-channel currents from diethylpyrocarbonate-modified NMDA receptors in cultured rat brain cortical neurons. J. Gen. Physiol. **105:** 837–859.
6. GARDNER, D., Ed. 1992. Forty Years of Membrane Current in Nerve. Physiological Reviews (Suppl.) **72**(4): S1–S190.
7. COLQUHOUN, D. & A. G. HAWKES. 1995. The principles of the stochastic interpretation of ion-channel mechanisms. *In* Single-Channel Recording, 2nd edition. B. Sakmann & E. Neher, Eds.: 397–482. Plenum Press. New York.
8. COLQUHOUN, D. & B. SAKMANN. 1985. Fast events in single-channel currents activated by acetylcholine and its analogues at the frog muscle end-plate. J. Physiol. **369:** 501–557.
9. LABARCA, P., J. A. RICE, D. R. FREDKIN & M. MONTAL. 1985. Kinetic analysis of channel gating. Application to the cholinergic receptor channel and the chloride channel from *Torpedo* californica. Biophys. J. **47:** 469–478.
10. MCMANUS, O. B., A. L. BLATZ & K. L. MAGLEBY. 1985. Inverse relationship of the durations of adjacent open and shut intervals for Cl and K channels. Nature **317**(6038): 625–627.

11. MAGLEBY, K. L. & D. S. WEISS. 1990. Identifying kinetic gating mechanisms for ion channels by using two-dimensional distributions of simulated dwell times. Proc. R. Soc. Lond. Ser. B Biol. Sci. **241:** 220–228.
12. KLECKNER, N. W. & R. DINGLEDINE. 1988. Requirement for glycine in activation of NMDA-receptors expressed in *Xenopus* oocytes. Science **241:** 835–837.
13. BENVENISTE, M., J. CLEMENTS, L. VYKLICKY, JR. & M. L. MAYER. 1990. A kinetic analysis of the modulation of *N*-methyl-D-aspartic acid receptors by glycine in mouse cultured hippocampal neurones. J. Physiol. **428:** 333–357.
14. LESTER, R. A. J. & C. E. JAHR. 1992. NMDA channel behavior depends on agonist affinity. J. Neurosci. **12:** 635–643.
15. JAHR, C. E. & C. F. STEVENS. 1990. A quantitative description of NMDA receptor-channel kinetic behavior. J. Neurosci. **10:** 1830–1837.
16. HOWE, J. R., S. G. CULL-CANDY & D. COLQUHOUN. 1991. Currents through single glutamate receptor channels in outside-out patches from rat cerebellar granule cells. J. Physiol. **432:** 143–202.
17. GIBB, A. J. & D. COLQUHOUN. 1992. Activation of *N*-methyl-D-aspartate receptors by L-glutamate in cells dissociated from adult rat hippocampus. J. Physiol. **456:** 143–179.
18. MAYER, M. L., L. VYKLICKY, JR. & J. CLEMENTS. 1989. Regulation of NMDA receptor desensitization in mouse hippocampal neurons by glycine. Nature **338:** 425–427.
19. SHIRASAKI, T., T. NAKAGAWA, M. WAKAMORI, N. TATEISHI, A. FUKUDA, K. MURASE & N. AKAIKE. 1990. Glycine-insensitive desensitization of *N*-methyl-D-aspartate receptors in acutely isolated mammalian central neurons. Neuroscience Letters **108:** 93–98.
20. CHIZHMAKOV, I. V., N. I. KISKIN & O. A. KRISHTAL. 1992. Two types of steady-state desensitization of *N*-methyl-D-aspartate receptor in isolated hippocampal neurones of rat. J. Physiol. **448:** 453–472.
21. CULL-CANDY, S. G. & M. M. USOWICZ. 1987. Multiple-conductance channels activated by excitatory amino acids in cerebellar neurons. Nature **325:** 525–528.
22. JAHR, C. E. & C. F. STEVENS. 1987. Glutamate activates multiple single channel conductances in hippocampal neurons. Nature **325**(6104): 522–525.
23. JACKSON, M. B. 1993. Activation of receptors directly coupled to channels. *In* Thermodynamics of Membrane Receptors and Channels. M. B. Jackson, Ed.: 249–293. CRC Press. Boca Raton, FL.
24. SIGWORTH, F. J. & S. M. SINE. 1987. Data transformations for improved display and fitting of single-channel dwell time histograms. Biophys. J. **52:** 1047–1054.

The Classification of Bioamine Receptors

How Helpful Are Molecular Phylogenies?

PHILIPPE VERNIER,[a] BRUNO CARDINAUD,[a]
HERVÉ PHILIPPE,[b] AND JEAN-DIDIER VINCENT[a]

[a]*Institute Alfred Fessard*
CNRS, UPR-2212
91198 Gif-sur-Yvette, France

[b]*Biologie Cellulaire-4*
University Paris XI
91405, Orsay, France

Monoamine receptors coupled to G proteins constitute a gene family (*i.e.*, genes encoding proteins with similar structure and functions) which exhibits a large molecular diversity. The concept of gene family implies that the origin of its multiple members is primarily gene duplications. These genes derived from a common ancestor gene by successive steps of duplications followed by mutations and selection by physiological constraints in animal species and exhibit ancestor-descent relationships. Two kind of diversification mechanisms should be considered. Firstly, gene duplications spreading in large populations yield receptor subtypes (paralogous genes), and secondly, speciation events multiply species populations which evolve on their own and homologous receptors in these species (orthologous receptors) genetically diverge to the same extent than the species themselves. The pharmacological differences between receptors reflect in part this evolutionary diversification, but pharmacology alone cannot unravel the nature of receptor homologies. "Natural" classifications based on phylogenetic methods use sequences of genes and proteins and provide clues to the genetic events which generated molecular diversity. They easily distinguish between paralogous receptors (true subtypes) and orthologous receptors (species homologues), and help to homogenize receptor nomenclature. To clarify the diversification of bioamine receptor family close to the origin of vertebrates we have cloned a dozen of dopamine receptors in "lower" vertebrates (lampreys, hagfish, European eel) and used various phylogenetical methods to classify all the known sequences of vertebrate bioamine receptors (FIG. 1). Important results are: i) Classes, subtypes, species homologues and polymorphic variants of the receptors are clearly individualized on the structural basis. ii) Carefully controlled tree topologies suggest that the main classes of receptors were generated before the divergence of vertebrates from invertebrates (D1, D2, α1, α2, β, 5HT1A, 5HT2, *etc.*), and that the diversification of these receptor classes into subtypes (D1a, D1b, D1c, α2a, α2b, α2c, *etc.*) was contemporary to vertebrate emergence and is likely to depend on two steps of genome tetraploidization. However, the use of molecular phylogeny methods has also severe limitations: the rate of receptor evolution is highly irregular in this protein family (but proportionally similar for all the domains of the receptors). iii) "Saturation" of sequence similarity occurs at the basis of the tree and the relationships of the main classes of bioamine

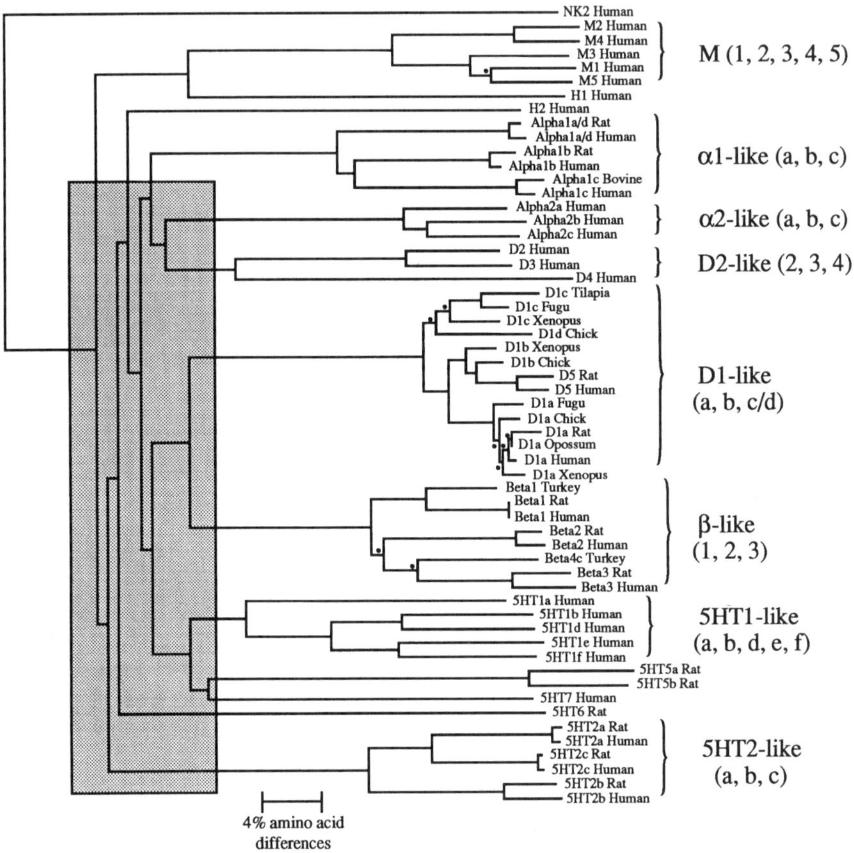

FIGURE 1. Except for the D1 dopamine receptor class, only the human orthologous genes of the various monoamine receptor subtypes were used to construct the tree. Entry of sequences, alignments, choice of the domains to be analyzed (mainly the transmembrane segments and surrounding regions), distance calculations and tree constructions were carried out using the MUST package (Philippe[1]). The presented tree was constructed with the Neighbor Joining Method of Saitou and Nei, and completed by a Bootstrap analysis which calculates the probability of occurrence of the presented branching for 1000 possible trees calculated by the program. The branching at the basis of this receptor tree, included in the grey box have little if any significance and cannot be used to infer the phylogenetic relationships of these receptor classes. The other nodes not strongly supported by the Bootstrap analysis (>80%) are marked by spots. Note the existence of at least three subtype of D1 receptors, although it is not known at present if the D1c/d subtype has homologues in mammals.

receptors cannot be resolved with confidence. The phylogenetic analysis gives essential informations on the nature of the functional constraints of this receptor family and provides a clearer understanding of the functions of bioamine receptor in the context of the physiology of organisms. The tissue-specific expression and functional peculiarities of the bioamine receptor subtypes have to be understood primarily in terms of evolutionary adaptation, according to the functions of an organism taken as a whole and not only as a sum of cells. Finally, pharmacological, biochemical and anatomical data can be integrated into the phylogenetical analysis which could be used as a framework for a more comprehensive classification integrating most of the operational data of receptor pharmacology and biochemistry.

REFERENCE

1. Philippe, H. 1993. Nucl. Acid. Res. **21:** 5264–5272.

Neurokinin Receptors

Comparison of Data from Classical Pharmacology, Binding, and Molecular Biology

D. REGOLI,[a] K. NGUYEN,[b] AND G. CALÓ[a]

[a]*Institute of Pharmacology*
Medical School
University of Ferrara
Ferrara 44100 Italy

[b]*Department of Pharmacology*
University of Sherbrooke
Sherbrooke J1H5N4 Canada

INTRODUCTION

Tachykinin pharmacology in the early 1980s was in precarious state, since no correlation could be demonstrated between results obtained in biological (guinea-pig ileum) and binding (rat brain membranes using [^{125}I]BH-SP) assays[1] (TABLE 1). We undertook a three-point program: i) utilization of isolated vessels, ii) search for monoreceptor systems, and iii) preparation and testing of selective agonists.

RESULTS

Among the most sensitive and selective blood vessels we tried for NK-1 response, the dog carotid artery (which responds with concentration-dependent relaxation to neurokinins (NKs)[2]) and the rabbit vena cava[3] (which responds with contractions to neurokinins) were identified as pure NK-1 systems. A good correlation could be demonstrated between the results of biological and binding assays, as shown in TABLE 1.[4] A similarly good correlation could be demonstrated between the biological responses of the rabbit pulmonary artery (NK-2 system[3,5]) and the binding of [^{3}H]NKA to rat duodenum membranes,[3,6] as well as between data from biological assays in the rat portal vein (NK-3 system[7]) and those of the binding of [^{125}I]BH-ELE to rat brain synaptosomes.[8]

Work on selective agonists started from the observation that DiMeC$_7$ ([pGlu^6, MePhe^8,Sar^9]SP$_{(5-11)}$) is active on NK-1 and NK-3 but not on NK-2 receptors (pD2 values: dog carotid artery = 7.5; rabbit pulmonary artery = <4.0; rat portal vein = 6.3).[3] A structure activity study in which Sar was used in position 9 of SP and MePhe in position 7 of NKB led to the discovery of [Sar^9, Met(O)$_2$^{11}]SP and Ac[Arg^6, Sar^9, Met(O)$_2$^{11}]SP$_{(6-11)}$, two potent and selective NK-1 receptor agonists and of [MePhe^7]NKB, the potent and selective NK-3 stimulant.[4,9] Indeed, these two compounds as well as [Nle^{10}]NKA$_{(4-10)}$, the selective NK-2 receptor agonist, identified in another study,[4] are extremely selective for their own receptor in both the functional and the binding assays[3,4] (see bottom of TABLE 1).

Use of "monoreceptor systems", namely biological assays whose response is contributed by a single receptor and the use of selective agonists has provided solid

TABLE 1. Relative Activities (SP = 100) of SP, Some Fragments, Tachykinins and Selective Agonists on Various Tissues and on Rat Brain Membranes

	GPI	RVD	Rat Brain ([^{125}I]BH-SP)	DCA	RbVC
SP	100	100	100	100	100
$SP_{(2-11)}$	190	80	30	47	ND
$SP_{(4-11)}$	90	230	5	4	ND
Physalaemin	160	60	190	100	178
Eledoisin	90	8600	30	32	12
Kassinin	110	18–100	10	15	14
Ac[Arg^{6}Sar^9,Met(O)$_2$^{11}]$SP_{(6-11)}$	135	<0.1	31	282	110
[Nle^{10}]$NKA_{(4-10)}$	16	40	<0.1	<0.1	<0.1
[MePhe^7]NKB	96	<0.1	<0.1	<0.1	0.1

GPI; guinea-pig ileum; RVD: rat vas deferens; DCA: dog carotid artery; RbCV: rabbit vena cava. The results concerning the effects of SP, its fragments and tachykinins are taken from ref. 1 except for the data obtained in the DCA and in the RbVC which are from ref. 3. The data regarding the effects of selective agonists are from ref. 4. ND: not determined.

and valid biological data that can be used for comparison with findings of binding assays and molecular biology experiments. One such comparison is shown in TABLE 2, using the naturally occurring peptides SP, NKA and NKB. The data summarized in TABLE 2 indicate that the order of potency of agonists is the same when measured with the three approaches and such an order defines the NK-1 receptor. Similar results have been found for NK-2 and NK-3 receptors (see review article by Regoli *et al.*[3]).

DISCUSSION

An important difference of apparent affinity is however observed between molecular biology, binding, and biological data, these last being much lower than the others. Lower accessibility of receptors in a complex tissue such as the RbVC can be invoked to explain the difference: in fact, all receptors of the cell surface of cell suspensions are readily accessible to agonists whereas only a fraction of receptors

TABLE 2. Comparison of NK-1 Receptor Affinities for the Three Neurokinins Measured with Biological (pD_2), Binding ($-logIC_{50}$) Assays and with Molecular Biology ($-logIC_{50}$) on COS Cells

Ligand (Ref. n°)	Biological Activity Rabbit Vena Cava – (3)	Binding Affinity Rat Brain Membranes [^{125}I]BH-SP (3)	Molecular Biology COS Cells [^{125}I]BH-SP[a] (10, 11)
SP	8.6	9.3	9.8
NKA	7.3	6.3	7.7
NKB	7.2	5.3	7.3

[a] Binding of [^{125}I]BH-SP was measured in plasma membranes of COS cells transfected with the cloned cDNA of the rat NK-1 receptor.

located in the smooth muscle multiple layers of the RbVC are presumably occupied and activated. Another limitation of the comparison presented in TABLE 2 is that the sets of data belong to two different species, rabbit and rat: species differences between NK-1 receptors have been demonstrated[3,12] and can in part account for the differences. A better and more precise comparison will be possible when data obtained with neurokinins, selective agonists and antagonists become available from biological and binding assays in human tissues.

REFERENCES

1. LEE, C. M., L. L. IVERSEN, M. R. HANLEY & B. E. B. SANDBERG. 1982. The possible existence of multiple receptors for substance P. N. S. Arch. Pharmacol. **318:** 281–287.
2. COUTURE, R., P. GAUDREAU, S. ST. PIERRE & D. REGOLI. 1980. The dog common carotid artery: A sensitive bioassay for studying the vasodilator effects of substance P and of Kinins. Can. J. Physiol. Pharmacol. **58:** 1234–1244.
3. REGOLI, D., A. BOUDON & J. L. FAUCHERE. 1994. Receptors and antagonists for substance P and related peptides. Pharmacol. Rev. **46:** 551–599.
4. REGOLI, D., G. DRAPEAU, S. DION & R. COUTURE. 1988. New selective agonists for neurokinin receptors: Pharmacological tools for receptor characterization. Trends Pharmacol. Sci. **9:** 290–295.
5. D'ORLEANS-JUSTE, P., S. DION, G. DRAPEAU & D. REGOLI. 1986. Different receptors are involved in the endothelium-mediated relaxation and the smooth muscle contraction of the rabbit pulmonary artery in response to substance P and related neurokinins. Eur. J. Pharmacol. **125:** 37–44.
6. BERGSTROM, L., J. C. BEAUJOUAN, Y. TORRENS, M. SAFFROY, J. GLOWINSKI, S. LAVIELLE, G. CHASSAING, A. MARQUET, P. D'ORLEANS-JUSTE, S. DION & D. REGOLI. 1987. ^{3}H neurokinin A labels a specific tachykinin binding site in the rat duodenal smooth muscle. Mol. Pharmacol. **32:** 764–771.
7. MASTRANGELO, D., R. MATHISON, H. J. HUGGEL, S. DION, P. D'ORLEANS-JUSTE, N. E. RHALEB, G. DRAPEAU, P. ROVERO & D. REGOLI. 1987. The rat isolated portal vein: A preparation sensitive to neurokinins, particularly neurokinin B. Eur. J. Pharmacol. **134:** 321–326.
8. BEAUJOUAN, J. C., Y. TORRENS, A. VIGER & M. J. GLOWINSKI. 1984. A new type of tachykinin binding site in the rat brain characterized by specific binding of a labelled eledoison derivative. Mol. Pharmacol. **26:** 248–254.
9. DRAPEAU, G., P. D'ORLEANS-JUSTE, S. DION, N. E. RHALEB, N. ROUISSI & D. REGOLI. 1987. Selective agonists for substance P and neurokinin receptors. Neuropeptides **10:** 43–54.
10. SHIGEMOTO, R., Y. YOKODA, K. TSUCHIDA & S. NAKANISHI. 1990. Cloning and expression of a rat neuromedin K receptor cDNA. J. Biol. Chem. **265:** 623–628.
11. INGI, T., Y. KITAJIMA, Y. MINAMITAKE & S. NAKANISHI. 1991. Characterization of ligand binding properties and selectivities of three rat tachykinin receptors by transfection and functional expression of their coded cDNAs in mammalian cells. J. Pharmacol. Exp. Therap. **259:** 968–975.
12. GITTER, B. G., D. C. WATERS, R. F. BRUNS, N. R. MASON, J. A. NIXON & J. J. HOWBERT. 1991. Species differences in affinities of nonpeptide antagonists for substance P receptors. Eur. J. Pharmacol. **197:** 237–238.

The α and ω of G Protein–Coupled Receptors

A Novel Method for Classification

HENRI MOEREELS, PAUL J. LEWI, LUC M. H. KOYMANS, AND PAUL A. J. JANSSEN

Janssen Research Foundation
B-2340 Beerse, Belgium

For comparison and classification of G protein–coupled receptors we used their primary structures instead of their pharmacological profiles. With the use of receptor identity profiles and spectral map analysis this resulted in a "receptor map" which represents a graphical receptor classification.[1] Here, a novel way for classification is presented for 113 neurotransmitter and opioid receptors based on counting amino acid residues. With the use of sequence alignment,[2] 10 conserved key residues were identified: α (the first Met residue of thc sequence), Asn in the tentative first transmembrane domain (TM1), Asp (TM2), Cys (top of TM3), Arg (bottom of TM3), Trp (TM4), Cys (in the loop between TM4 and TM5), Pro (TM5), Pro (TM6) and Pro (TM7). The key residue ω represents the last (non-conserved residue of the sequence. These key residues define 10 segments which are indicated in the insert of FIGURE 1. Consecutive segments form a partition. The number of residues between various key residues has already been found to be characteristic for each receptor type or subtype.[3] The surprising and striking new finding is that all 113 sequences can be classified uniquely by means of only three partitions. These are N_1-$C_{4/5}$, $C_{4/5}$-P_6 and P_6-ω, respectively. Hence, each sequence can be characterized by three numbers which represent the number of residues in the corresponding partitions. These numbers are used as the coordinates of a diagram which we have called BIN-map and which is shown in FIGURE 1. In this diagram a bin is regarded as a three-dimensional box which encloses all the sequences that belong to one and only one particular receptor type or subtype. Let us consider as an example the number of residues between the conserved key residues for the human-β_2 receptor sequence: α-49-$\mathbf{N}_1$-27-$\mathbf{D}_2$-26-$\mathbf{C}_3$-24-$\mathbf{R}_3$-26-$\mathbf{W}_4$-32-$\mathbf{C}_{4/5}$-19-$\mathbf{P}_5$-76-$\mathbf{P}_6$-34-$\mathbf{P}_7$-90-ω. This results in N_1-$C_{4/5}$ = 135 (27 + 26 + 24 + 26 + 32), $C_{4/5}$-P_6 = 95 (19 + 76) and P_6-ω = 124 (34 + 90). The sequences of the rat, mouse and hamster only differ from the human sequence in P_6-ω having a value of 129. As such the β_2-receptor sequences are located in the bin 124–129 on the P_6-ω axis. The BIN-map reveals that the combination of a long $C_{4/5}$-P_6 with a long P_6-ω is not observed in the sequences. The same holds for a short $C_{4/5}$-P_6 in combination with a short P_6-ω. An interesting and intriguing observation is that the BIN-map, based on counting amino acid residues, shows a remarkable resemblance to the "receptor map"[1] resulting from spectral map analysis obtained from receptor identity profiles. This resemblance and the observation that different receptors can be described by counting residues between conserved key residues are still not explained.

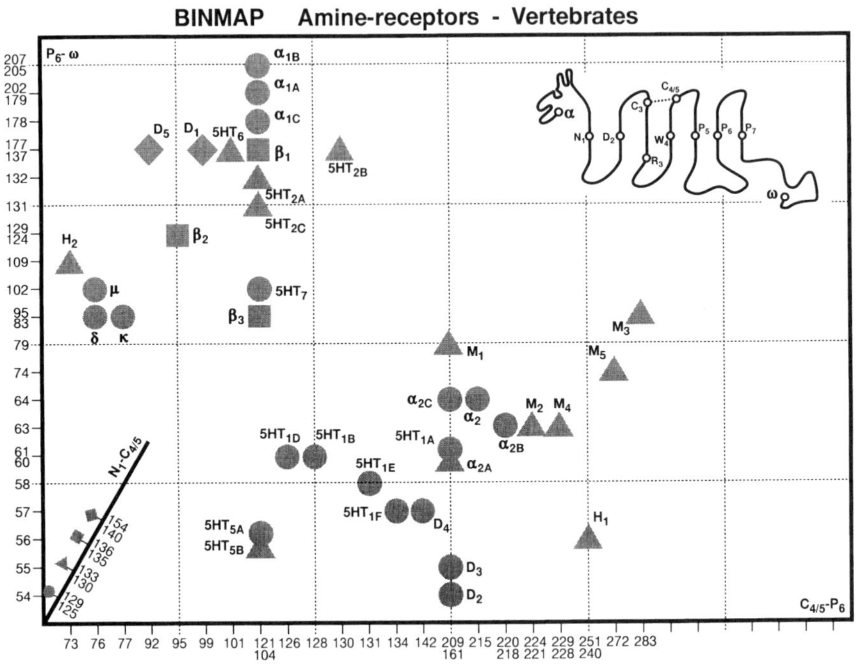

FIGURE 1. BIN-map for 113 neurotransmitter and opioid receptors. The key residues are indicated in the right hand top corner. Each subtype indicated in the map covers a number of sequences; β_1: 4 sequences, β_2: 4, β_3: 3, α_{1A}: 2, α_{1B}: 3, α_{1C}: 2, α_{2A}: 4, α_{2B}: 4, α_2C: 3, α_2: 1, D_1: 4, D_2: 7, D_3: 4, D_4: 1, D_5: 2, H_1: 4, H_2: 3, $5HT_{1A}$: 3, $5HT_{1B}$: 4, $5HT_{1D}$: 3, $5HT_{1E}$: 1, $5HT_{1F}$: 3, $5HT_{2A}$: 4, $5HT_{2B}$: 3, $5HT_{2C}$: 3, $5HT_{5\alpha}$: 2, $5HT_{5\beta}$: 2, $5HT_6$: 1, $5HT_7$: 4, M_1: 4, M_2: 4, M_3: 3, M_4: 4, M_5: 2, μ: 2, δ: 3, κ: 3 (© Janssen Research Foundation, 1994).

REFERENCES

1. LEWI, P. J. & H. MOEREELS. 1994. Receptor mapping and phylogenetic clustering. *In* Advanced Computer-Assisted Techniques in Drug Discovery, **3:** 131–162. H. van de Waterbeemd, Ed. VCH Weinheim.
2. MOEREELS, H., L. DE BIE & J. P. TOLLENAERE. 1990. CGEMA and VGAP: A colour graphics editor for multiple alignment using a variable GAP penalty. Application to the muscarinic acetylcholine receptor. J. Comp.-Aided Mol. Design **4:** 131–145.
3. MOEREELS, H., P. J. LEWI, L. M. H. KOYMANS & P. A. J. JANSSEN. 1996. The α and ω of G-protein coupled receptors: A novel method for classification, Part 1. Receptors and Channels. **4:** 19–30.

Evolutionary Conservation of the 5-HT$_{2B}$ Receptors

J.-F. COLAS,[a] D.-S. CHOI,[a] J.-M. LAUNAY,[b]
AND L. MAROTEAUX[a,c]

[a]*Université de Strasbourg*
IGBMC CNRS-INSERM
BP 163-67404 Illkirch Cedex, France

[b]*FRA C. Bernard*
Faculté de Pharmacie
4 Av. de l'Observatoire
75006 Paris, France

We recently characterized the 5-HT$_{2B}$ subfamily of serotonin receptors from human, mouse, and *Drosophila* species. The protein sequences and the pharmacology of these receptors are related, indicating a possible evolutionary conservation.

The 5-HT$_2$ family of serotonin (5-hydroxytryptamine, 5-HT) receptors mediates many central and peripheral physiological functions of serotonin in vertebrates. Central nervous system effects include neuronal sensitization to tactile stimuli and mediation of the hallucinogenic effects of lysergic acid diethylamide and related phenylisopropylamine hallucinogens; cardiovascular effects include contraction of blood vessels and alteration of platelet shape. The 5-HT$_2$ receptor (5-HT$_2$R) family comprises i) the ubiquitous 5-HT$_{2A}$R (formerly 5-HT$_2$R); ii) the 5-HT$_{2C}$R (formerly 5-HT$_{1C}$R), mainly expressed in the choroid plexus; and iii) the rat fundic 5-HT$_{2F}$R and the mouse 5-HT$_{2C}$R which have recently been assigned as to the new 5-HT$_{2B}$R subclass.[1] In addition, we recently cloned a 5-HT$_2$-like R in *Drosophila.*

Alignment of the mouse 5-HT$_{2B}$ sequence[2] with that of various aminergic receptors, shows less than 30% homology with rat 5-HT$_{1A}$R and mouse 5-HT$_{1B}$R. This places this mouse cDNA within the 5-HT$_2$R family.[3] However, the levels of sequence identity with both the mouse 5-HT$_{2A}$R (60%) and 5-HT$_{2C}$R (57%) are low. By comparison, homologies between 5-HT$_{2A}$R and 5-HT$_{2C}$R of the same species are around 70%, and those between the 5-HT$_{2A}$R or the 5-HT$_{2C}$R from different species are over 95% and 85% respectively. The low similarity of the rat 5-HT$_{2B}$R with the rat 5-HT$_{2A}$R and 5-HT$_{2C}$R (68%, and 70%, respectively) compared to its 88% homology with the mouse 5-HT$_{2B}$R protein over the transmembrane (TM) region, supports the classification of this rat receptor within the 5-HT$_{2B}$R subfamily. The human 5-HT$_{2B}$R,[4–6] displays only 79% and 82% identity to the rat and mouse 5-HT$_{2B}$ protein, respectively. The lower homologies observed with human 5-HT$_{2A}$R (71%) and 5-HT$_{2C}$R (73%) justify its classification as the third member of the 5-HT$_{2B}$R subfamily. Interestingly, we have cloned a distantly related *Drosophila* receptor cDNA that codes for a larger protein of 869 amino acids,[7] which is more similar (40% homology over the TM region) to the mammalian 5-HT$_{2A}$R, 5-HT$_{2B}$R, or 5-HT$_{2C}$R than to any other receptor so far known including *Drosophila* 7TMRs. However, original characteristics distinguish this receptor, such as the presence of

[c] Author to whom correspondence should be addressed.

TABLE 1. Competition Experiment for [^{125}I] DOI-Labeled Membrane from 5-HT_2Dro and 5-HT_2Dro N-terminus Deleted (N) COS-1 Transfected Cells: Comparison with the Rat 5-HT_{2A}, Mouse (M) and Human (H) 5-HT_{2B} and Pig 5-HT_{2C} Receptors

Drug		pKi					
		5-HT_{2A}	5-HT-$_{2B}$M	5-HT_{2B}H	5-HT_{2C}	5-HT_2DroN	5-HT_2Dro
Agonists							
1	*N*-Acetyl-5-HT	5.5	8.1	8.1	6.1	7.2	7.3
2	Tryptamine	6.0	6.9	6.8	7.2	5.8	5.9
3	5-CT	3.5	6.7	7.5	5.7	4.4	5.5
4	(±)α-Me-5-HT	7.3	6.5	6.8	7.3	6.8	6.7
5	5-HT	5.5	5.8	7.9	7.5	7.8	6.7
6	1-Me-5-HT	6.3	5.6	5.7	8.4	5.5	5.6
7	5-MeOH-Tryptamine	5.5	5.6	7.6	7.6	5.3	5.4
8	8-OH-DPAT	5.0	5.3	5.5	5.2	4.9	5.0
9	Quipazine	6.2	5.2	6.6	6.7	5.2	5.1
10	2-Me-5-HT	5.2	5.1	7.0	5.8	5.8	5.7
11	5-Me-Tryptamine	5.2	5.0	4.7	8.1	4.9	5.0
12	NNdiMe-5-MeOHTrypt.	6.2	4.9	5.2	7.0	4.2	4.3
28	RU 24969	6.0	4.9	7.2	6.5	4.3	4.4

Antagonists							
13	Ritanserin	9.3	8.4	7.6	8.6	8.3	8.4
14	Pizotifen	7.8	8.3	8.1	8.1	8.1	8.2
15	Methysergide	8.6	7.9	7.2	8.6	7.1	7.2
16	Mesulergine	8.4	7.7	7.4	8.8	7.2	7.3
17	Methiothepine	8.8	7.5	7.5	7.6	7.1	7.2
18	Cyproheptadine	8.5	7.6	7.3	7.9	7.2	7.4
19	Ketanserin	8.9	7.2	6.4	7.0	8.2	7.2
20	Spiperone	8.8	7.2	4.9	5.9	7.2	7.3
21	Setoperone	8.6	7.2	6.9	7.3	7.7	7.6
22	TFMPP	6.6	7.1	7.2	7.2	7.2	7.3
23	Rauwolscine	6.1	6.7	7.0	5.8	6.7	6.8
24	Buspirone	6.1	6.6	5.8	5.1	6.5	6.6
25	Yohimbine	6.0	6.4	7.4	4.4	6.4	6.4
26	Bufotenine	6.4	5.8	6.2	7.2	6.0	6.8
27	ICS 205-930	5.3	5.0	4.8	4.6	4.9	5.0
29	Iodocyanopindolol	4.9	4.7	4.3	5.0	4.8	4.6
30	MDL 72 222	6.8	4.7	<4	6.3	4.7	4.5
31	Mianserin	8.0	6.4	7.1	8.1	5.7	5.8
32	*Cis*-flupenthixol	7.6	6.6	6.8	6.1	5.4	5.5
33	Haloperidol	6.6	6.5	4.8	4.8	5.3	5.5
34	Clozapine	7.2	6.4	6.3	5.9	5.5	5.6
35	Sulpiride	4.5	4.9	4.3	4.1	6.0	5.9

Each value is the mean of at least 3 independent trial run in triplicate. 5-CT, 5-carboxamidotryptamine; 8-OH-DPAT, (±)-8-hydroxy-2-(di-n-dipropylamino)tetralin; TFMPP, N-(3-trifluoromethylphenyl)piperazine.

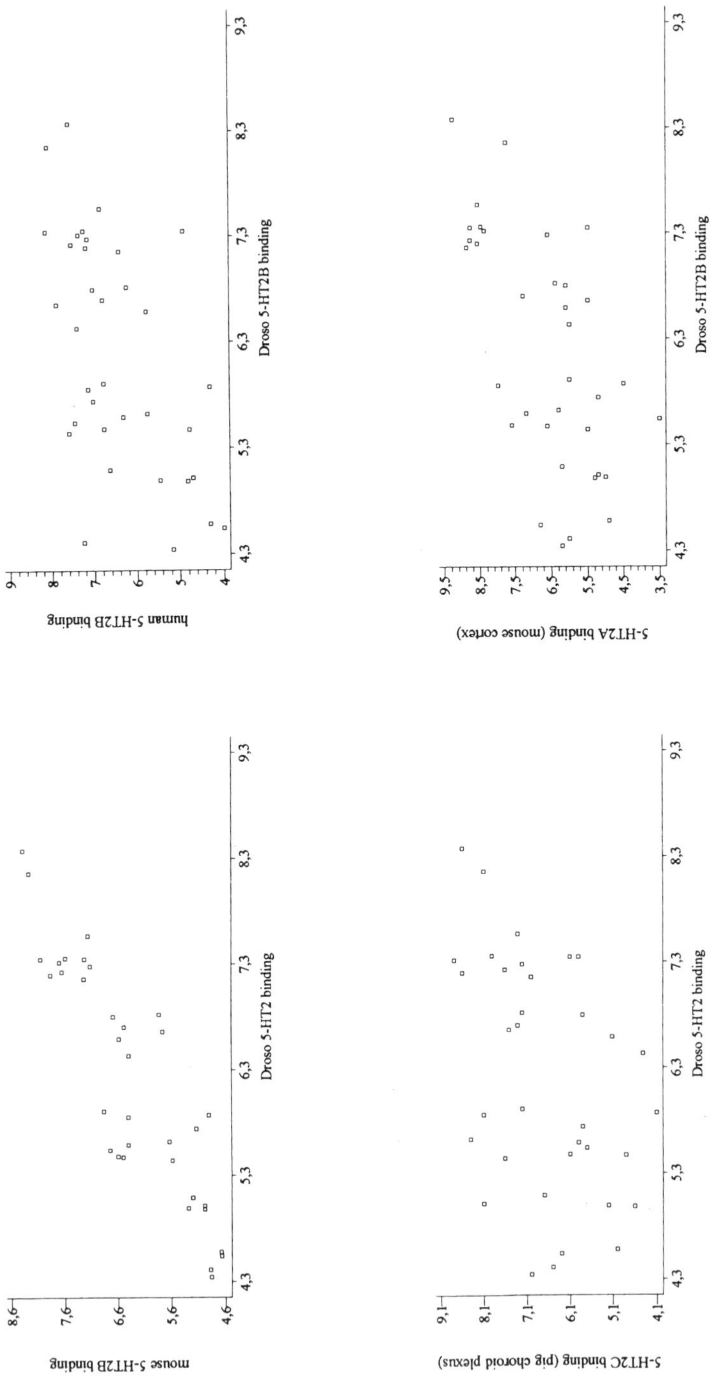

FIGURE 1. Comparison of pKi values of *Drosophila* 5-HT$_2$R with other 5-HTR.

a long amino-terminus also described in the *Drosophila* muscarinic acetylcholine, serotonin 5-HTDro$_1$, and octopamine receptors.[8]

Confirming the sequence comparisons, the pharmacology of the 5-HT$_{2B}$Rs differs from that of other 5-HTRs. After transfection of the cDNAs into the monkey kidney COS-1 cells, the pharmacological profile of the 5-HT$_{2B}$Rs were analyzed. The pharmacological values of the mouse 5-HT$_{2B}$R correlate with those of 5-HT$_{2A}$R (rs = 0.64, p <0.001) or 5-HT$_{2C}$R (rs = 0.42, p <0.05), but not with other 5-HT$_2$R subtypes. Furthermore, like 5-HT$_{2A}$R and 5-HT$_{2C}$R, the mouse 5-HT$_{2B}$R transfected into fibroblast cell line is coupled to the PLC second messenger system as well as in the serotoninergic cell line 1C11*.[9] The pharmacological profile of the human 5-HT$_{2B}$R correlates with that of the mouse 5-HT$_{2B}$R (rs = 0.60, p <0.001) and pig 5-HT$_{2C}$R (rs = 0.44, p <0.01) but not rat 5-HT$_{2A}$R. Moreover, the pharmacology published for the rat 5-HT$_{2B}$R, either transiently expressed in COS cells[10] or stably expressed,[11] does not correlate with that of the mouse; however, strong correlation is obtained with the human 5-HT$_{2B}$R (not shown). Thus, despite their structural differences, the pharmacology of the rat 5-HT$_{2B}$R is closer to that of the human than to that of the mouse. The comparison of *Drosophila* 5-HT$_2$R pharmacological properties (FIG. 1, TABLE 1) with those of other 5-HTR shows a good correlation with those of mouse 5-HT$_{2B}$R (rs = 0.87, p <0.001), pig 5-HT$_{2C}$R (rs = 0.64, p <0.001), rat 5-HT$_{2A}$R (rs = 0.60, p <0.001), or human 5-HT$_{2B}$R (rs = 0.55, p <0.01),[1,9] but no significant correlation is seen for any other serotonin receptor (not shown). Therefore, it appears closer to the mammalian 5-HT$_2$Rs than to the *Drosophila* 5-HT$_1$DroR, 5-HTDro$_{2A}$R, and 5-HTDro$_{2B}$R.[8] In accordance with the new mammalian nomenclature,[1] we propose that this receptor be called 5-HT$_2$Dro and that 5-HTDro$_{2A}$, 5-HTDro$_{2B}$, and 5-HT$_1$Dro, be renamed 5-HT$_{1A}$Dro, 5-HT$_{1B}$Dro, and 5-HT$_7$Dro, respectively.

The original long amino-terminus of the 5-HT$_2$DroR, when deleted affects the pharmacology only slightly (TABLE 1): The most important effects are seen for the 5-HT, 5-CT, and ketanserin which cause a 10-fold change in affinity. This indicates that the external N-terminus affects the conformation of the receptor TM regions involved in interactions with these compounds, information which may be of importance for receptor modeling, which usually uses deleted N-terminus receptor sequence.

REFERENCES

1. HOYER, D. *et al.* 1994. Pharmacol. Rev. **46:** 157.
2. LORIC, S. *et al.* 1992. FEBS Lett. **312:** 203.
3. PEROUTKA, S. J. 1992. Neuropharmacology **31:** 609.
4. SCHMUCK, K. *et al.* 1994. FEBS Lett. **342:** 85.
5. KURSAR, J. D. *et al.* 1994. Mol. Pharmacol. **46:** 227.
6. CHOI, D.-S. *et al.* 1994. FEBS Lett. **352:** 393.
7. COLAS, J.-F. *et al.* 1995. Proc. Natl. Acad. Sci. USA **92:** 5441.
8. HEN, R. 1992. Trends Pharmacol. Sci. **13:** 160.
9. LORIC, S. *et al.* 1995. Mol. Pharmacol. **47:** 458.
10. FOGUET, M. *et al.* 1992. EMBO J. **11:** 3481.
11. WAINSCOTT, D. B. *et al.* 1993. Mol. Pharmacol. **43:** 419.

General Discussion

J. BLACK: I have had two wonderful days listening to some of the best papers on receptor pharmacology and biochemistry which I have heard in a long time and I congratulate the organizers for bringing all these expert people together. This symposium is being held under the auspices of IUPHAR and the European Federation of Pharmacological Societies (EPHAR). When we are dealing with something which is organized on behalf of these Societies, we should keep at the back of our minds that pharmacology is about drugs, and drugs are about sick people.

Pharmacology has been for a long time and still is, concerned with the description of the effects of drugs on living systems usually in relationship to physiological or pathophysiological consequences. However, since the discovery of enzymes in the 1920s and progressively onwards, there has been the development of a pharmacological activity which attempts to analyze the actions of drugs in terms of underlying chemical systems. This analytical aspect of pharmacology is a dominant theme among pharmacologists today. The descriptive aspect of pharmacology is related to the *selectivity* of drugs and the analytical aspect has got to do with the *specificity* of drugs.

Since the 1940s pharmacologists have been using biochemical information, methods and metaphors to classify drugs in order to define expectations and outcomes particularly in relationship to their clinical use. Enzymes provided the first basis for the effective specification of drug actions. This was possible because in the 1920s and 1930s biochemists got together to deal with an explosion of information in their field. From a very early stage, they set about ordering and defining the new types of enzymes which were being described. Thus, pharmacologists had available to them an underlying orderly substrate against which they could measure and classify the action of drugs.

When I began in pharmacology, receptors were only used as a word by pharmacologists and as a mathematical operator by a small subgroup interested in the algebra developed to describe pharmacological experience. Subsequently, the concept of receptor developed into an explanatory model of drug action. I think the first successful example of this was Ahlquist's definition of the α- and β-adrenoceptor subtypes as an explanation for the behavior of both natural (noradrenaline, adrenaline) and foreign substances (isoprenaline). Ahlquist considered his work as an intellectual ordering exercise. He did not regard the receptors as physical entities but as intellectual constructs to explain the different behavior of the ligands he was investigating. Clearly, there is still an enormous intellectual element involved and the challenge of this meeting is how it is we are going to use certain chemical ideas and information to classify these receptors and provide a basis for describing drug actions.

In the latest developments receptors are now defined as chemical entities and we all know that the development of ideas about receptors has been associated with a quite remarkable explosion of selective and specific drugs. What comes out of it is a fascinating circularity: on the one hand we use the definition of receptors to classify drugs, on the other hand we use drugs to expose complexity in receptor classification.

The question that arises now is the definition of receptor. Our colleagues in physiology have been using the word receptor for years in terms such as chemoreceptor, mechanoreceptor or baroreceptors. The word "receptor" seems to me to be meaningless without a descriptive suffix. The one I have been proposing for a long while is "hormone receptor," based on the fact that hormones are nothing else other than chemical communicators, that is, molecules involved in intercellular communication. In the first 60 years of this century, the endocrinologists talked about hormones and never about receptors, whereas the pharmacologists talked about receptors and never about hormones. It is only comparatively recently that the two aspects have come together, with hormone receptors being recognized as the interface between the hormone and its effect. However, it is the hormone which has traditionally been the instrument for illuminating the existence of the cellular transduction machinery. The traditional pharmacological method for doing this was to start from the physiological end of systems using for example organ bath techniques and even whole animals. The only thing which the pharmacologists could do was to vary the dose or the concentration of the hormone or its surrogates and so pharmacologists have been obsessed with dose response curves and with their interpretation. Indeed I would say a major part of pharmacology today is involved with hermeneutics, that is, methods of interpreting the finer points of dose-response curve data.

With the development of new analytical methods and biochemistry, the transduction machinery has been progressively exposed, adding a new layer of complexity to the problem of receptor classification. Finally, the latest stages of research have been the molecular cloning and expression of receptors.

What has been exposed during this meeting is a massive amount of analytical data and concepts, revealing the immense complexity and variation of the receptor proteins and of the G-proteins, channels and enzymes coupled to them. The challenge thrown to us by IUPHAR is how to handle this complexity in a way which will be suitable for the classification of drugs.

Biochemists use an enzyme classification system that ignores structure. However, it is becoming evident that the least ambiguous part of our knowledge about receptors is the structure of proteins, once they have been isolated and cloned. Therefore, structural data are important by themselves.

How can we handle the massive amount of structural, biochemical and pharmacological information that we have and that we are adding to every day? The answer could be by setting up a relational database, which can be addressed in many different ways, only limited by our imagination. Besides the structural information, all the biochemical data can be tucked away in the database, ready for being queried. How then can we query the database? The way could be to start from the hormone. The hormone could be used to identify a group of related receptor proteins. Fortunately, only a relatively small number of proteins has been so far allied to one hormone. We might number the identified proteins from 1 to n which will make communications unambiguous. The unambiguity of chemical structure and the selection of the hormone should be the basis for this system. Thus, what we require is a database which we address with the hormone and then ask for related information. Whatever the final solution is, we have to think very hard to get a system that will be useful for all disciplines.

E. Barnard: I fully support the idea of a database as a component of receptor classification, since it will allow the elimination of a great deal of enquiry. However, once it has been set up, we still need a system of classification according to which new types or subtypes of receptors can be defined and named prior to inserting it into the database.

J. Black: Can't we simply number the receptor subtypes in the order they show up? This is what the enzymologists have been doing for many years. We have no problem with the classification of complement proteins. We have now 78 different

CD proteins and we can simply look them up. Those which are accessed frequently we will soon know; the others are still there in the database.

P. HUMPHREY: I think that for simplicity we have to maintain trivial names, which convey immediate information about the drugs or hormones that interact with a certain type of receptor. The view that is emerging in the IUPHAR Nomenclature Committee is to use both a colloquial and an intelligent numbering system, based on a series of reference numbers giving the corresponding key information. Arguably, the first number could refer to the hormone (for example, indicated with a three letter code), the second could refer to the structure of the receptor (which gives information also about the transduction mechanism). The question is then: do we number the receptor(s) as we find them? Do we follow the current classification schemes? And do we use, as I was hoping, transduction systems as a means of classification? I believe in the importance of the transduction pathway as a key characteristic of receptor activation, which is able to differentiate between receptor classes as we have done for 5-HT receptors.

J. BLACK: I don't believe that we can have a method of paraphrasing meaningfully all the different aspects of receptor classification. The question is are we dealing with entities (according to the genomic point of view) or with systems? I think we are dealing with systems. Probably, we cannot develop a method of paraphrasing in a few numbers the terrific complexity of possibility in the systems we are dealing with, but we can insert all available knowledge in the database and extract the information we need.

E. BARNARD: I think we cannot just number receptors in the order they are found, but we need some guiding principle to define and classify them. It is much more informative to present a newly cloned receptor as a $5HT_1$-type than as receptor "752." In this respect, it is important to underline that, at the moment, receptor structure is the only definitive element we have, whereas transduction pathways show too many uncertainties. Furthermore, nowadays we can get the sequence of a new receptor very quickly, whereas it may take some while to identify the transduction pathway. On the other hand, receptor structure and function do go hand-in-hand (for example, G-protein-coupled receptors which are characterized by 7 transmembrane domains).

J. BLACK: So far there seems to be no indication that we will have more than, say, 20 receptor subtypes in each class. So we get to the small number through the hormone and then get to the database. The database would not tell you how you have to read it, rather it can be read any way you want to read it.

S-E. DAHLEN: The enormous expansion of knowledge in the receptor field makes the use of a database necessary. At the same time, we should not forget the importance of communication. This means that two orders of receptor classification could be necessary, one within the specialist field and the other for "marketing." Another point that is emerging in this meeting is that receptor classification cannot anymore be based on one single aspect, but should be the result of integrating information from various disciplines (pharmacology, biochemistry, molecular biology), combining and relating all available data on receptor structure, transductional pathways, membrane protein interactions, *etc.*

N. SHANKLEY: In the database, receptors could be classified with a code corresponding to the name of the hormone, followed by a code corresponding to the protein structure and by a third code indicating the related primary transduction pathway. However, one problem with this process of classification was illustrated by Professor Milligan who reviewed the emerging evidence of "receptor promiscuity," that is, the fact that the same receptor protein may be coupled to two (or more) transduction mechanisms. Thus, we may no longer be able to classify receptors as entities. Nevertheless, a specific receptor-transducer system could be addressed in a database by labeling the receptor with an "intelligent" accession number made up of a combination of these codes. These combinations could be linked to existing trivial names.

D. HOYER: Regarding information on transduction pathways, we should not go over the top; the point is not to indicate which specific G-protein a receptor is coupled to, but whether a receptor is coupled to a G-protein or to an ion channel or whether it contains a protein kinase sequence within its structure. In other words, it is only important to have a general understanding of the function of transductional elements. Although receptor protein structure contains the information on transduction pathways, some further limited indication may be needed to be given for those not familiar with a particular field.

The way we structure a database may not be too important. Certainly, the hormone will be taken as a starting point, the order of the rest of the information may not be too important as long as we can access all the information.

G. MILLIGAN: I agree with Dr. Hoyer on the fact that extensive and detailed information on transductional mechanisms could produce almost a cell-type specific designation. At this stage of setting up the system, what is important is to have a general understanding of the function of transductional elements. Thus, although the knowledge of the primary signaling mechanisms is important, I do feel that structural elements are going to play a key role in receptor nomenclature. At the same time, the history of developments in receptor classification should be maintained, keeping a number of the trivial names by which we can easily access the information.

T. SCHWARZ: The number of receptors will certainly increase in the future, but probably not so dramatically, since we already notice today that the rate of discovery of new structures is leveling off. The importance of receptor nomenclature is communication, which implies the definition of reliable access into the system. I would like to focus on the importance of a more detailed numbering system encoding the receptor structures that would make it possible to communicate data about mutations, interactions, points of contact with G-proteins, *etc.* A residue number, counting from the start of the receptor is not informative when comparing receptors, whereas "residue x" in "TM domain y" is rather stable and informative among receptors. Otherwise there is no need for a very detailed numbering system in receptor classification.

P. HARTIG: I think that in setting up the system we want to be looking to the far future. The point of focusing a nomenclature on one gene, one receptor subtype concept allows us to deal with the issues that will be coming. Presumably, as new peptide hormone systems are discovered in the brain, there will be an explosion of hormone receptor subtypes (consider, for example, the neurokinins and somatostatin). In a data system prepared for the way that science will probably operate 10–15 years from now, the basic set of proteins and receptors in the brain as defined by the human genome will be our reference, at a molecular level, for understanding brain function and we can use this concept as an organizing principle. In addition, I think that in no way should we lose sight of the absolute critical importance of the physiological system. Some of the phenomena that we are seeing may not ever be seen in physiological systems but only in our artificial systems. One other point I wish to make is one made by Professor J. P. Green a number of years ago. He said "please think of the students to come in the future; do not make unnecessary changes to what you have already said; do not reuse names to mean something different than you used to mean by them." I think that a lot of the obvious logical things to do (and not to do) might flow from consideration of this view.

Concerning multiunit receptors (*e.g.*, the nicotinic cholinoceptor), many practitioners in this field have chosen to define a single gene name for each polypeptide making up a particular subunit and then naming the combinations (*e.g.*, alpha7 beta3). Thus, rather than being concerned about naming all possible combinations, the simpler and less confusing way to approach the question is to name each of the genes of each subunit and then, through a physiological enquiry, to find out what combinations occur in natural systems.

P. HUMPHREY: One of the aspects concerning transduction mechanisms is the fact that different profiles are observed in different systems. In particular, recombinant systems may not reflect the physiological condition of endogenous receptors, thus providing misleading information that cannot be extrapolated to the whole tissues. I think that with proper circumspection we could be more confident about transduction pathways, especially second messenger systems, with regard to receptor characterization. Thus, we might agree to broad transduction classes for grouping receptors (as we have done for 5-HT_1, 5-HT_2, and 5-HT_3).

D. REGOLI: I agree with the idea of considering the receptor as an entity. It is possible, with a code, to identify the entity with the name of the hormone for which it is made and then by numbering the different receptor subtypes. I work with a scheme in which the entity reflects the genetics and the subtypes reflect pharmacological behavior.

T. RIMELE: In setting up a relational database, the question is to establish a numbering system for the genes. For example, each gene could be identified with a number and then each receptor could be identified with a further number that points to the entire signaling pathway. If we are considering access via the Internet, then we can easily define paths to all available information behind each receptor or gene, also for every cell system and signaling pathway. Does the human genome initiative have some type of numbering system for the genes?

P. HARTIG: The genes that come out of the genome project have a sequential accession number, which is all that is needed in a database.

A. NORTH: With regard to the suggestion of a new database, what has not been raised is the fact that a considerable database does exist in the form of Genbank. There the receptors are classified by the hormone then the species, so we must be careful about redoing a classification that already exists and works well.

T. BONNER: Very little of the information that this audience wants is contained within Genbank. There is simply the correlation between the name that this community has given for the receptor and the sequence.

J. BOCKAERT: I agree with the idea that we should start with hormones, which are easy to remember, and not with numbers. It also is important to remember that, besides hormones, other receptor "ligands" exist like photons and calcium ions, so perhaps ligand is a better term than hormone. Concerning transduction mechanisms, these are too variable from one cell to the other to be used for a logical receptor classification. As far as receptor structure is concerned, current knowledge on its relation to receptor function is still incomplete and may sometimes lead to wrong conclusions. I would therefore prefer the ligand and then the gene as the building blocks of the database.

J. LOMASNEY: We do not seem to mind that various ligands can bind to an individual receptor, so why should we be bothered by the fact that an individual receptor is able to couple to different G-proteins? The way coupling is achieved is presumably determined by the affinity of the G-protein for the receptor, a parameter which is still not accessible. Probably, in the future it will be possible to determine a rank order of the interactions of an individual receptor with various G-proteins. Since G-proteins are natural ligands for the receptor, in that sense transduction is important, although as yet not practically useful for receptor classification.

D. HOYER: We recently wrote "receptors are cellular macromolecules that are concerned directly and specifically in chemical signaling between and within cells; combinations of the hormone, neurotransmitter, drug, intracellular messenger with its receptor initiate a change in cell function" [Jenkinson *et al.* (1995). Pharmacol. Rev. **47:** 255–266]. Does anyone disagree with this?

J. BLACK: I do!

J. SHAW: I too disagree. Not because I disagree with the concept of what a receptor is, but a large proportion of the pharmacological community are working on things like L-channels and a whole range of binding sites which don't have a natural ligand. I think if we exclude them we risk the danger of parallel systems developing or of the whole thing falling into disuse because of confusion.

J. BLACK: Dr. Janssen, is a photon a ligand?

P. JANSSEN: No it isn't! I am of course 100% in favor of a database but there are at least 7 reasons why setting up a reliable and complete database is going to be very difficult:

1. Polymorphism in receptors is much more common than expected.
2. There are receptor mutations which are important for the transmission of many genetic diseases.
3. There may be species-dependent differences in receptor structure; for example, the muscarinic receptor in the fruit fly Drosophila is different from the muscarinic receptor in vertebrates.
4. In the published sequences of receptors there are many errors (approximately 5%) that are difficult to detect. A reliable database should take this finding into account.
5. Most receptors are glycosylated and at present the chemical structure of the sugars is not well known. Looking to the future, information on glycoproteins would not be simple to insert into the database.
6. Most receptors coupled to G-proteins contain one or two fatty acids attached to a cysteine in the C-terminal. This is how the receptors are anchored to the membrane, the knowledge of which is still very incomplete.
7. 7-TM receptors are phosphorylated after activation: information on the exact site of phosphorylation, that should be part of the database, but may also be difficult to insert.

D. TRIST: What we need is an instrument which is going to give flexibility, in order to be able to change the system or let it evolve. At the moment, a flexible database seems to be a very good way to start, although it is going to be difficult, as emphasized by Dr. Janssen.

E. BARNARD: I wonder if I could try to summarize. I think everybody agrees with the suggestion that the hormone or chemical messenger should be named first. The following three suggestions for nomenclature have arisen:

1. First, all numbers after the hormone code are arbitrary. This means that one does not have to worry about what we have got, one simply assigns the next available number in the database.
2. The second possibility is to include some information which relates to the subclasses of that type of receptor.
3. The third possibility is to encode as much structural information as possible.

Concerning the first point, I would argue that systematic numbers have advantages over arbitrary numbers. Systematics in itself has an additional value, that is to say, the systematic classification of receptors implies the introduction of concepts of relationship (*cf.* systematic classification in animal and plant phylogeny). As far as subclasses of hormone receptors (point 2) are concerned, it would be possible to devise a simple code indicating receptor subtypes and these should be consecutively numbered. However, the question is more complex in the case of receptors with many subclasses (see for example 5-HT receptors, for which 7 or more subtypes have been identified). There is a good argument for coding each subcategory to include both structural and functional information which as I was arguing in my presentation ultimately converge. The third possibility, although it would be an extreme, implies a starting code indicating the type of receptor we are dealing with (*e.g.*, 7-TM structure), followed by numbers coding as much available information as possible.

J. BLACK: It is important to remember that IUB and IUPAC put real money into the business of classifying enzymes. Should we send a message to IUPHAR that we require the same?

M. SPEDDING: Establishing a database is an immense undertaking. Obviously, if you get it wrong at the start, it remains wrong for a long period afterwards. Setting up a database a few years ago probably would have resulted in a catastrophe. For example, splice variants would have been inappropriately coded and we would have had a very different view on G-protein coupling. I believe that structure is a prime determinant for receptor classification, but also that transduction, which can entirely change the rank order of potency of agonists in a given tissue, is an important variable to be included. We now need a logical numbering system which is robust. At present, the IUPHAR Nomenclature Committee are publishing a small series of recommendations for the use of trivial names for receptors and splice variants, *etc.*

M. LEW: I don't think we should hold back from building a database based on a current knowledge simply because we might get it wrong. We are not trying to create something which is fixed and immutable. Moreover, in a relational database it probably does not matter how we name the receptors. We can have 3 or 4 different nomenclatures used by different groups that will be drawn together by the database. Thus, it may be convenient for the medicinal chemist to deal with receptors on the basis of which drugs activate or inactivate them, whereas the molecular biologist might refer to how the genes are related. It is the database that pulls it all together.

A. NORTH: Would IUPHAR be interested in having a commercial publishing operation running the database for them? So much information is already made available through electronic media published by several companies.

P. HUMPHREY: This has been considered. It would be a costly operation and first of all we have to determine if we want to do it and also how we want it to be done before we commission anything. There are logistical problems. In established molecular databases (*e.g.*, GenBank and Swissprot) you also access pharmacological information that comes from us, pharmacologists. These databases are rapidly updated with functional information because the database managers read the pharmacological literature. Clearly we are going to have an interrelationship with these kinds of databases. As far as names are concerned, you can search by any term on computers today so it doesn't really matter what you call the receptor. I think the point about the numbering system is that it must provide a registry code which relates unambiguously to a definitive protein, but also as a systematic number provides information apropos classification.

R. BOND: There are receptors which are activated by more than one hormone (*e.g.*, adrenaline and noradrenaline for α- or β-receptors and peptides which activate the same receptor). How do you deal with that?

M. LEW: I suggest that both transmitters/hormones should be put in the database. People interested in NPY or in PYY would look at their own preferred entry but would get to the same information through the database.

M. SPEDDING: IUPHAR has contacted approximately 100 journals with regard to current classification proposals and the response has been excellent. Currently, IUPHAR works with 20 different receptor nomenclature subcommittees each with 10–15 members. Consensus has to be reached and this can be a slow process but usually leads to a classification that is accepted with time.

[At this point the meeting considered written questions submitted by delegates]

Can we still regard antagonist dissociation equilibrium constants as a "cornerstone" for receptor classification?

G. MARTIN: My confidence in using agonist information for receptor classification purposes has fallen. My fall-back position is to return to simple competitive antagonists. Can I take it then that pK_B estimates are still the way forward for drug receptor classification?

T. KENAKIN: Yes!

D. TRIST: As I understand it, if the two-state model is applicable then, in theory at least, it is possible that antagonist affinity estimates could be contaminated by an efficacy term. In practice, I believe the error would be small although it does question the reliability of pK_B estimates.

G. MARTIN: That was one observation that led to my question with particular reference to 7-TM GPCR receptors. Then there was information presented on ion channels which left me equally dubious that the approach was viable there. Dr. Pallotta reported that agonists could remain bound to the channel and could not be displaced by the addition of a competitive reversible antagonist. To me that means you cannot get to a steady-state or equilibrium situation and so pK_B estimation in this situation is also unreliable.

J. BOCKAERT: Another example comes from the observation that in cells transfected with mGlu 1 receptors, the antagonist 4-carboxylphenylglycine is able to competitively block the response of the agonist ACPC but not that of the natural agonist glutamate.

G. MARTIN: Does this mean that when we use antagonists as our cornerstone for classification and obtain two different answers with two agonists (*i.e.*, agonist-dependent pK_B values) that we have defined two receptors?

R. EGLEN: I think that antagonism is important, but it is very easy to get carried away with small receptor systems for which there are selective compounds available. In the case of muscarinic receptor subtypes, for which we have no highly selective antagonists, one is obliged to refer to the profile of antagonist affinity values. It is very difficult, in fact, to distinguish operationally prejunctional M_2 from M_4 receptors. This is also true for other receptors for which no antagonist is currently available, for example, 5-HT_7 receptors and also several peptide receptors. I would also like people to consider other receptor areas which have not been addressed at this conference, for example, olfactory receptors, which are G-protein coupled but which lack even an endogenous ligand that has been adequately described. However, there are exogenous ligands which can activate the receptor. Clearly antagonists are important but other features should be considered.

Concerning the dissociation of the ternary complex AR*G to R*G in the cube model, is it possible to observe agonism long after the agonist has left the receptor? Is there any evidence for such a receptor memory?

D. CLARK: Referring to the cube model, it is theoretically possible to observe responses even when the agonist, A, dissociates from the ternary complex AR*G. This means that we can observe agonism a long time after the agonist has left the receptor. There are, in fact, drugs with low affinity for the receptor that produce sustained responses which are not reversible by wash-out. Although I'm aware of the possibility of compartmentalization, I wonder whether an agonist may leave an "impression" on the receptor, so that it remains in a functional state long after the agonist is gone.

P. JANSSEN: In the case of rhodopsin, the photon isomerizes the 13-*cis*-retinal to the 13-*trans*-retinal and this process goes on for a long time after the photon is gone. This is a clear example that the presence of the so-called agonist is not needed.

Is constitutive activation a frequent phenomenon? How many and which constitutively activated systems have been described?

B. PALLOTA: The single channel recordings from the nicotinic acetylcholine receptor have shown what appears to be the spontaneous activity of a channel, with occasional brief msec openings, in the apparent absence of any known endogenous ligands for that receptor. Thus, under these conditions, what is recorded should represent an ion channel which is spontaneously active. We tried to look for similar activity in the case of NMDA receptors, but it is extremely difficult to ensure that background endogenous ligands have been eliminated.

Should a classification system attempt to encompass receptors with no known natural ligands, *i.e.*, orphan receptors?

S. FORD: Practically there has to be something that originally is just an entry in the database. Subsequently, as information increases it can be assigned to a class. Olfactory receptors exemplify this process. There are classified by homology and also in terms of the fact that they are found in the olfactory epithelium. I'm sure that some of those olfactory receptors are not in fact going to be olfactory; they have been misclassified. This is acceptable, at least we can look at them in a sort of manageable parcel. There are about a hundred new receptors being discovered as "orphans" every year and, in this respect, it would be better to have even a crude (and partially wrong) classification than none at all. So far, orphan receptors have been classified on homology and also on chromosome localization (which can work quite well because some receptor genes appear in clusters). If you have all the information available you may start to crack the code and I believe this will be a major challenge for the next ten years.

E. BARNARD: I think that orphan receptors are going to be a major player in this scene. Already two bacterial genomes are known (although they do not have many receptors in our sense) and the *C. elegans* genome is near completion and there are rumors that this is also true of the yeast gene. They certainly will give us final catalogues with a number of receptors that we do not know. Of course, when the human genome project is finished, there will be large numbers of still unidentified receptors. We will recognize that they are G-protein-linked receptors but we won't know what they are. I think we should have a policy towards them. I propose to ignore them until a specific ligand is identified. By definition you cannot do experiments on them until you have identified a ligand, natural or otherwise. We have enough problems with those which we can already experiment on! As far as olfactory receptors are concerned about 500 have been reported but the vast majority of those are known as sequences obtained by PCR which are homologous to the small number actually cloned and shown to respond to odorants. However, some of them may not be odorant receptors at all. Again, the best policy is ignore them until they are related to a known odorant when there will be no classification problem since those that are known as odorant receptors are very homologous in a group of their own within the main superfamily of G-protein-coupled receptors.

J. BLACK: The reason why I proposed the term "hormone receptors" rather than "ligand" is simply because of the relationship to physiology. We have been hugely successful in learning to control physiology because we control intercellular communication. That is not true in the case of odorant receptors or rhodopsin, which are not involved in intercellular communication.

How much does pathology influence receptor promiscuity and what are the implications for drug design?

G. MILLIGAN: This is a difficult question to approach in any detail, but there is a number of elements that we can consider. The first is that there are pathophysiological situations arising from mutational modification of a variety of receptors. Within the G-protein-coupled receptor field, there are examples such as retinitis pigmentosa due to a mutation in the rhodopsin receptor and diabetis insipidus due to mutations in various positions within the chain of the vasopressin V_2 receptor. To specifically address the question, I don't think that pathology changes the promiscuity of coupling, but undoubtedly many of these mutations affect the profile of receptor activation (as observed by studying the function of mutant receptors in reconstituted systems) with pathological consequences. One example that I am aware of is that of the TSH receptor for which there have been reports that mutations in position 623 change the profile between cAMP and inositide metabolism activation.

D. HOYER: In human prostate cancer muscarinic receptors seem to select another G-protein. This phenomenon can be observed even in the absence of receptor mutation. In the somatostatin field, for example, receptors are expressed in certain types of tumors which couple to phosphatases rather than adenylate cyclase.

R. EGLEN: I was particularly interested in terms of the ratio of receptors to G proteins as a function of disease and maybe even development. For example, in the case of heart failure, in addition to large changes in the population of β-adrenoceptors, changes in Gi protein levels have been reported. These findings may be relevant in terms of clinical application and maybe recombinant systems could provide potential disease models.

M. SPEDDING: This could be the subject of an entire conference. There are data on changes in receptor-G protein coupling with aging, although the importance of these alterations (in terms of pathophysiological consequences) is not completely known. Going just a little bit back from total promiscuity, there may be changes in R and R′, dependent on the membrane properties during aging or in pathological conditions such as diabetes. Many drugs may show selectivity for diseased tissues on the basis of this phenomenon.

E. BARNARD: Another case in which pathology is important in relation to receptors is the cystic fibrosis protein, CFTR. Mutations, discovered because of the progress in understanding the human genetics of the disease, definitely affect the transduction properties of the receptor in a specific way. A common mutation affects the airway CFTR more than the pancreatic cell CFTR. Probably, we will discover many other cases.

G. MARTIN: Concerning the expression "receptor promiscuity" as far as receptor-G protein coupling is concerned. I wonder whether the receptor is promiscuous or is it the G proteins which are promiscuous. This may sound like a semantic issue, but it might help to shape the way we define a receptor. If we think about the G-proteins as being promiscuous, then the receptor protein is nothing more than a sorting house for information coming in and going out and that determines the sort of taxonomy that you would want to put in place. It would make sense to have a structurally based taxonomy and thereafter we know we are dealing with an array of potential systems. At this moment, it might be an impossible task to try and order these systems, while we are still discovering what happens beyond receptor occupation.

Is intrinsic efficacy a useful parameter to measure?

T. COSTA: I believe so. If the question relates to intrinsic efficacy as defined by Furchgott then it is still the best we can use for classification.

Why are so few drugs directly targeted at G proteins?

G. MILLIGAN: Traditionally, there have been two reasons for this. The first is that there has not been a particularly easy screen to measure an agent able to regulate the G-proteins directly, not in the same way as we've had for ligand binding studies at the level of the receptor. The second reason, more trivial is that people have steered clear on the basis that an agent directed at G proteins has to cross the plasma membrane. A third element is that the guanine nucleotide binding pocket would be, by definition, relatively similar amongst a series of closely related G-proteins. The guanine nucleotide binding pocket, which would be one likely target, is clearly not only going to be in what we call G-proteins, but it will be very similar in the whole host of small molecular weight binding G-proteins (ras protooncogenes, proteins involved in protein trafficking and proteins such as tubulin). Nevertheless, we are starting to consider the possibility that it is possible to obtain agents with some selectivity for particular G proteins. For example, there are reports of up to 100-fold Gs versus Gi G-protein selectivity for drugs based on the molecular structure of suramin. At present, however, the degree of selectivity is still not high and the widespread distribution of identical G-proteins across tissues continues to pose problems.

Concluding Remarks

J. BLACK: We do seem to agree that with such a huge amount of information to which we want to have readier access; a new database dedicated to our purposes would be attractive to all of us. The problems only come in how we want to talk to each other about receptors and communicate with each other in writing. This is an editorial problem as far as writing is concerned. As far as conversation is concerned, we can do anything we like, as long as we understand each other!

Agonist Potency Ratios at m1 and m3 Muscarinic Receptors Expressed in A9L and CHO Cells

M. H. RICHARDS

Marion Merrell Dow
16, rue d'Ankara
Strasbourg, France

Expressing a given receptor in different cell lines which appear to differ in their complement of G proteins and/or effectors for a particular second messenger system provides a means to compare the effects of these components on receptor activation. A comparison was made of the formation of inositol-monophosphates (IP_1) elicited by full and partial agonists activating m1 and m3 human muscarinic receptors expressed in A9L and CHO cells at similar receptor densities.[1] In CHOm1, A9Lm1 and A9Lm3 cells, carbachol-induced IP_1 formation was approximately linear for up to 60 minutes. In CHOm3 cells preincubated with [^{3}H]-*myo*-inositol for 0.5, 24, 48 or 72 hours, IP_1 steady-state formation occurred at about 30 minutes in the presence of oxotremorine-M and carbachol and about 45 minutes with pilocarpine. Thus, pilocarpine was a weak partial agonist ($<$ 50% of maximum response) after 20 minutes stimulation and a strong partial agonist ($>$ 50% of maximum) after 60 minutes stimulation. Log occupancy curves were calculated for four full agonists (FIG. 1). At m3 receptors, the shifts in log occupancy curves between A9L and CHO cells were similar for the four agonists and the agonist potency ratios at m3 receptors did not differ between cell lines suggesting that these agonists activated the same complement of second messenger components in each cell line. In contrast, agonist log occupancy curves at m1 receptors displayed greater dispersal in CHO than A9L cells and agonist potency ratios for carbachol and methacholine were significantly different in the two cell lines.[2] This suggests that agonist structure may determine the particular isoforms of G proteins/effectors activated by m1 receptors in A9L and CHO cells.

REFERENCES

1. RICHARDS, M. H. & P. VAN GIERSBERGEN. 1995. Br. J. Pharmacol. **114:** 1241.
2. RICHARDS, M. H. & P. VAN GIERSBERGEN. 1995. Life Sci. **57:** 397.

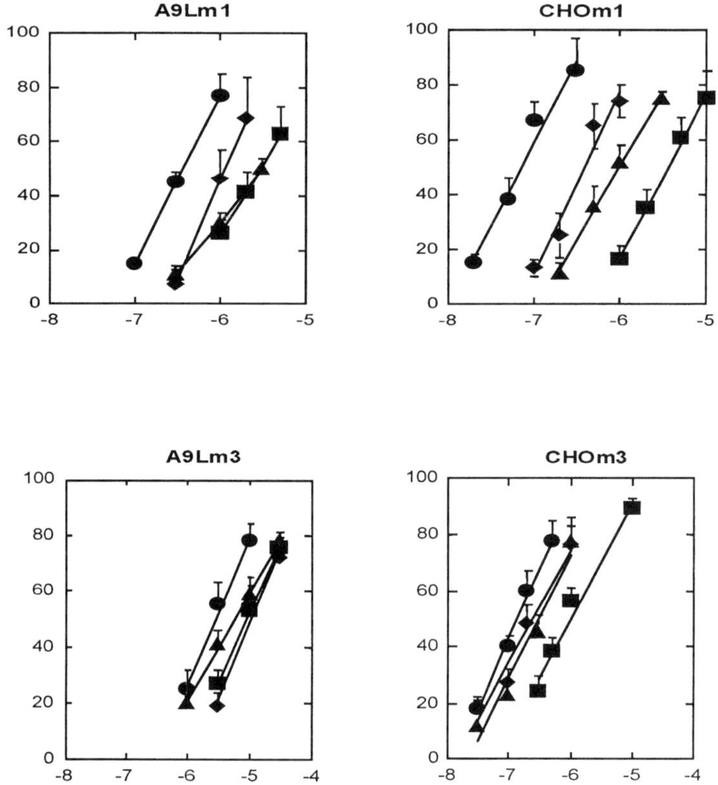

FIGURE 1. Log occupancy curves at m1 and m3 receptors expressed in A9L or CHO cells: oxotremorine-M (●), carbachol (■), methacholine (▲), muscarine (◆), n = 3 − 9. X axes = log molar concentrations, Y axes = % maximal response.

[^{3}H]5-HT Binding to 5-HT$_{1nonA\text{-}nonB}$ Receptors in Rat Hypothalamus Is Not Representative of 5-HT$_7$ Receptors

MARCO GOBBI, LUCA PAROTTI,
AND TIZIANA MENNINI

Istituto di Ricerche Farmacologiche "Mario Negri"
Via Eritrea 62
20157 Milano, Italy

A novel 5-HT receptor subtype, termed 5-HT$_7$, has recently been cloned from rat brain.[1,2] When expressed in mammalian cells it showed distinct pharmacological properties, *e.g.*, high affinity for both 8-OH-DPAT and ritanserin.[1,2] In the rat brain, the highest mRNA expression was found in the hypothalamus.[2]

In the present study, [^{3}H]5-HT binding in rat hypothalamic membranes was assayed to confirm the possibility of measuring 5-HT$_7$ receptors.[3] The binding was performed as described by Sleight *et al.*,[3] with the only difference that 3 μM ($\pm$)pindolol (instead of 0.1 μM) was used to fully saturate 5-HT$_{1A\text{-}1B}$ receptors, according to the data shown in FIGURE 1. At this concentration pindolol is inactive on 5-HT$_7$ receptors.[2] In these conditions, [^{3}H]5-HT binding could be further inhibited by 10 μM methiothepin (used to provide non-specific binding, FIG. 1). To characterize these 5-HT$_{1nonA/nonB}$ binding sites we tested 5-HT, 8-OH-DPAT, ritanserin, and mianserin. The inhibition curve of 5-HT had an IC$_{50}$ of 1.4 nM and a slope of 1.1 (FIG. 2) whereas the inhibition curves of all the other compounds (FIG. 2) had slopes significantly lower than one (0.41 $\pm$ 0.04, 0.35 $\pm$ 0.15 and 0.58 $\pm$ 0.11, respectively, means $\pm$ SD, n = 3 experiments). These results indicate that, in our hands, the pindolol-insensitive [^{3}H]5-HT binding sites are heterogeneous and, mainly, that this binding assay does not allow to *selectively* label 5-HT$_7$ receptors in rat hypothalamus, challenging the previous report.[3] They also suggest that the proposed down-regulation of 5-HT$_7$ receptors after fluoxetine treatment[3] should be considered with caution and confirmed using more selective and sensitive binding assays.[4]

REFERENCES

1. RUAT, M. *et al.* 1993. Proc. Natl. Acad. Sci. USA. **90:** 8547.
2. SHEN, Y. *et al.* 1993. J. Biol. Chem. **268:** 18200.
3. SLEIGHT, A. J. *et al.* 1995. Mol. Pharmacol. **47:** 99.
4. TO, Z. P. *et al.* 1995. Br. J. Pharmacol. **115:** 107.

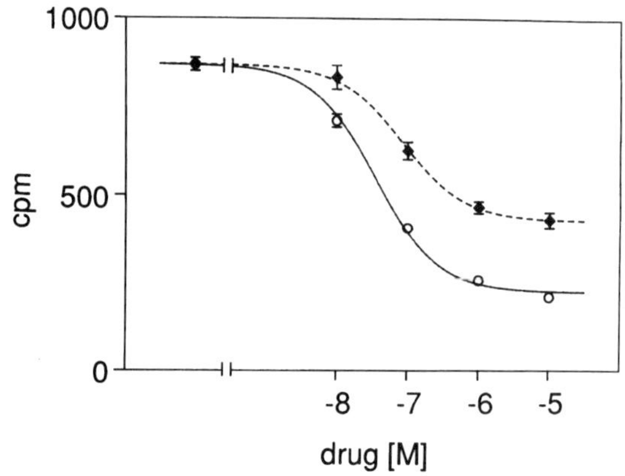

FIGURE 1. Inhibition of [^{3}H]5-HT binding in rat hypothalamic membranes by methiothepin (*open circles*) or (±)pindolol (*filled diamonds*).

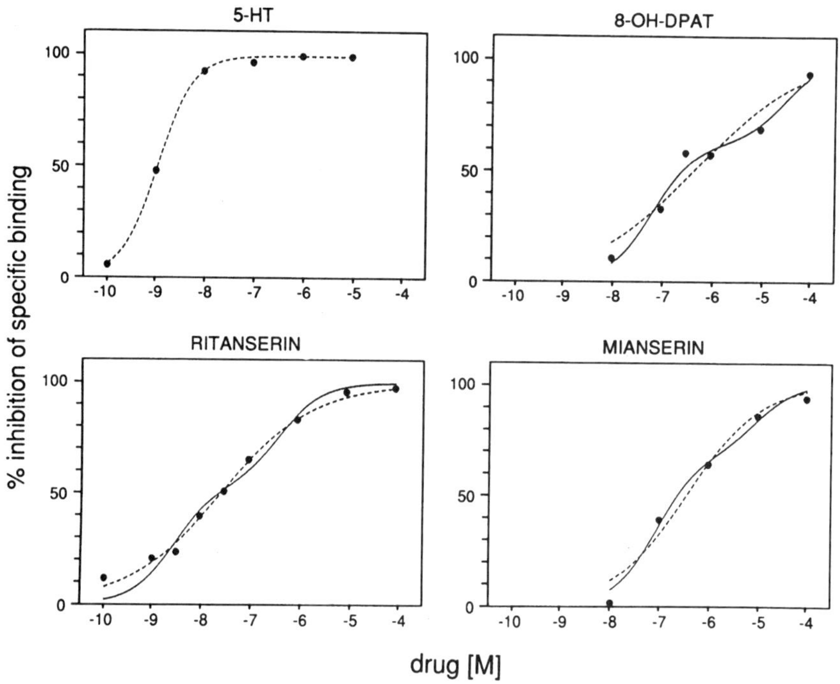

FIGURE 2. Inhibition of [^{3}H]5-HT binding in rat hypothalamic membranes by different drugs. 3 μM (±)pindolol was present in all assays, and non-specific binding was defined in the presence of 10 μM methiothepin.

Working Hypothesis on the Classification of Cys-Leukotriene Receptors in Airways

V. CAPRA, G. E. ROVATI,
AND S. NICOSIA

Institute of Pharmacological Sciences
University of Milan
v. Balzaretti 9
20133 Milan, Italy

Cysteine containing leukotrienes (Cys-LTs) have been shown to be potent smooth muscle constrictors, cause mucus hypersecretion, plasma extravasation and contribute to the onset of inflammation in the airways, therefore playing a fundamental role in asthma. Cys-LT receptors (Cys-LTR) have not yet been cloned or purified, but by integrating pharmacological and radioligand binding studies it is possible to attempt a classification. On the basis of the classical Cys-LT antagonists (FPL55712, SKF 104353, ICI 198,615, MK-571) two different receptors have been identified in guinea pig trachea, which are now named *Cys-LT$_1$* (sensitive to classical antagonists and BAY u9773) and *Cys-LT$_2$* (insensitive to classical antagonists but sensitive to BAY u9773) according to IUPHAR.[1] More recently, the existence of *Cys-LT$_3$*, resistant also to BAY u9773, has been proposed.[2] Functional data obtained in human bronchi suggested that this tissue does not contain different Cys-LTR for LTC_4 and LTD_4.[3] Identification of a true LTC_4 receptor by means of binding studies has been so far impeded, possibly because LTC_4 binds with relatively high affinity to abundant non-receptor proteins involved in its synthesis and transport. We performed binding studies with ^{3}H-LTC_4 in membranes from human lung (HL) parenchyma: in the presence of S-decyl-GSH, which binds to non-receptor LTC_4 sites, computerized analysis of binding data revealed the presence of two different classes of binding sites, while in its absence only one low affinity (non-receptor?) class was predominant in most of the experiments. The high affinity sites labeled by ^{3}H-LTC_4 seem different from those labeled by ^{3}H-LTD_4 on the basis of agonist and antagonist affinities, and the existence of two distinct binding proteins for LTC_4 and LTD_4 has been confirmed by photoaffinity labeling. In human pulmonary vein, BAY u9773 indicated the presence of a *Cys-LT$_2$* receptor,[4] which does not distinguish between LTD_4 or LTC_4. On the basis of binding, photoaffinity labeling and functional data from our and other laboratories, we propose the existence of two distinct Cys-LTR in HL parenchyma, one for LTD_4 (*Cys-LT$_4$* ?, *i.e.*, sensitive to classical antagonists, insensitive to Bay u9773), and one for LTC_4 (*Cys-LT$_3$*) and a single Cys-LTR in human bronchi (TABLE 1).

TABLE 1

	Cys-LT₁		*Cys-LT₂*		*Cys-LT₃*		*Cys-LT₄*
Sensitive to: Classical antagonists	Yes		No		No		Yes
BAY u9773	Yes		Yes		No		No ?
Agonists	LTD_4 = LTC_4	LTD_4	LTD_4 = LTC_4	LTC_4	LTD_4 = LTC_4	LTC_4	LTD_4
Distribution	Rat lung GP lung Human bronchi Human lung[a]	GP trachea	Ferret airways Human pulmonary vein Sheep bronchi	GP trachea	GP lung	Human lung[b]	Human lung[b] ?

[a] Data from literature.
[b] Data from our laboratory.

REFERENCES

1. COLEMAN, R. A., R. M. EGLEN, R. L. JONES, S. NARUMIYA, T. SHIMIZU, W. L. SMITH, S. E. DAHLÉN, J. M. DRAZEN, P. J. GARDINER, W. T. JACKSON, T. R. JONES, R. D. KRELL & S. NICOSIA. 1995. *In* Advances in Prostaglandin, Thromboxane, and Leukotriene Research, Vol. 23: 283. B. Samuelsson, *et al.*, Eds. Raven Press. New York.
2. TUDHOPE, S. R., N. J. CUTHBERT, T. S. ABRAM, M. A. JENNINGS, R. J. MAXEY, A. M. THOMPSON, P. NORMAN & P. J. GARDINER. 1994. Eur. J. Pharmacol. **264:** 317.
3. BUCKNER, C. K., R. D. KRELL, R. B. LARAVUSO, D. B. COURSIN, P. R. BERNSTEIN & J. A. WILL. 1986. J. Pharmacol. Exp. Ther. **237:** 558.
4. LABAT, C., J. L. ORTIZ, X. NOREL, I. GORENNE, J. VERLEY, T. S. ABRAM, N. J. CUTHBERT, S. R. TUDHOPE, P. NORMAN, P. J. GARDINER & C. BRINK. 1992. J. Pharmacol. Exp. Ther. **263:** 800.

Cloning and Functional Expression of Guinea Pig Atrial α_2-Adrenergic Receptor Subtypes[a]

SAMUEL P. S. SVENSSON, AMY C. PORTER,
AND JOHN W. REGAN

*Departments of Pharmacology & Toxicology, Physiology,
and the Program in Neuroscience
College of Pharmacy
University of Arizona
Tucson, Arizona 85721*

Functional studies have shown that the α_2-adrenergic receptor (α_2-AR) antagonist, SKF 104078, has very low affinity for presynaptic α_2-ARs in the guinea pig (gp) atrium. In this study, we have cloned the gp homologues of the human α_2A, α_2B, and α_2C subtypes and have studied them in isolation by heterologous expression in cultured mammalian cells. Oligonucleotide primers, designed from conserved areas of the human α_2-ARs were used in a polymerase chain reaction (PCR) with template cDNA synthesized from gp atrial mRNA. Three PCR products were obtained which shared identity with the three human α_2-AR subtypes. A gp genomic library was screened with a cDNA clone encoding a portion of the gp-α_2A and genes containing the complete coding sequences of the: gp-α_2A (450 a.a.); gp-α_2B (448 a.a.); and gp-α_2C (455 a.a.) subtypes were obtained. The amino acid sequence identity between the gp-α_2ARs and their corresponding human homologues was: 87% for the gp-α_2A; 85% for the gp-α_2B; and 89% for the gp-α_2C. The nucleotide sequences have been submitted to GenBank under accession numbers: gp-α_2A, U25722; gp-α_2B, U25723; and gp-α_2C, U25724. The gp genes were subcloned into a eukaryotic expression plasmid and were expressed transiently in COS-7 cells. The binding of the α_2-selective antagonist, [^{3}H]MK-912, to membranes prepared from these cells was specific and of high affinity with K_d of 810 pM for the gp-α_2A, 2,700 pM for the gp-α_2B and 110 pM for the gp-α_2C. Competition for the binding of [^{3}H]MK-912 by SKF 104078 indicated that it was of moderately high affinity (~100 nM) but that it was not selective for any of the gp α_2-AR subtypes (TABLE 1). The fact that SKF 104078 was not selective for any gp α_2-AR subtypes suggests that an additional subtype may still exist or that there is additional complexity with respect to the functional activation of the guinea pig atrial α_2-ARs. Co-expression of gp α_2-AR subtypes with a cAMP responsive reporter gene resulted in agonist-dependent modulation of the reporter gene activity (CAT; chloramphenicol acetyl-

[a] This work has been supported by grants from Arizona Affiliate of American Heart Association, R. W. Johnson Pharmaceutical Research Institute, NIH (EY09355), MFR and the Swedish Society for Medical Research.

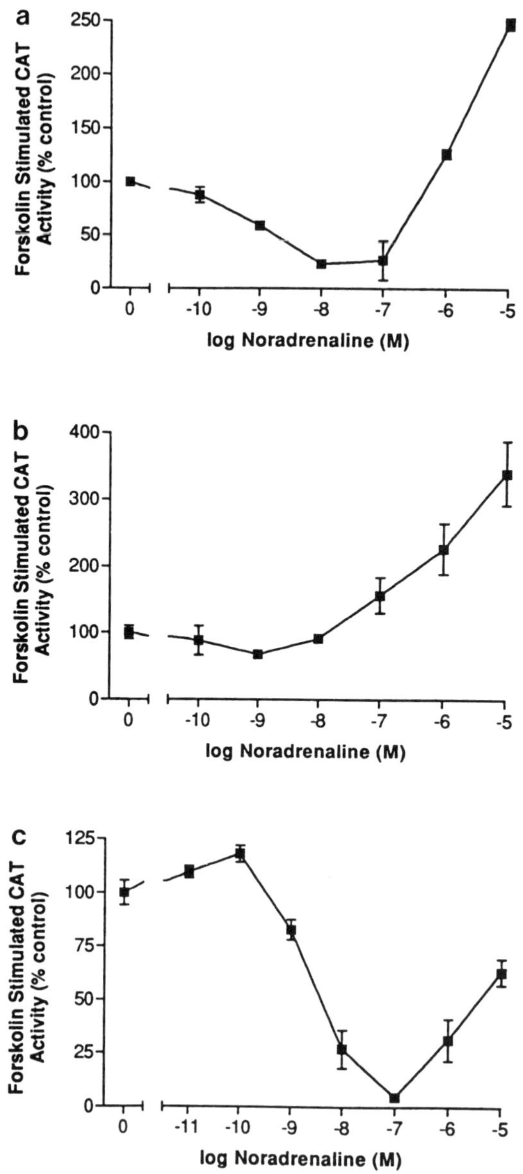

FIGURE 1. Functional interaction of the cloned guinea pig α_2AR subtypes with a cAMP-responsive reporter gene in transfected JEG-3 cells. cAMP-dependent reporter gene activity was monitored using a plasmid containing a cAMP response element (CRE) linked to the gene encoding chloramphenicol acetyl transferase (CAT). CAT activity was initially stimulated with 1 μM forskolin (FSK) and the dose-response curves were generated for each of the subtypes using noradrenaline. **a**: gp-α_2A; **b**: gp-α_2B; **c**: gp-α_2C. Data are the means ± S.E.M. for duplicate plates of cells and representative of three experiments.

transferase) (FIG. 1). Overall, the pharmacology of the cloned gp subtypes was in agreement with data obtained for the native gp receptors and was functionally similar to that of the cloned human α_2-AR subtypes. The availability of the cloned guinea pig α_2-ARs and the knowledge that they may all be expressed in the heart will help towards understanding the effects of α_2-adrenergic agents on the heart and their role in cardiac function.

TABLE 1. Binding Affinities (K_i's, nM) of Adrenergic Ligands at the Cloned Guinea Pig α_2-Adrenergic Receptor Subtypes

Agonists	α_2A	α_2B	α_2C
noradrenaline	783 ± 82	86 ± 14.5	254 ± 49
clonidine	9.5 ± 0.6	35 ± 4.5	95 ± 16
oxymetazoline	25 ± 3	250 ± 16	106 ± 43
dexmedetomidine	0.55 ± 0.10	0.83 ± 0.31	4.6 ± 0.8
Antagonists			
yohimbine	20.5 ± 4.4	81 ± 6	3.4 ± 0.2
SKF104078	79 ± 70	132 ± 52	104 ± 26
atipamezole	0.04 ± 0.02	1.8 ± 0.4	0.6 ± 0.1
prazosin	1443 ± 201	152 ± 42	163 ± 40
chlorpromazine	157 ± 76	13 ± 5.5	10 ± 2.5

K_i's were determined by nonlinear regression analysis (Prism™ software, GraphPAD Inc., San Diego, USA) with data obtained from radioligand binding competition studies using [^{3}H]MK-912. Data are the means ± S.E.M. of three separate experiments.

Apparent Heterogeneity of 5-HT_{2A} Receptors Observed in Functional Studies with Antagonists May Reflect Species-related Differences

E. J. MYLECHARANE,[a] J. LOBB,[a] L. TYNDALL,[a]
AND G. R. MARTIN[b]

[a]*Department of Pharmacology*
University of Sydney
Sydney NSW 2006 Australia

[b]*Receptor Pharmacology Group*
Wellcome Research Laboratories
Beckenham, Kent BR3 3BS, United Kingdom

Published pA_2 (or pK_B) values for ketanserin, trazodone and LY 53857 at the 5-HT_{2A} receptor mediating contraction in rat jugular vein, caudal artery, femoral artery and uterus are generally higher than those in other rat tissues (portal vein and aorta) as well as those in rabbit, guinea-pig and dog smooth muscle preparations, whereas spiperone pA_2 values are consistently higher in the rat.[1] These differences raise the possibility of intra- as well as interspecies 5-HT_{2A} receptor heterogeneity; however, variations in experimental conditions and analysis in different laboratories could either mask genuine or reveal spurious differences in pA_2 values.

Antagonist activity has now been assessed using a standardized protocol[2] in rat jugular vein, caudal artery, portal vein and aorta, and in rabbit aorta, common carotid artery and trachea. The data obtained for ketanserin, trazodone, spiperone and LY 53857 are summarized in TABLE 1. In some of these tissues, the available evidence indicates that pA_2 values may be influenced by a cocaine-sensitive 5-HT uptake system, thus in the present investigation, pK_B data have been obtained in the rat aorta in both the absence and presence of cocaine (30 μM). In the rabbit common carotid artery and trachea, competitive antagonism could not be obtained in the absence of cocaine; in the rat portal vein, however, cocaine enhanced the myogenic activity present in this tissue, resulting in non-surmountable antagonism. In the other tissues tested, cocaine did not influence antagonist pK_B values, thus the pK_B data in TABLE 1 are those obtained in the absence of cocaine.

As shown in TABLE 1, trazodone pK_B values were comparable in rat jugular vein, rat caudal artery and rat portal vein, as well as in rat aorta (provided that cocaine is present). Those for spiperone in these tissues were also comparable. These values were consistently higher than those in the rabbit tissues. Ketanserin pK_B values were similar in rat jugular vein and rat aorta (provided that cocaine is present), but higher than in the rabbit aorta; it produced non-surmountable antagonism in the other rat tissues. LY 53857 was competitive in only the rat and rabbit aorta preparations, thus intraspecies comparisons are not possible. The higher pK_B values within rat tissues, relative to those in rabbits, suggest that the apparent heterogeneity of 5-HT_{2A} receptors is species-related.

TABLE 1. 5-HT_{2A} Receptor Antagonist pK_B Values in Rat and Rabbit Smooth Muscle Preparations

Tissue	Ketanserin	Trazodone	Spiperone	LY 53857
Rat				
Jugular v. (− cocaine)	9.05 ± 0.16[a] (1.12 ± 0.10)	8.02 ± 0.07[a] (0.90 ± 0.08)	9.66 ± 0.11[a] (0.98 ± 0.14)	NSA
Caudal a. (− cocaine)	NSA[a]	8.32 ± 0.12[a] (0.91 ± 0.08)	9.63 ± 0.09[a] (0.91 ± 0.07)	NSA
Portal v. (− cocaine)	NSA	8.56 ± 0.25 (1.24 ± 0.30)	9.75 ± 0.27 (1.40 ± 0.30)	NSA
Aorta (+ cocaine)	9.22 ± 0.12 (0.90 ± 0.07)	7.86 ± 0.10 (1.12 ± 0.07)	9.64 ± 0.21 (1.23 ± 0.13)	7.78 ± 0.16 (0.96 ± 0.21)
Aorta (− cocaine)	8.81 ± 0.17 (0.82 ± 0.10)	7.58 ± 0.10 (0.90 ± 0.07)	9.27 ± 0.10 (0.98 ± 0.07)	7.75 ± 0.17 (0.88 ± 0.22)
Rabbit				
Aorta (− cocaine)	8.56 ± 0.09[a] (0.91 ± 0.09)	7.23 ± 0.19[a] (1.00 ± 0.13)	9.28 ± 0.10[a] (1.10 ± 0.06)	7.43 ± 0.14 (0.81 ± 0.11)
Carotid a. (+ cocaine)	NT	7.22 ± 0.17 (1.25 ± 0.23)	8.75 ± 0.32 (1.11 ± 0.20)	NT
Trachea (+ cocaine)	NT	7.10 ± 0.17 (0.97 ± 0.12)	NSA	NT

pK_B values and Schild plot slopes (in parentheses). Values shown are means ± S.E.; n = 3 − 7 at each antagonist concentration tested.
a. = artery; v. = vein; NSA = non-surmountable antagonism; NT = not tested.
[a] Data from Leff and Martin[2] (1986).

REFERENCES

1. MYLECHARANE, E. J. 1990. Agonists and antagonists of 5-HT_2 receptors. *In* Cardiovascular Pharmacology of 5-Hydroxytryptamine. P. R. Saxena, D. I. Wallis, W. Wouters & P. Bevan, Eds.: 81–100. Kluwer. Dordrecht.
2. LEFF, P. & G. R. MARTIN. 1986. Peripheral 5-HT_2-like receptors. Can they be classified with the available antagonists? Br. J. Pharmacol. **88:** 585–593.

[^{3}H]-RS-79948-197, a High Affinity Radioligand Selective for α_2-Adrenoceptor Subtypes

C. M. MILLIGAN,[a] C. J. LINTON,[a] L. PATMORE,[a]
N. GILLARD,[a] G. J. ELLIS,[b] AND P. TOWERS[b]

[a]*Quintiles Scotland Ltd*
Heriot-Watt Research Park
Edinburgh, EH14 4AP, United Kingdom

[b]*Amersham International plc.*
Forest Farm, Whitchurch
Cardiff, United Kingdom

At least three subtypes of α_2-adrenoceptors have now been identified on the basis of differences in tritiated antagonist binding, amino acid sequence and chromosomal location (see TiPS Receptor & Ion Channel Nomenclature Supplement, 1995). This paper describes the binding characteristics of [^{3}H]-RS-79948-197 ([8aR, 12aS, 13aS]-5, 6, 8a, 9, 10, 11, 12, 12a, 13, 13a-decahydro-12-ethanesulfonyl-3-methoxy-6H-isoquino [2,1-g] [1, 6] naphthyridine HCl (Syntex US patents No. 4,791,108 and 4,886,798).

METHODS

Membranes from rat cortices or neonate lungs were prepared as described by MacKinnon *et al.*[1] α_2C10 and OK cells (2×10^7 cells/ml) were homogenized in 10 volumes buffer (50 mM Tris HCl, 0.5 mM EDTA, pH 7.4) before centrifuging (15 min at 48,000 $\times$ *g*) and washing. In saturation assays 20–200 μg protein and [^{3}H]-RS-79948-197 (0.01–2 nM; 65 Ci mmol^{-1}) were incubated to equilibrium in a final volume of 1 ml assay buffer for 90 min at 25°C. NSB was determined in the presence of 1 μM (α_2C10 cell, lung and cortex assays) or 10 μM (OK cell assay) phentolamine. In kinetic assays association kinetics were carried out under the same conditions as above. Binding was initiated by the addition of 0.1 nM [^{3}H]-RS-79948-197 and terminated at timed intervals. At equilibrium, a saturating concentration of phentolamine was added to dissociate bound [^{3}H]-RS-79948-197. In competition assays 0.1 nM [^{3}H]-RS-79948-197 was incubated with various drugs (1 pM – 0.1 mM) and 20–200 μg protein in 1 ml assay buffer for 90 min at 25°C. NSB represented <5% in α_2C10 cells and rat cortex and approximately 10% in the rat neonate lung and OK cell assays. Data analysis was performed using 'Ligand'[2] or GraFit (Erithacus Software Ltd). The kinetic equilibrium dissociation constants were determined from the ratio $K_d = K_2/K_1$ (where $K_1 = (K_{obs} - K_2)/[\text{ligand}]$, K_{obs} is the observed initial rate constant, K_1 the association and K_2 the dissociation rate constant).

TABLE 1. Kinetic and Saturation Data for [^{3}H]-RS-79948-197

α_2-Subtype	α_{2A}	α_{2B}	α_2	α_{2D}
Source	α_2C10 cells	rat lung	OK cells	rat cortex
Kinetic expts:				
K_d (pM)	230 ± 50	30 ± 2	20 ± 2	80 ± 10
Saturation expts:				
K_d (pM)	130 ± 40	40 ± 2	20 ± 2	60 ± 10
B_{max} (fmol/mg protein)	1438 ± 140	136 ± 42	150 ± 6	231 ± 10

Values = mean ± S.E. mean (n ≥ 3).

RESULTS

Binding of [^{3}H]-RS-79948-197 to α_2C10 cells, neonatal lung, OK cells and rat cortex in kinetic assays was rapid, reversible and of high affinity and in saturation assays it bound with high (pM) and approximately equal affinity to a single population of binding sites in all four tissues (TABLE 1). In competition assays unlabeled RS-79948-197 showed low affinity for α_1 (α_2/α_1 selectivity ratio > 10,000), $5HT_{1A}$, $5HT_2$, β_1, β_2, D_1, D_2, I_2, M_1, M_2 and dihydropyridine binding sites ($pK_i < 6$). The affinities of subtype selective ligands at the [^{3}H]-RS-79948-197 site in these assays agreed well with previous studies,[1,3] confirming that it is indeed α_2 subtypes that are being labeled.

DISCUSSION

[^{3}H]-RS-79948-197 labels α_{2A}, α_{2B}, α_{2C} and α_{2D} (possible species variant of α_{2A}—see MacKinnon *et al.*[3]) adrenoceptor subtypes with very high (20–130 pM) and approximately equal affinities (2–7 fold selective for the α_{2C} subtype). This profile has significant advantages over other α_2-adrenoceptor radioligands, *e.g.*, [^{3}H]-yohimbine which has only moderate affinity (~8 nM) and poor selectivity over α_1-adrenoceptors (selectivity ratio ~100), [^{3}H]-idazoxan also labels imidazoline I_2 receptors and [^{3}H]-MK 912 which, while exhibiting high (0.46–2 nM) affinity for α_2 subtypes, is some 10–50 fold selective for the α_{2C} subtype.[4,5]

REFERENCES

1. MACKINNON, A. C., A. T. KILPATRICK, *et al.* 1992. Br. J. Pharmacol. **106,** 1011–1018.
2. MUNSON, P. J. & D. RODBARD. 1980. Analyt. Biochem. **107:** 220–239.
3. MACKINNON, A. C., M. SPEDDING & C. M. BROWN. 1994. TiPS **15**(4): 119–123.
4. UHLEN, S., X. YUN, V. CHHAJLANI, *et al.* 1992. Br. J. Pharmacol. **106:** 986–995.
5. UHLEN, S., A. C. PORTER & R. R. NEUBIG. 1994. J. Pharm. Exp. Ther. **271:** 1558–1565.

Acylation Differentiates Two Forms of Agonist Binding to Rat 5-HT$_{1A}$ Receptors

Palmitoylated G Protein—Agonist States?

CATHERINE DACQUET, CHRISTELLE MACIA, AND MICHAEL SPEDDING

Division of Molecular Biology and Screening
Institut de Recherches Servier
125 Chemin de Ronde
78290 Croissy sur Seine, France

Palmitoylation of some G proteins is a major event implicated in the coupling between receptors and G proteins. S 14506, a very strong 5-HT$_{1A}$ agonist with high affinity for the receptor (pK_1 = 9[1]) was compared with the classical 5-HT$_{1A}$ agonist: 8-OH-DPAT. We have previously defined two distinct types of agonist binding at rat hippocampus 5-HT$_{1A}$ receptors where 8-OH-DPAT and S 14506 bind to the receptors,[2] but S 14506 binding is resistant to GppNHp, whereas 8-OH-DPAT binding is reduced; binding of the two agonists is inversely modified by cations, allowing the possibility that S 14506 binds to a distinct agonist state of the receptor. [^{3}H]-8-OH-DPAT in the presence of GppNHp was used as a decoupling agent between the G protein and the receptor. When palmitoyl coenzyme A (PCoA 1 μM) was added to the hippocampus membranes, the 8-OH-DPAT competition curve became monophasic despite the presence of GppNHp (100 μM) in the incubation medium (Hill coefficient values significantly changed from 0.66 ± 0.06 to 0.96 ± 0.11, n = 3). No change in the 8-OH-DPAT K_i values was observed (2.96 ± 1.12 nM and 2.48 ± 0.64 nM, n = 3). On the contrary, addition of PCoA induced a biphasic competition curve with S 14506, suggesting the presence of high and low affinity binding sites (Hill coefficient values equal to 0.70 ± 0.03 and 0.52 ± 0.05, n = 3). S 14506 K_i values were not modified by PCoA (1.46 ± 0.43 nM and 3.16 ± 1.48 nM, n = 3). In the absence of GppNHp, no effect of PCoA on 8-OH-DPAT and S 14506 binding was observed suggesting a stronger action when the 5-HT$_{1A}$ receptors were preferentially in the decoupled state. We therefore propose that the model proposed by Bond *et al.* may be modified to:

$$\begin{array}{ccc} \text{R (inverse agonists)} & \rightleftharpoons & \text{R* (agonists)} \\ & \nwarrow\!\!\searrow \quad \swarrow\!\!\nearrow & \\ & \text{R* palm} & \end{array}$$

REFERENCES

1. LIMA *et al.* 1994. Br. J. Pharmacol. **112:** 492–498.
2. DACQUET, C. *et al.* 1994. Can. J. Physiol. Pharmacol. **72:** 454.
3. BOND, R. A. *et al.* 1995. Nature **374:** 272–274.

An Improved Method for Analysis of Competitive Agonist/Antagonist Interactions by Non-linear Regression

MICHAEL J. LEW[a] AND JAMES A. ANGUS

Department of Pharmacology
The University of Melbourne
Parkville, Victoria 3052, Australia

Functional experiments designed to determine antagonist pK_B values are almost invariably analyzed by methods based on the formula and linear regression devised by Schild.[1] This type of analysis has two potential problems: i) it requires that the protocol for each tissue includes a control curve, so a single concentration-response curve (CRC) per tissue design cannot be used; and ii) the use of concentration ratios leads to a bias towards underestimation of the pK_B with low degrees of antagonism. Methods that overcome these problems have been published,[2,3] but are not really straightforward in application, and have not been widely adopted. Here we report a new method that overcomes problems associated with Schild analysis, and is easy to apply with analysis tools employed by most pharmacologists.[4]

This method, like Schild's equation, is based on the Gaddum formula, but rather than using CRC shift as an index of antagonist effect, it simultaneously compares the locations of all of the CRCs to theoretical predictions using a nonlinear regression. Application of this nonlinear regression method involves first determination of agonist pEC_{50} values and comparisons of CRC slopes and maxima, all exactly as would be done for a Schild analysis. Next the pEC_{50}s are plotted against the appropriate antagonist concentrations. A linear scale for antagonist concentration allows the control pEC_{50}s to be plotted at [antagonist] = zero. The pK_B is obtained by a nonlinear regression using the formula

$$pEC_{50} = c - \log([B] + 10^{-pK_B}) \tag{1}$$

where [B] is the antagonist concentration and c is simply a fitting constant. Most charting applications for personal computers will provide explicitly an estimated standard error for the pK_B obtained in this way. The competitivity of the agonist/antagonist interaction can be tested by a comparison of the goodness of fit (F-test, see ref. 5) of the previous formula with two more flexible formulae that allow deviations equivalent to Schild plots that are either non-linear

$$pEC_{50} = c - \log([B](1 + n[B]/10^{-pK_B}) + 10^{-pK_B}) \tag{2}$$

or have a non-unity slope

$$pEC_{50} = c - \log([B]^n + 10^{-pK_B}). \tag{3}$$

[a] Fax: + 61 3 9347 1452; E-mail: (Michael_Lew@MUWAYF.unimelb.edu.au).

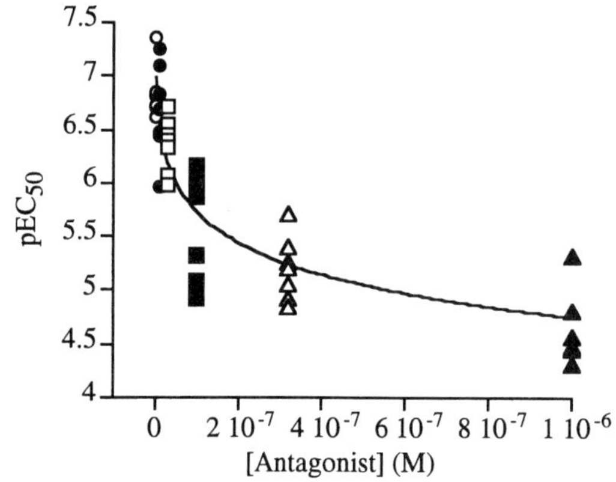

FIGURE 1. Plot of pEC_{50} vs [B] for the interaction of methysergide and 5-HT in the dog coronary (data from ref. 4). Each point is the pEC_{50} of an individual 5-HT concentration-response curve, and the curve shown is the fit of Equation 1 with the resulting pK_B of 8.2 (95% ci 7.9, 8.6). The fit was not significantly improved by equations 2 and 3.

FIGURE 1 shows an example of the plot and regression that provide the pK_B estimate, and the suggested display of the experimental results is the Clark plot, as shown in FIGURE 2.

This method treats all of the data identically, with the control CRC being represented simply with [B] = 0, rather than being assigned a special meaning as it is in the Schild methods. Because the non-linear regression method and the Schild methods are based on the same theory they yield identical values from ideal data. However, statistical bootstrap simulations[6] indicate that with non-ideal data the new method yields a slightly higher mean pK_B with a tighter and more symmetrical distribution than the Schild method. This difference is primarily the result of a bias towards underestimation of pK_B by the Schild method at low antagonist concentrations because of the log(r-1) term.

This new method is preferred over the Schild method mainly for its advantage of allowing single CRC experimental protocols, but it probably provides slightly more accurate values and is applicable to any experimental design where the Schild methods might otherwise be used.

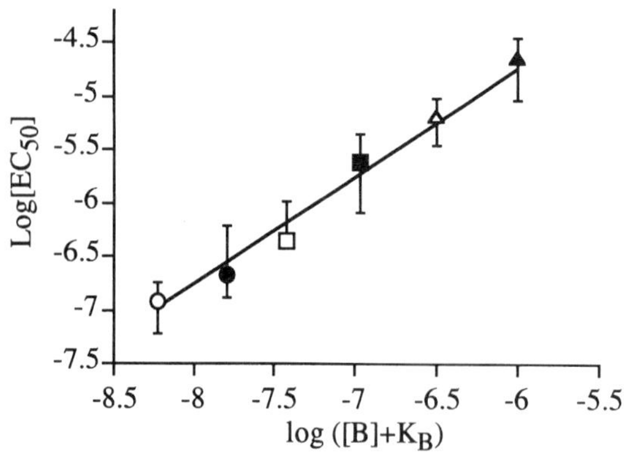

FIGURE 2. Clark plot display of the results. Each point represents the mean location of the agonist concentration-response curves at a particular antagonist concentration. The line indicates the expected ideal interaction at the determined pK_B and the error bars are the 95% ci of the line. The pK_B is indicated by the x-location of the first data point (*i.e.* [B] = 0). For a full explanation of the Clark plot see refs. 3 and 4.

REFERENCES

1. Arunlakshana, O. & H. O. Schild. 1959. Some quantitative uses of drug antagonists. Br. J. Pharmacol. **14:** 48–58.
2. Waud, D. R. 1975. Analysis of dose-response curves. *In* Methods in Pharmacology. E. E. Daniel & M. Paton, Eds.: 471–506. Plenum Press. New York.
3. Stone, M. & J. A. Angus. 1978. Developments of computer-based estimation of pA_2 values and associated analysis. J. Pharmacol. Exp. Ther. **207:** 705–718.
4. Lew, M. J. & J. A. Angus. 1995. An improved method for analysis of competitive agonist/antagonist interactions by non-linear regression. Trends Pharmacol. Sci. **16:** 328–337.
5. Kenakin, T. P. 1993. Pharmacologic Analysis of Drug-Receptor Interaction. Raven Press. New York.
6. Efron, B. & R. J. Tibshirani. 1993. An Introduction to the Bootstrap. Chapman & Hall. New York.

ERK Activation and Cellular Proliferation in Response to Muscarinic Acetylcholine Receptor Agonists[a]

PING WANG, N. DHANASEKARAN, AND GARY R. LUTHIN[b]

Allegheney University of the Health Sciences and Temple University, Fels Institute Philadelphia, Pennsylvania 19102

In cells transfected with plasmids encoding cloned muscarinic receptors, a cell proliferative response is seen in response to challenge by muscarinic receptor agonists. This is dependent on the level of receptor expression, and often can only be shown when receptors are expressed at relatively high levels. We have shown that the human prostate cancer cell lines PC3 and DU145 express both m1 and m3 receptors, at levels seen in the normal human prostate gland. These receptor subtypes are typically linked to phosphoinositide (PI) metabolism, and are therefore candidates for inducing cell proliferative responses by activation of protein kinase C (PKC). It is unclear whether activation of PKC is required for cell proliferation, and it is possible that PKC-independent responses, such as the MAP kinase (ERK) pathways, are activated directly by neurotransmitter receptors. We chose these cells to determine whether the endogenous receptors could induce cell proliferation, and to attempt to define biochemically the mechanism of signal transduction associated with the proliferative response.

We measured whole cell PI metabolism, cAMP accumulation, and the cellular mass of phospholipids (diacylglycerol, PI, PS, PE, and water-soluble metabolites) in response to carbachol. In PC3, but not DU145 cells, PI metabolism was increased by a short (5–10 minute) incubation with carbachol. cAMP levels were not altered in either cell line under the same conditions, nor was the total mass of any phospholipid measured in the absence of lithium. This is in contrast to cells expressing cloned receptor subtypes, where the m3 and m4 receptors were shown to increase PI and decrease cAMP metabolism, respectively. In both PC3 and DU145 cells ERK activity was shown to increase dramatically following carbachol treatment, and in both cell lines thymidine uptake was increased in the next 24-hour period. We have shown that in specimens of benign prostate hyperplasia (BPH), the m1 receptor was the predominant receptor subtype that was expressed. Using a co-immunoprecipitation assay, we showed that the m1 receptor interacted with the guanine nucleotide binding protein (G protein) G_{16} in addition to G_{i1} and $G_{q/11}$ in BPH but not in cerebral cortex, where the m1 receptor is also expressed. In PC3 and DU145 cells, the m1 and m3 receptors were shown to couple directly to the G proteins G_i, $G_{q/11}$, and G_{12}, but not G_{16}. BPH and metastatic prostate cancer are derived from different zones of the prostate, and is tempting to speculate that the propensity of the different zones to become metastatic relates to the coupling of receptors to

[a] This research was supported by Grant NIH DK43333 to G.R.L.

[b] Author to whom correspondence should be addressed.

potentially oncogenic G proteins. This is stated in the context that constitutively active G_{16} has been shown to suppress cell growth, while G_i, $G_{q/11}$, and G_{12} have been shown to promote cell proliferation and transformation. G_i and $G_{q/11}$ are thought to be responsible for cAMP and PI metabolism, respectively, while the function of G_{12} is not clear at this time. Of interest in these studies is that while the DU145 cells express muscarinic receptors that are believed to be linked to PI metabolism, we did not see activation of this pathway in these cells. We have therefore shown that a selectivity of cell signaling occurs at the level of receptor-G protein coupling, cellular metabolism of cAMP and PI, and activation of cell proliferative pathways in these cell lines. We are currently attempting to integrate these concepts in the context of cell proliferative responses to muscarinic cholinergic agonists, given that we now have established apparently different signaling pathways in the cells that lead to activation of ERK and cellular proliferation.

Co-Agonism between Glutamate and Glycine on the NMDA Receptor

Does It Influence the Estimation of Antagonist Dissociation Constant?

P. FINA, M. CORSI, C. CARIGNANI,
AND D. G. TRIST

Pharmacology Research Department
GlaxoWellcome S.p.A.
Via A. Fleming 4
37135 Verona, Italy

INTRODUCTION

The NMDA receptor is probably unique in that it requires simultaneous stimulation by two agonists, glutamate (Glu) and glycine (Gly). NMDA receptor activation leads to an increase of permeability of the channel to Ca^{2+}, Na^{+} and K^{+}. The co-agonism between Glu and Gly has previously been described by Marvizon and Baudry in terms of a ternary complex model.[1] This study extends their work to an operational model and tries to analyze the influence of co-agonism on the estimation of the dissociation constant for antagonists.

RECEPTOR THEORY FOR THE AGONISTS AND ANTAGONISTS RECEPTOR INTERACTION

The effect is allowed only when both Glu and Gly are bound to the receptor.

$$
\begin{array}{ccccccc}
\mathrm{RB} & \overset{\mathrm{KB}}{\Leftrightarrow} \mathrm{B}\ + & \mathrm{R} & +\ \mathrm{Gly} & \overset{\mathrm{Kgly}}{\Leftrightarrow} & \mathrm{RGly} & \\
+ & & + & & & + & \\
\mathrm{Glu} & & \mathrm{Glu} & & & \mathrm{Glu} & \\
\Updownarrow \gamma\mathrm{Kglu} & & \Updownarrow \mathrm{Kglu} & & & \Updownarrow \alpha\mathrm{Kglu} & \\
\mathrm{GluRB} & \overset{\gamma\mathrm{KB}}{\Leftrightarrow} \mathrm{B}\ + & \mathrm{RGlu} & +\ \mathrm{Gly} & \overset{\alpha\mathrm{Kgly}}{\Leftrightarrow} & \mathrm{GluRGly} & \Rightarrow \mathrm{Effect}
\end{array}
$$

R is the NMDA receptor, Kgly and Kglu are the dissociation constants for Gly and Glu, respectively. The parameter, α, is introduced in the model to account for the modification of the affinity of one agonist by the binding of the other. If Effect is defined as the response obtained in the presence of both [Glu] and [Gly], Max the maximal response obtained by [Gly] in the presence of a defined amount of [Glu], Em the maximal tissue response, τ the transducer constant, [Gly_{50}] the glycine concentration that induces 50% of the maximal response and n as the slope parameter, then:

$$\text{Effect} = \frac{\text{Max} \cdot [\text{Gly}]^n}{[\text{Gly}_{50}]^n + [\text{Gly}]^n} \quad \text{with} \quad \text{Max} = \frac{\text{Em} \cdot \left(\tau \cdot \dfrac{[\text{Glu}]}{\alpha \cdot \text{Kglu} + [\text{Glu}]}\right)^n}{1 + \left(\dfrac{\tau \cdot [\text{Glu}]}{\alpha \cdot \text{Kglu} + [\text{Glu}]}\right)^n}$$

$$[\text{Gly}_{50}] = \frac{\text{Kgly} \cdot \left(\dfrac{\alpha\text{Kglu} + \alpha[\text{Glu}]}{\alpha\text{Kglu} + [\text{Glu}]}\right)}{\left[2 + \left(\dfrac{\tau \cdot [\text{Glu}]}{\alpha\text{Kglu} + [\text{Glu}]}\right)^n\right]^{1/n} - 1}$$

The estimated τ value depends on [Glu]/Kglu and on α. The equations may be simplified only when [Glu] $\gg$ Kglu to:

$$\text{Max} = \frac{\text{Em} \cdot \tau^n}{1 + \tau^n} \qquad [\text{Gly}_{50}] = \frac{\alpha \cdot \text{Kgly}}{(2 + \tau^n)^{1/n} - 1}$$

The antagonist [B] for the glycine binding site (where KB is the dissociation constant) can interact with two receptor forms, R and RGlu. The parameter, γ, is introduced, to describe the possibility of the antagonist to bind to the two different forms of the receptor. Then, the Effect is

$$\text{Effect} = \frac{\text{Max} \cdot [\text{Gly}]^n}{\left([\text{Gly}_{50}] \cdot \left(1 + \dfrac{\left(\dfrac{\gamma\text{Kglu} + [\text{Glu}]}{\gamma\text{Kglu} + \gamma[\text{Glu}]}\right)[\text{B}]}{\text{K}_\text{B}}\right)\right)^n + [\text{Gly}]^n}$$

with [Glu] $\gg$ Kglu the equation can be simplified to:

$$\text{Effect} = \frac{\text{Max} \cdot [\text{Gly}]^n}{\left([\text{Gly}_{50}] \cdot \left(1 + \dfrac{[\text{B}]}{\gamma\text{K}_\text{B}}\right)\right)^n + [\text{Gly}]^n} \qquad \text{and} \qquad \gamma\text{KB} = \text{KB} \quad \text{when } \gamma = 1$$

AGONIST AND ANTAGONIST RECEPTOR INTERACTION: DATA SIMULATION

Simulated concentration response curves (CRC) for Gly indicate that there is an increase of the maximal effect increasing τ. Gly_{50} increases changing the α value from 0.1 to 10. CRCs for Gly in the absence and presence of a glycine antagonist (FIG. 1) have been simulated. The γ value has been changed from 1 (similar affinity of the antagonist for the two receptor forms) to 0.1 or 100 (different affinity of the antagonist for the two receptor forms). The dose ratio (FIG. 2) in the absence and presence of the antagonists can be over estimated ($\gamma < 1$) or under estimated ($\gamma > 1$).

CONCLUSIONS

The co-agonism between Glu and Gly on NMDA receptors involves that: a) τ represents the intrinsic efficacy of the system bound to both Glu and Gly, b) Max

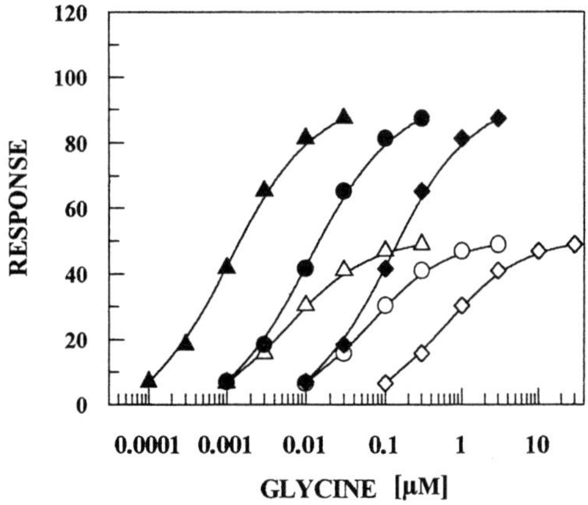

FIGURE 1. Graphic simulation of CRCs for glycine according to the proposed receptor model. The following values were assigned to the parameters: Em = 100; n = 1; [Glu] = 30; Kglu = 0.3; $\tau = 1$ for the empty symbols or $\tau = 10$ for the filled symbols. α was set equal to 1 (no allosterism between Glu and Gly, *circle*), equal to 0.1 (+ allosterism between Glu and Gly, *triangle*) and = 10 (− allosterism between Glu and Gly, *square*).

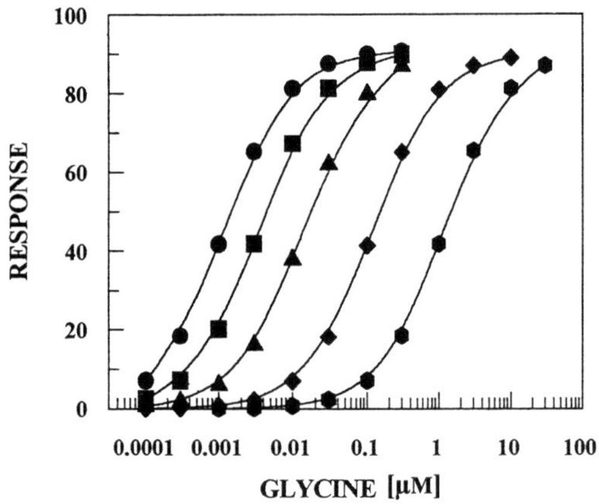

FIGURE 2. Graphic simulation of CRCs for glycine in the absence and presence of a glycine antagonist according to the proposed receptor model. The following values were assigned to the parameters: [Glu] = 30 μM; $\alpha = 0.1$; Kglu = 0.3 μM; Kgly = 0.1 μM; n = 1; Em = 100; $\tau = 10$; [B] = 10 μM and KB = 0.1 μM. Control (●), antagonist with $\gamma = 1$ (◆), antagonist with $\gamma = 0.1$ (⬢), antagonist with $\gamma = 10$ (▲) and antagonist with $\gamma = 100$ (■).

and Gly_{50} are function of both τ and α, c) a glycine antagonist can bind differently to the two forms of the NMDA receptor (R and GluR), so leading to the estimation of an apparent KB. In conclusion, differences in KBs in different tissues may suggest a NMDA receptor heterogeneity as well as diversities in the allosterism played by Glu across tissues, *i.e.*, different γ values.

REFERENCE

1. Marvizon, J. C. & M. Baudry. 1993. Anal. Biochem. **213:** 3–11.

Estimation of the Relative Potency of Two Glycine Antagonists on NMDA-induced Depolarizations in BRSC and ARCS

M. CORSI, L. BETTELINI, P. FINA, AND D. G. TRIST

Pharmacology Research Department
GlaxoWellcome S.p.A.
Via A. Fleming 4
37135 Verona, Italy

INTRODUCTION

The NMDA receptor requires simultaneous stimulation by two agonists glutamate and glycine. The resulting co-agonism may, then, affect not only the agonists receptor interaction but also the estimation of the dissociation constant for antagonists of one binding site. In this work, the potency of two glycine antagonists have been studied against NMDA-induced depolarizations in hemisected baby rat spinal cord (BRSC) and adult rat cortical slices (ARCS).

METHODS

Hemisected spinal cord of 6–11 old day rats and cortical slices (wedge-like form, 3) from adult rats (150–250 g), were placed through a greased slot in the central barrier of a two compartment bath as previously described.[1,2] When the effect of Acea-1021 (ACEA)[3] and 7-chlorokynurenic acid (CKY) were tested, they were added to the tissues 15 minutes before the concentration response curve (CRC) to NMDA. Responses were quantified as percentage response in comparison to KCl 10 mM (BRSC) or kainate 30 μM (ARCS).

DATA ANALYSIS

NMDA CRCs in BRSC and ARCS were fitted to a logistic equation. Considering the amount of endogenous glycine to be much greater than its dissociation constant for the receptor, then $[NMDA_{50}]$ and Max are equal to:

$$[NMDA_{50}] = \frac{\alpha \cdot K_{NMDA}}{(2 + \tau^n)^{1/n} - 1} \qquad \text{Max} = \frac{Em \cdot \tau^n}{1 + \tau^n}$$

α is the allosteric parameter of the co-agonism between NMDA and glycine, K_{NMDA} is the receptor dissociation constant of NMDA, τ is the transducer parameter and Em is the maximal tissue response. Antagonist IC_{50}s were estimated by fitting a logistic equation to the response obtained by NMDA at 30 or 100 μM. When

TABLE 1. pIC_{50}s (±S.E.) Obtained for ACEA and CKY in BRSC and ARCS

NMDA (30 μM)	pIC_{50} ACEA	pIC_{50} CKY	ACEA/CKY
BRSC	5.04 ± 0.06	5.17 ± 0.14	1.35
ARCS	5.67 ± 0.01	4.67 ± 0.01	0.10
NMDA (100 μM)			
BRSC	4.94 ± 0.06	4.92 ± 0.06	0.95
ARCS	5.63 ± 0.08	4.02 ± 0.06	0.024

$[NMDA] \gg K_{NMDA}$, the ratio of IC_{50}s for the two antagonists in one tissue can be written as:

$$\frac{IC_{50\,ACEA}}{IC_{50\,CKY}} = \frac{\gamma_{acea} \cdot KB_{acea}}{\gamma_{CKY} \cdot KB_{CKY}}$$

γ is the allosteric factor that describes the capability of the antagonist to bind to the two receptor forms (R and R_{NMDA}), KB is the antagonist dissociation constant for the free form. When [NMDA] is not much greater than K_{NMDA} the ratio $[NMDA]/K_{NMDA}$ may influence the IC_{50}s ratio but only if γ_{acea} is $\neq \gamma_{CKY}$.

RESULTS

ACEA and CKY antagonized in a dose-dependent manner the NMDA-induced contractions in BRSC and ARCS. The antagonism at the highest dose of antagonists tested was almost complete. IC_{50}s were estimated for both ACEA and CKY and they are reported in TABLE 1. The ratio between the IC_{50}s for ACEA/CKY in the BRSC was greater than that observed in the ARCS when estimated at the level of NMDA response equal to 30 μM. This difference was maintained when the analysis was repeated at the level of NMDA response equal to 100 μM.

CONCLUSIONS

The IC_{50}s ratio of ACEA/CKY in the BRSC was consistently greater than in ARCS at both NMDA concentrations used. This suggests that the influence of $[NMDA]/K_{NMDA}$ is of minor importance over the other parameters. The ratio of IC_{50}s in the same tissue for the two antagonists mirrors the ratio of the dissociation constants and the allosteric factors. Hence, the results obtained are expected only if BRSC and ARCS contain a heterogeneous NMDA receptor population (different KBs) or if the allosteric factors for the two glycine antagonists, γ, in the two tissues are different.

REFERENCES

1. BIRCH, P. J. *et al.* 1988. Eur. J. Pharmacol. **156:** 177–180.
2. HARRISON, H. L. & M. A. SIMMONDS. 1985. Br. J. Pharmacol. **84:** 281–391.
3. WOODWARD, R. M. *et al.* 1995. Mol. Pharmacol. **47:** 568–581.

Functional Study of the Excitatory α-Adrenoceptor in Guinea Pig Ileum, in Relation to the Hypothesis of an Atypical Receptor

ALICE BRICOLA,[a] NEIDE H. JURKIEWICZ,
AND ARON JURKIEWICZ[b]

Department of Pharmacology
Escola Paulista de Medicina
Federal University of São Paulo
Caixa Postal 20372
São Paulo 04034-970, Brazil

INTRODUCTION

According to functional studies, the α_{1A}-adrenoceptor subtype has a high affinity for phentolamine, WB-4101, 5-methyl-urapidil and (+)niguldipine, and is not inactivated by the irreversible antagonist chloroethylclonidine (CEC). On the contrary, the α_{1B}-adrenoceptor subtype is blocked by CEC, but has a lower affinity for the other antagonists.[1]

It has been previously suggested that there are several populations of excitatory α-adrenoceptors in guinea pig terminal ileum.[2] In contrast, it has been recently advanced that the guinea pig ileum has a single population of atypical excitatory α_1-adrenoceptors, because of the lack of inhibitory effect of CEC and low affinity for 5-methylurapidil and WB-4101, tested against phenylephrine.[1]

Our objective was to verify if concentration-response curves for epinephrine (Epi), norepinephrine (NE), and phenylephrine (Ph) are shifted by the antagonists prazosin (Pz), indoramin (In), piperoxane (Pip) and WB-4101 according to the expectations for an interaction with a single population of adrenoceptors.

MATERIALS AND METHODS

Dose-response curves for drug-induced isotonic contractions were made in three different parts of the ileum: 1) terminal ileum, close to the ileo-caecal valve; 2) 20–50 cm from the ileo-caecal valve; and 3) more than 50 cm from the ileo-caecal valve. Antagonists were only used in the terminal ileum.

The shifts induced by competitive antagonists on cumulative dose-response curves were analyzed by means of Schild plots. Experiments were performed in Tyrode solution, at pH 7.2–7.4, 0.5 g load, 37°C, after blocking β-adrenoceptors with

[a] Permanent address: Pontifícia Universidade Católica de Campinas, São Paulo, Brazil.
[b] Address correspondence to Prof. Aron Jurkiewicz; Fax: (55-11) 549-2127 or (55-11) 576-4569.

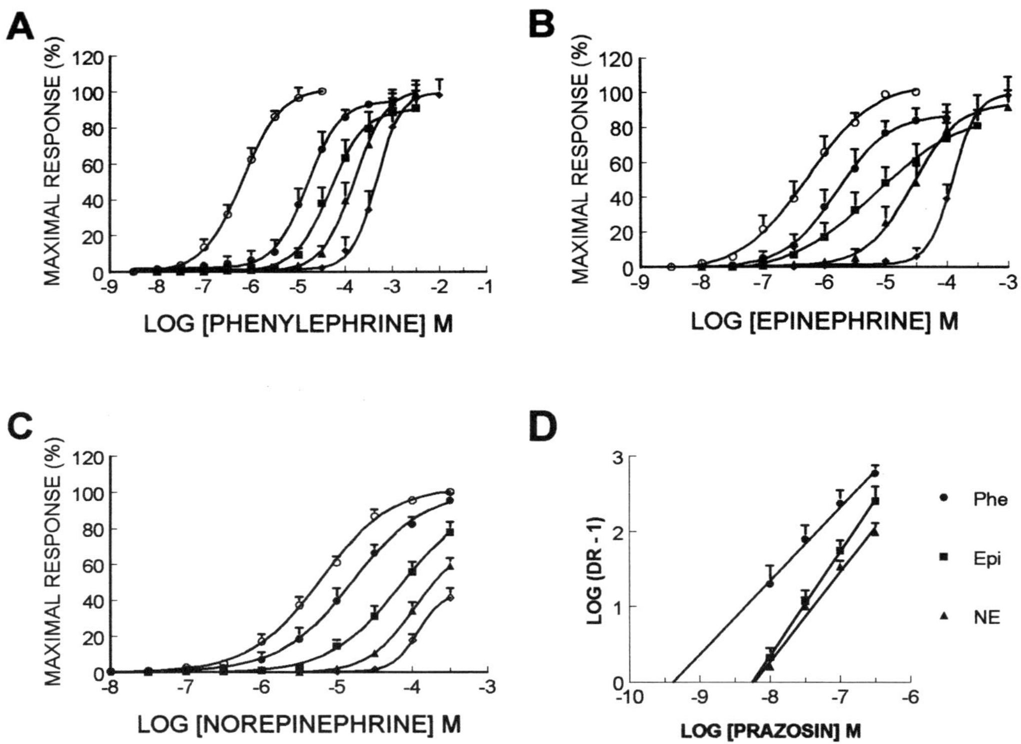

FIGURE 1. Guinea pig ileum. Responses to phenylephrine **(A)**, epinephrine **(B)** and norepinephrine (C—only first part of the curve), in absence (○) and presence of prazosin 10^{-8} M (●); 3×10^{-8} M (■); 10^{-7} M (▲); 3×10^{-7} M (◆). n = 7–9 experiments. Vertical bars show S.E.M. **D**: Schild regression lines for the corresponding curves in **A**, **B**, and **C**.

TABLE 1. Values of pA_2 and Slopes of Schild Lines for Competitive Antagonists on Contractions Induced in the Guinea Pig Terminal Ileum by α-Adrenoceptor Agonists[a,b]

	Phenylephrine		Epinephrine		Norepinephrine	
	pA_2	slope	pA_2	slope	pA_2	slope
Prazosin	9.65 ± 0.37*	0.97 ± 0.10	8.25 ± 0.06	1.37 ± 0.09**	8.28 ± 0.07	1.17 ± 0.09
WB-4101	9.26 ± 0.07	1.45 ± 0.08**	9.63 ± 0.28	1.27 ± 0.17	9.31 ± 0.13	1.10 ± 0.09
Indoramin	8.74 ± 0.11	1.55 ± 0.08**	8.45 ± 0.10	1.71 ± 0.16**	8.34 ± 0.13	1.20 ± 0.13
Piperoxane	6.38 ± 0.08	1.28 ± 0.10	6.16 ± 0.14	1.03 ± 0.18	6.43 ± 0.21	0.62 ± 0.24

[a] In the presence of cocaine (10^{-5} M) and propranolol (4.10^{-6} M) for epinephrine and norepinephrine experiments.
[b] Number of experiments was 6–9 for each pair of antagonist × agonist.
*Significantly different from the corresponding values for the same antagonist; **significantly different from 1.0.

propranolol (4×10^{-6} M), neuronal uptake with cocaine (10^{-5} M) and extraneuronal uptake with corticosterone (10^{-5} M).

RESULTS

(a) Single doses of agonists induced dose-dependent contractions in terminal ileum. These contractions were preceded by a relaxation when segments distant from the ileo-caecal valve were used, indicating the presence of inhibitory adrenoceptors.

(b) The dose-response curves for norepinephrine, but not for the other agonists, were biphasic, since when a plateau was reached, a second curve was obtained starting at 10^{-3} M.

(c) The rank order of potencies measured from pD_2 values for the three agonists were compatible with an interaction with α-adrenoceptors and were not significantly changed after block of neuronal or of extraneuronal uptake, or of β-adrenoceptors, indicating that these factors play a minor role in ileum.

(d) Even after block of uptake and of β-adrenoceptors, the slopes of at least four Schild regression lines, obtained from the shifts induced by antagonists (FIG. 1) were higher than 1.0, as shown in TABLE 1.

(e) In these conditions, the pA_2 values for Pz varied according to the agonist being Ph, Epi, or NE (TABLE 1).

DISCUSSION AND CONCLUSIONS

Taken together, the four findings: a) inhibitory effect that preceded contractions for norepinephrine, mainly in segments far from the ileo-caecal valve; b) double curve for norepinephrine; c) slopes higher than 1.0 in many of the Schild plots; and d) variation of pA_2 values for prazosin according to the agonist, indicate that more than a single population of α-adrenoceptors are functionally present in the terminal guinea-pig ileum.

Although the hypothesis of an atypical adrenoceptor in ileum[1] cannot be ruled out at present, our results represent a cautionary finding in relation to functional classification or subclassification of receptors.

REFERENCES

1. ABEL, P. W., W. ZENG, J. E. PORTER, M. A. SCOFIELD, F. LIU, I. GONZALEZ-CABRERA, F. J. DOWD & W. B. JEFFRIES. 1995. Pharmacol. Commun. **6:** 29–38.
2. BRICOLA, A., N. H. JURKIEWICZ & A. JURKIEWICZ. 1984. Braz. J. Med. Biol. Res. **17:** 549 (Abs).

Functional Change of the Balance between α_{1A} and α_{1B} Adrenoceptor Populations after Transplantation of the Vas Deferens to the Intestine

ANDRE S. PUPO, NEIDE H. JURKIEWICZ, AND ARON JURKIEWICZ[a]

Department of Pharmacology
Escola Paulista de Medicina
Federal University of São Paulo
Caixa Postal 20372
São Paulo 04034-970, Brazil

INTRODUCTION

The α_1-adrenoceptor involved in contractions to norepinephrine (NE) in the rat vas deferens has been identified as α_{1A} by means of both transductional and operational methods.[1,2] Our objective was to verify, by means of the determination of apparent affinity constants of antagonists (pA_2 values) and of the effects of chloroethylclonidine (CEC) and of the calcium antagonist isradipine, if this receptor population is changed after transplantation of the vas to the intestinal wall for seven days.[3]

MATERIALS AND METHODS

Transplantation of rat vas deferens was made as previously described.[3] Seven days after transplantation, the transplanted organ was mounted under the same conditions as the normal vas deferens, for isometric contraction experiments.

Experiments in normal organs were made in the presence of cocaine (6 μM), corticosterone (10 μM) and propranolol (0.1 μM) to block neuronal plus extraneuronal uptake and β-adrenoceptors, respectively. In transplanted vas deferens only corticosterone (10 μM) and propranolol (0.1 μM) were used, since this organ behaves as a denervated vas.[3]

The selective α_1-adrenoceptor antagonists prazosin (1–10 nM), 5-methylurapidil (5-MU, 10–100 nM), WB-4101 (1–30 nM) were incubated before (45 min) and during cumulative addition of norepinephrine. The irreversible antagonist CEC (100 μM) was washed out before norepinephrine. The L-type calcium channel blocker isradipine (0.1 μM) was incubated for 30 min under sodium light. The pA_2 values were calculated from the shifts induced on dose-response curves, according to the method of Arunlakshana and Schild.[4]

[a] Address correspondence to Prof. Aron Jurkiewicz; Fax: (55-11) 549-2127 or (55-11) 576-4569.

RESULTS AND DISCUSSION

We have observed that in normal vas deferens, 5-methylurapidil (5-MU), prazosin and WB-4101 antagonized the contractions induced by NE showing high pA_2 values (TABLE 1). In addition, CEC was ineffective against NE contractions, while isradipine caused a complete block of NE curve (FIG. 1A). These results corroborate the finding that the responses to NE in the normal vas are mediated by α_{1A} adrenoceptors. In transplanted vas, the pA_2 values for antagonists remained high (TABLE 1), indicating that norepinephrine-induced contractions are also mediated predominantly by α_{1A}-adrenoceptors. However, contrary to the results in normal vas deferens, CEC induced a shift to the right on dose-response curves to NE (42 ± 8 times) added to an inhibition of 20% of the maximal response, while isradipine was unable

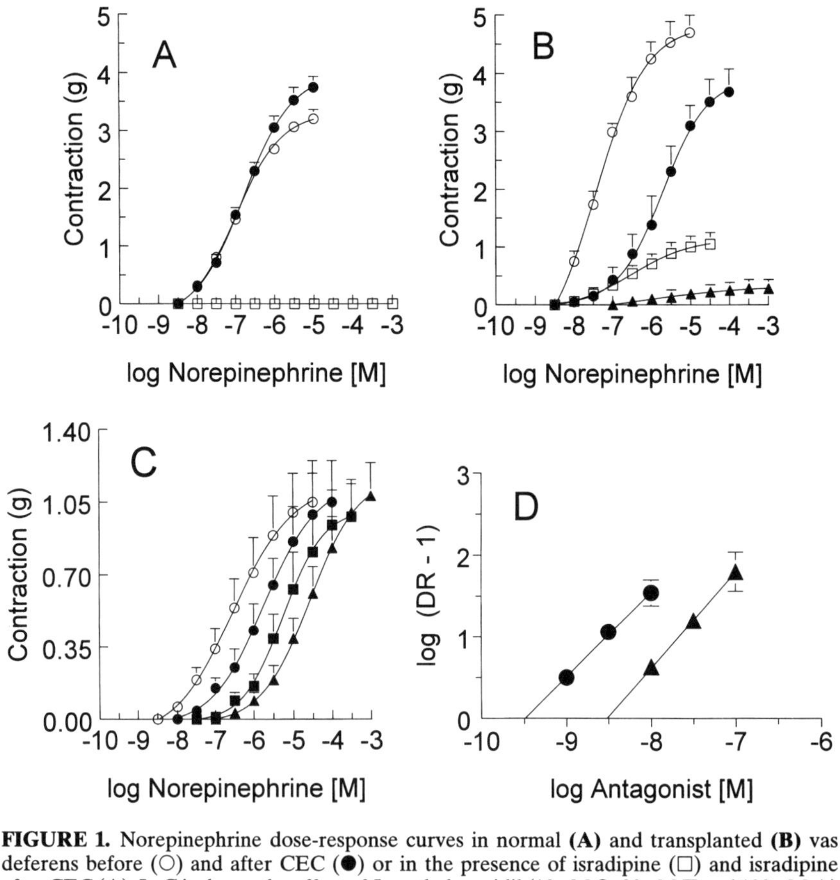

FIGURE 1. Norepinephrine dose-response curves in normal **(A)** and transplanted **(B)** vas deferens before (○) and after CEC (●) or in the presence of isradipine (□) and isradipine after CEC (▲). In **C** is shown the effect of 5-methylurapidil (10 nM ●, 30 nM ■ and 100 nM ▲) on isradipine-resistant norepinephrine contractions (○) in a transplanted vas deferens and in **D** the respective Schild plot (▲). A Schild line for prazosin (●), obtained in the same conditions as for 5-methylurapidil, is also shown in **D**. Number of experiments = 6–8 for each dose-response curve or Schild regression line.

TABLE 1. Values of pA_2 and Slopes of Schild Lines for Prazosin (PZS), WB-4101 and 5-Methylurapidil (5-MU) against Norepinephrine in Normal and Transplanted Vas Deferens in the Absence or Presence of Isradipine (I)

	Normal		Transplanted	
	pA_2	Slope	pA_2	Slope
PZS	9.52 ± .12	1.06 ± .16	9.65 ± .07	1.03 ± .04
WB-4101	9.57 ± .11	0.99 ± .03	9.68 ± .05	0.92 ± .03
5-MU	8.76 ± .11	1.07 ± .03	8.70 ± .09	0.85 ± .10
5-MU + I	not determined		8.56 ± .08	1.17 ± .19
PZS + I	not determined		9.49 ± .05	1.04 ± .08

n = 6–8.

to cause a complete block of NE, leaving a residual effect of about 20% of the maximum (FIG. 1B). This remaining contraction to NE after CEC was abolished by isradipine, indicating a participation of α_{1B}-adrenoceptors as well (FIG. 1B). However, when we have further investigated this isradipine-resistant component (*i.e.*, the component expected to be related to α_{1B}-adrenoceptors) by means of 5-MU (FIGS. 1C and 1D) and prazosin, the respective pA_2 values were high (TABLE 1), indicating an interaction with α_{1A}-adrenoceptors.

It is concluded that: 1) as judged from its functional characteristics, the α_1-adrenoceptor population in rat vas deferens shows some degree of plasticity, and can vary as an adaptive phenomenon to perturbations, such as transplantation, and 2) there is an apparent incompatibility between operational and transductional criteria for adrenoceptor classification in the transplanted rat vas deferens, since besides α_{1A}-adrenoceptors, we have identified an adrenoceptor that is blocked by CEC and mediates isradipine-resistant effects (characteristic of α_{1B}-adrenoceptors), but shows high affinity for 5-MU (characteristic of α_{1A}-adrenoceptors).

REFERENCES

1. HAN, C., P. W. ABEL & K. P. MINNEMAN. 1987. Nature **329:** 333–335.
2. BURT, R. P., C. R. CHAPPLE & I. MARSHALL. 1995. Br. J. Pharmacol. **115:** 467–475.
3. JURKIEWICZ, N. H., A. JURKIEWICZ, C. B. GOMES & J. G. AUCELIO. 1977. J. Pharmacol. Exp. Ther. **203:** 112–119.
4. ARUNLAKSHANA, O. & H. O. SCHILD. 1959. Br. J. Pharmacol. **14:** 48–58.

Functional Classification of Non-competitive Antagonism in Vas Deferens

LUCIENNE C. MARTINI, LUCIA GARCEZ-DO-CARMO, AND ARON JURKIEWICZ[a]

Department of Pharmacology
Escola Paulista de Medicina
Federal University of São Paulo
Caixa Postal 20372
São Paulo 04034-970, Brazil

In recent years, thanks to the advance with respect to drug transduction mechanisms, and to the discovery of a number of second messengers, several non-competitive-like antagonists have been used in functional studies in smooth muscle. Our purpose was to reinvestigate the functional characteristics of non-competitive antagonism, by using a modified Schild plot[1,2] and the determination of pD'_2 values,[2] based on experiments performed in rat vas deferens.

THEORY

According to receptor theory, non-competitive antagonists are drugs expected to modify the effects of agonists by interfering with transduction mechanisms (*i.e.*, the chain of events leading from drug-receptor interaction to the effect). The non-competitive antagonist cannot be displaced by the agonist, unlike competitive antagonism, because agonist and antagonist interact with different sites. Mainly from the classical theoretical and quantitative experimental work of Schild and of Ariens and co-workers,[1,2] non-competitive antagonists are known to cause a dose-dependent decay of the height of dose-response curves for agonists, without shifts to the right, according to the equation:

$$E_A/E_{AB} = ([B]/K_B) + 1$$

where E_A/E_{AB} is the ratio between the effect of an agonist A in the absence (E_A) and presence (E_{AB}) of a non-competitive antagonist B, and [B] and K_B are the respective concentration and equilibrium dissociation constant for B. The effect ratio ($ER = E_A/E_{AB}$) can be used in the Schild equation modified for non-competitive antagonism,[2] in the form:

$$\log (ER - 1) = \log [B] - \log K_B$$

[a] Address correspondence to Prof. Aron Jurkiewicz; Fax: (55-11) 549-2127 or (55-11) 576-4569.

Experimentally, if log (ER − 1) is plotted against log [B] one should expect a straight regression line with a slope of 1.0, from which the negative logarithm of K_B, or pD'_2 value, can be interpolated as an estimate of antagonist affinity.

EXPERIMENTS

FIGURE 1 shows that although the maximal heights of dose-response curves for norepinephrine in rat vas deferens were decreased by different non-competitive antagonists, two of the slopes of regression lines on the modified Schild plot were different from 1.0, contrary to the expectations from receptor theory (see TABLE 1). In spite of the described distortions on slopes, the values of pD'_2 were interpolated from regression lines. As expected, each antagonist presented a particular value of pD'_2, expressing its respective affinity (TABLE 1). Similar results were obtained by using the same antagonists against acetylcholine, KCl or $BaCl_2$ as agonists, except that a shift to the right (0.2–1.0 log units) was observed for the curves of KCl and $BaCl_2$, in addition to the decrease of maximal effect, whichever the antagonist (TABLE 1).

Nifedipine and verapamil that block calcium voltage-dependent channels, antagonized the effects of several agonists, with slopes not significantly different from unity. The pD'_2 values for nifedipine were the same, whichever the agonist (TABLE 1). However, for the three other antagonists, the pD'_2 values were lower when obtained against KCl or $BaCl_2$ (TABLE 1).

DISCUSSION

Except for nifedipine, the so-called non-competitive antagonists caused changes on agonist dose-response curves that were not in accord with the expectations from

TABLE 1. Values of pD'_2, Slope of Modified Schild Regression Lines (±S.E.M.) and Shift to the Right of Dose-Response Curves of Agonists Induced by Non-competitive Antagonists

Antagonists		Agonists: Norepinephrine	Acetylcholine	KCl	$BaCl_2$
Papaverine	pD'_2	5.10 ± 0.03	5.02 ± 0.08	4.92 ± 0.07**	4.69 ± 0.13**
	Slope	2.33 ± 0.21*	1.40 ± 0.10*	1.54 ± 0.14*	0.87 ± 0.22
	Shift***	no	no	0.5	1.0
Caffeine	pD'_2	2.80 ± 0.03	2.80 ± 0.06	2.30 ± 0.02**	2.30 ± 0.06**
	Slope	1.90 ± 0.10*	1.60 ± 0.10*	2.81 ± 0.40*	2.05 ± 0.22*
	Shift***	no	no	0.2	0.6
Verapamil	pD'_2	7.20 ± 0.06	7.00 ± 0.10	6.30 ± 0.04**	6.70 ± 0.06**
	Slope	0.78 ± 0.04	0.92 ± 0.05	1.00 ± 0.10	1.20 ± 0.13
	Shift***	no	no	0.2	1.0
Nifedipine	pD'_2	7.50 ± 0.04	7.60 ± 0.03	7.20 ± 0.20	7.30 ± 0.10
	Slope	0.80 ± 0.06	1.40 ± 0.10	0.79 ± 0.16	0.75 ± 0.10
	Shift***	no	no	0.7	0.6

*, Significantly different from 1.0; **, significantly different from the pD'_2 value for norepinephrine; ***, maximal measurable shift to the right (in log units) of mean dose-response curve.

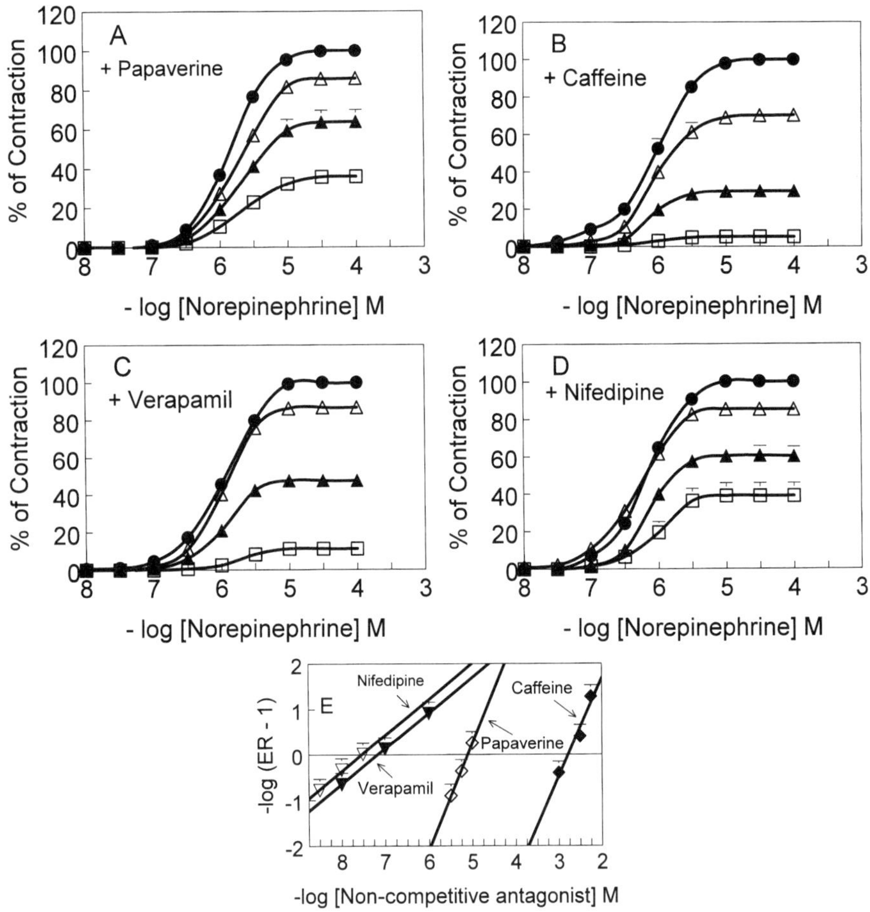

FIGURE 1. Dose-effect curves for isotonic contractions induced by norepinephrine in the absence and presence of antagonists. **A**: Papaverine: control (●), 3 μM (△), 6 μM (▲) and 10 μM (□); **B**: Caffeine: control (●), 1 mM (△); 3 mM (▲) and 6 mM (□); **C**: Verapamil: control (●), 0.01 μM (△), 0.1 μM (▲) and 1 μM (□); **D**: Nifedipine: control (●), 3 nM (△), 10 nM (▲) and 30 nM (□). In **(E)**, Schild plot corresponding to the shifts shown in **A**, **B**, **C**, and **D**: Nifedipine (▽), Verapamil (▼), Papaverine (◇) and Caffeine (◆).

receptor theory, namely: a) some pD_2' values were different from other for the same antagonist, depending on the agonist; b) some decays of the curves, expressed as slopes on the Schild modified plot, were different from 1.0; and c) some dose-response curves were shifted to the right.

Very few quantitative studies of this kind are available and it is not known how far the principle of classifying non-competitive antagonists by pD_2' values, as compared to the use of pA_2 values for classifying competitive antagonists, can be useful for the study of receptor mechanisms. However, on the basis of the present data, it is suggested that the modified Schild plot can be used to select drugs that show a typical non-competitive behavior, indicating a single interaction place. Although several situations can generate atypical Schild plots for competitive antagonism,[1,3] no such studies have been made for the present case. However, it could be advanced, for instance, that the receptor is coupled to multiple signal transduction pathways, or that the non-competitive antagonist or the agonist interact with more than one receptor site, depending on the type of result.

REFERENCES

1. KENAKIN, T. P. 1970. Can. J. Physiol. Pharmacol. **60:** 249–265.
2. ARIENS, E. J. 1957. Arch. Int. Pharmacodyn. **110:** 275–299.
3. FURCHGOTT, R. F., A. JURKIEWICZ & N. H. JURKIEWICZ. 1973. *In* Frontiers in Catecholamine Research. E. Usdin and S. H. Snyder, Eds.: 295–299. Pergamon Press. New York.

M_3 Receptor Mobilizes Intracellular Calcium in Rat Stomach Fundus

SORAYA S. SMAILI,[a] MARIA ETSUKO OSHIRO,[b]
ALICE T. FERREIRA,[b] AND ARON JURKIEWICZ[a,c]

[a]*Department of Pharmacology*
[b]*Department of Biophysics*
Escola Paulista de Medicina
Federal University of São Paulo
Caixa Postal 20372
São Paulo 04034-970, Brazil

Stimulation of muscarinic receptors by carbachol (CCh) in rat stomach fundus produces a contraction[1] that is not affected by Ca^{2+} withdrawal (0Ca) or dihydropyridine L-type Ca^{2+}-channel blockers,[2-4] indicating an involvement of intracellular sources of calcium. These intracellular pools can be depleted by caffeine, thapsigargin,[3] and ryanodine (unpublished data). Our objective was to study which subtypes of muscarinic receptors are involved in this intracellular calcium ($[Ca^{2+}]_i$) translocation.

METHODS

Dose response curves for CCh-induced contractions in rat stomach fundus strips were obtained in absence (control) and presence of increasing concentrations of the muscarinic antagonists atropine, pirenzepine, hexahydrosiladifenidol (HHSiD), and 4-diphenyl acetoxy-methyl piperidine methiodide (4-DAMP), and pA_2 values were obtained from the corresponding Schild plots.[5] Experiments with fundus dispersed cells were also made for the study of intracellular calcium translocation.

RESULTS AND DISCUSSION

Results of contraction experiments are shown in TABLE 1. All the Schild slopes[5] were compatible with the expectations from receptor theory, except for 4-DAMP, which showed a slope greater than unity. The rank order of potencies, obtained from pA_2 values, suggests an interaction with M_3 receptors.[6]

In dispersed cells loaded with Fura-2 in normal Tyrode, stimulation of muscarinic receptors with CCh increased $[Ca^{2+}]_i$ from 281 to 310 nM (not shown). When the experiment was repeated in a calcium-free (0Ca) solution, CCh induced a similar increase, from 145 to 168 nM, indicating the use of internal pools of calcium. The subsequent addition of Ca^{2+} (1.8 and 3.6 mM) produced a further increase (from 168 to 287 nM and 369 nM, respectively) of intracellular calcium (FIG. 1A). In control

[c] Address correspondence to Prof. Aron Jurkiewicz; Fax: (55-11) 549-2127 or (55-11) 576-4569.

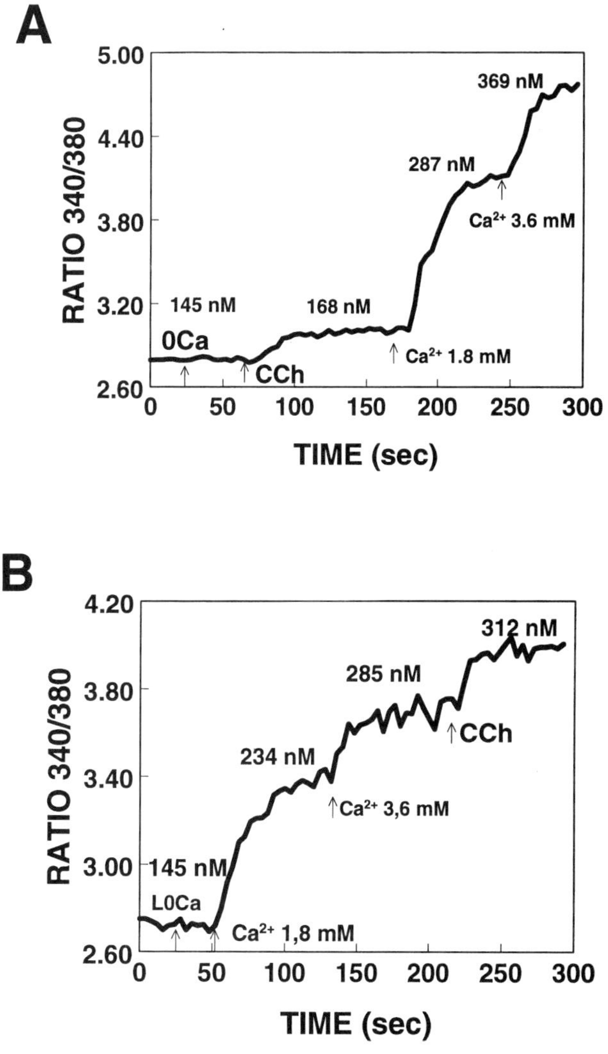

FIGURE 1. Typical experiment showing CCh-induced increase of intracellular calcium in stomach fundus. Fresh dispersed cells were obtained by incubation of mucosa-free fragments with collagenase (0.5 mg/ml), pronase (0.3 mg/ml) and bovine serum albumine (2 mg/ml), in calcium-free Tyrode solution (0Ca), for 45 min at 37°C, followed by filtration and centrifugation at 2,000 rpm, for 4 min. The pellet was suspended in Tyrode solution with albumin (1%) and Ca^{2+}, for one hour. Fluorescence was measured by alternating dual excitation wavelengths between 340 nm and 380 nm and emission wavelength at 510 nm in a SPEX fluorometer. The fluorescence ratios (340/380) were used to calculate changes in $[Ca^{2+}]_i$ according to Grynkiewicz *et al.*[7] Ratios of maximum (R_{max}) and minimum (R_{min}) fluorescence of Fura-2 were determined by addition of digitonin (R_{max}) followed by Mg^{2+} and EGTA (R_{min}) at pH 8. **A**: The cells were incubated in 0Ca Tyrode solution and activated by CCh (300 μM) followed by the addition of Ca^{2+} (1.8 and 3.6 mM). **B**: The cells were incubated in 0Ca Tyrode solution followed by the addition of Ca^{2+} (1.8 and 3.6 mM) and CCh (300 μM). The concentration of intracellular calcium at equilibrium (nM) is shown inside each figure.

TABLE 1. Values of pA_2 and Slopes Obtained from Schild Plots from the Shifts Induced by the Muscarinic Antagonists Atropine, Pirenzepine, Hexahydrosiladifenidol (HHSiD) and 4-Diphenyl Acetoxy-Methyl Piperidine Methiodide (4-DAMP) on Dose-response Curves for CCh

Antagonists	pA_2	Slope
Atropine	9.91 ± 0.33	0.92 ± 0.18
Pirenzepine	6.78 ± 0.07	1.20 ± 0.03
HHSiD	8.32 ± 0.11	1.22 ± 0.05
4-DAMP	9.45 ± 0.15	2.05 ± 0.14*

Values represent mean ± S.E.M. from eight experiments.
* Slope significantly greater than 1.0.

experiments, Ca^{2+} introduced after 0Ca incubation without CCh also produced an increase in $[Ca^{2+}]_i$ (from 145 to 285 nM), that was smaller than that in the presence of CCh (FIG. 1B). This result corroborates previous findings that muscarinic agonists release Ca^{2+} from intracellular sources.[2–4,8–10]

Two possibilities should be put forward to explain the rapid increase of $[Ca^{2+}]_i$ after reintroduction of Ca^{2+} in cells preincubated in 0Ca without CCh (FIG. 1B): one is that Ca^{2+} withdrawal produces depolarization that opens a voltage-operated channel and other is that 0Ca induces a depletion of the Ca^{2+} internal stores that causes mobilization of a messenger that opens Ca^{2+} channels to refill intracellular pools.[11,12]

We conclude that in fundus there is a predominance of M_3 muscarinic receptors that induce a release of intracellular calcium from ryanodine and thapsigargin-sensitive stores. The possible influence of a M_2 muscarinic subtype is not excluded, but our results strongly suggest a M_3 subtype, because this receptor is linked with intracellular Ca^{2+} mobilization via phosphoinositide breakdown.[6]

REFERENCES

1. HYPPOLITO-JURKIEWICZ, N. & A. JURKIEWICZ. 1967. Acta Physiol. Pharmac. Neerl. **14:** 223 (Abs).
2. SMAILI, S. S., N. H. JURKIEWICZ, A. G. GARCIA & A. JURKIEWICZ. 1991. Braz. J. Med. Biol. Res. **24:** 953–956.
3. SMAILI, S., S. M. T. CARVALHO & A. JURKIEWICZ. 1994. Can. J. Physiol. **72**(Suppl. 1): 579 (Abs).
4. SECREST, R., D. D. SCHOEPP & M. L. COHEN. 1989. J. Pharmacol. Exp. Ther. **250:** 971–978.
5. ARUNLAKSHANA, O. & M. O. SCHILD. 1959. Br. J. Pharmacol. **14:** 48–58.
6. CAUFIELD, M. P. 1993. Pharmacol. Ther. **58:** 319–379.
7. GRYNKIEWICZ, G., M. POENIE & R. Y. TSIEN. 1985. J. Biol. Chem. **260:** 3440–3450.
8. MURPHY, N. P., P. F. T. VAUGHAN, S. G. BALL & J. G. MCCORMACK. 1991. J. Neurochem. **57:** 2116–2123.
9. KAJITA, J. & H. YAMAGUCHI. 1993. Am. J. Physiol. **264:** L496–L503.
10. YANG, C. M., Y.-L. YO & Y.-Y. WANG. 1993. Br. J. Pharmacol. **110:** 983–988.
11. PUTNEY, J. W. JR. 1990. Adv. Pharmacol. **22:** 251–269.
12. RANDRIAMAMPITA, C. & R. Y. TSIEN. 1993. Nature **364:** 809–814.

Agonist Binding Properties for Recombinant Kappa Opioid Receptors Expressed in CHO-K1 Cells

L. CARBONI,[a] G. CAMPANA,[a] S. CACCIAGUERRA,[a] G. MURARI,[a] E. SPERONI,[a] M. S. PAPPALARDO,[b] G. RONSISVALLE,[b] AND S. SPAMPINATO[a,c]

[a]*Department of Pharmacology*
University of Bologna
Irnerio, 48-40126 Bologna, Italy

[b]*Institute of Medicinal Chemistry*
University of Catania,
Catania, Italy

Pharmacological studies have suggested that there are at least three major classes of opioid receptors, designated μ, δ and κ. These receptors differ in their affinity for various opioids, although there is a consensus quantitative rank order of ligand affinities across the μ, δ and κ binding sites.[1] Until recently, only tissues or cultured cells expressing endogenous receptors were available for evaluating opioid activities. Confounding factors included the necessity of using several tissues and/or cells as model systems for different receptor types. Recently, several groups have reported the molecular cloning of the μ, δ and κ receptors.[2,3] The availability of the cloned receptors now allows for studies of single receptors with regard to pharmacological profile, structure-function analysis and regulation of expression. The aim of this study was to characterize the pharmacological properties of the cloned κ opioid receptor and to investigate the receptor affinity of a novel benzomorphan derivative.

To pharmacologically characterize the cloned rat κ receptor, we expressed the cDNA encoding this protein stably in CHO-K1 cells. A $\approx$ 2 kilobase XbaI fragment of the cDNA coding for the rat kappa receptor (a kind gift of Dr. L. Yu) was cloned into the CMV promoter-based expression vector pRc/CMV (Invitrogen) and transfected into CHO-K1 cells using a calcium phosphate procedure. Cells were selected for 2 weeks in medium containing 500 μg/ml of the neomycin analogue G418 (GIBCO). Positive clones of transfected cells were used for binding assay as described.[2,4] The receptors were labeled with the selective κ agonist [^{3}H] U69,593 or with the non-selective opioid antagonist [^{3}H] naloxone. The binding of these radioligands was saturable, of high affinity, and superimposable to that observed for the native receptors expressed in guinea-pig cerebellum membranes obtained as previously reported.[4] Analysis of the saturation experiments demonstrated that [^{3}H] U69,593 bound to the cloned κ receptors with a K_D of 2.36 nM and a B_{max} of 19.5 fmol/mg protein (values are the mean of at least three experiments). Similarly, [^{3}H] U69,593 bound to the native κ receptor (K_D = 1.98 nM; n = 3) while the B_{max}, taken as an index of the receptor number, was higher (90.9 fmol/mg protein;

[c] Address correspondence to Prof. Santi Spampinato; Tel.: +39-51-243597; Fax: +39-51-248862; E-mail: (spampi@biocfarm.unibo.it).

TABLE 1. Affinity Constants (K_i, nM) of Opioid Agonists Determined by the Inhibition of the Specific Binding of [^{3}H] U69,593 to the Rat κ Opioid Receptor Expressed in CHO-K1 Cells

Agonist	K_i	Slope
U-69,553	7.40 ± 0.9	0.94 ± 0.07
U-50,488	18.70 ± 1.4	0.93 ± 0.03
DAMGO	>10,000	N.D.
DPDPE	>10,000	N.D.
CCB	4.22 ± 1.7	0.95 ± 0.04

Results are the mean ± S.E.M. (n = 3). N.D. = Not determined.

n = 3) than in CHO-K1 cells. The K_D for [^{3}H] naloxone binding to the cloned κ receptor was 4.22 nM and the B_{max} was 23.5 fmol/mg protein; [^{3}H] naloxone bound to the native κ receptor with a K_D of 10.86 nM and a B_{max} of 68.7 fmol/mg protein. No specific radioligand binding was detectable in appropriate non-transfected control cells.

To identify the pharmacological specificity of the cloned κ receptor, we evaluated a series of opioid agonists for their abilities to inhibit radioligand binding of [^{3}H] U69,593. As expected, U69,593 and U50,488 bound to the cloned receptor with K_i values in the low nanomolar range whereas the μ agent DAMGO and the δ agonist DPDPE bound with a very low potency to the κ cloned receptor (TABLE 1). These results correlate well with the values observed for native opioid receptors and reported in the literature.[2] In a previous study we have described the pharmacological profile of the new κ opioid agonist CCB (6,11-dimethyl-1,2,3,4,5,6-hexahydro-3{[2′-methoxycarbonyl-2′-(4-chlorophenyl)cyclopropyl] methyl}-2,6-methano-3-benzazocin-8-ol) and evaluated its binding profile to the native κ opioid receptor (K_i = 0.41 nM).[4] Interestingly, this κ opioid agonist did not significantly bind to μ and δ receptors (K_i > 25,000 nM). We confirmed that it displayed a high binding affinity to the cloned κ receptor (TABLE 1).

This study confirms the possibility of individually investigating the pharmacological properties of the cloned opioid receptor types in homogeneous cell lines with no native opioid receptors stably transfected with the cDNA for one of these receptors.

REFERENCES

1. GOLDSTEIN, A. 1987. Trends Pharmacol. Sci. **8:** 456–459.
2. RAYNOR, K., H. KONG, Y. CHEN, K. YASUDA, G. I. BELL & T. REISINE. 1994. J. Pharmacol. Exp. Ther. **45:** 330–334.
3. CHEN, Y., J. L. MESTECK & L. YU. 1993. Biochem. J. **295:** 625–628.
4. RONSISVALLE, G., O. PREZZAVENTO, L. PASQUINUCCI, A. MARRAZZO, F. VITTORIO, J. A. GOMEZ-VIDAL, L. CARBONI & S. SPAMPINATO. 1995. Life Sci. **57:** 1487–1495.

The Calcitonin Gene-related Peptide (CGRP) Receptor in Rat Smooth Muscles

F. WISSKIRCHEN AND I. MARSHALL

Department of Pharmacology
University College London
Gower Street
London WC1E 6BT, United Kingdom

Recently the CGRP receptor has been divided into type 1 (guinea pig atrium) and type 2 (rat vas deferens), partly based on differing affinities of the linear CGRP analogue [Cys(ACM)^{2,7}] hα CGRP and the antagonist hα CGRP$_{8-37}$.[1,2] While [Cys(ACM)^{2,7}] hα CGRP shows no activity (up to 10^{-6} M) at the CGRP$_1$ receptor, it is a weak agonist (1% of CGRP) at the CGRP$_2$ receptor.[1] The antagonist hα CGRP$_{8-37}$ has a higher affinity at the CGRP$_1$ receptor than at the CGRP$_2$ receptor.[2,3] The aim of the present studies was to establish the receptor subtype classification for CGRP in three rat isolated smooth muscles: prostatic vas deferens, pulmonary artery, and internal anal sphincter (IAS).

Tissues were set up (0.5 g tension) in Krebs solution at 37°C (95% O_2/5% CO_2). Cumulative dose response curves to human α CGRP (hα CGRP) and its analogues were produced on precontracted tissues. The antagonist hα CGRP$_{8-37}$ (up to 10^{-5} M) was equilibrated for 20 minutes and the peptidase-inhibitors (10^{-6} M of phosphoramidon, thiorphan, amastatin, bestatin and captopril) for 30 minutes.

In the vas deferens, hα CGRP, hβ CGRP and rat β CGRP were equipotent (pD$_2$ around 8.0), with rat amylin and [Cys(ACM)^{2,7}] hα CGRP being 10- and at least 1000-fold less potent, respectively. In the pulmonary artery and IAS, the CGRP peptides were 5- to 10-fold more potent (pD$_2$ around 8.4 and 8.8, respectively). Human adrenomedullin gave a pD$_2$ value of 8.5 $\pm$ 0.1, while rat amylin and [Cys(ACM)^{2,7}] hα CGRP were 30- and at least 1000-fold less potent than hα CGRP in the pulmonary artery. Salmon calcitonin showed no activity (up to 10^{-4} M) in the vas deferens and pulmonary artery. Hα CGRP$_{8-37}$ (10^{-5} M) gave pK$_B$ values against hα CGRP of 5.9 $\pm$ 0.1, 7.1 $\pm$ 0.1 and 5.7 $\pm$ 0.3 for vas deferens, pulmonary artery and IAS, respectively. Peptidase-inhibitors (10^{-6} M) did not modify the pD$_2$ values of hα CGRP. In the vas deferens the pD$_2$ values were 8.0 $\pm$ 0.1 and 7.9 $\pm$ 0.1 while in the pulmonary artery they were 8.4 $\pm$ 0.1 and 8.2 $\pm$ 0.1 in the absence and presence of peptidase-inhibitors, respectively. The affinity of hα CGRP$_{8-37}$ in the vas deferens and pulmonary artery was not significantly affected by the presence of peptidase-inhibitors.

Unlike Longmore *et al.*,[4] inhibition of peptidases did not alter either the potency of hα CGRP or the affinity of the antagonist hα CGRP$_{8-37}$. The subtype of the CGRP receptor in the pulmonary artery appears to be a CGRP$_1$ type on the basis of the relative potency of [Cys(ACM)^{2,7}] hα CGRP and the affinity of CGRP$_{8-37}$. However in the vas deferens and IAS the affinity of hα CGRP$_{8-37}$ is consistent with a CGRP$_2$ receptor subtype.

REFERENCES

1. DENNIS, T., A. FOURNIER, S. ST.-PIERRE & R. QUIRION. 1989. J. Pharmacol. Exp. Ther. **251**(2): 718–725.
2. DENNIS, T., A. FOURNIER, A. CADIEUX *et al.* 1990. J. Pharmacol. Exp. Ther. **254**(1): 123–128.
3. MIMEAULT, M., A. FOURNIER, Y. DUMONT *et al.* 1991. J. Pharmacol. Exp. Ther. **258**(2): 1084–1090.
4. LONGMORE, J., J. E. HOGG, P. H. HUTSON *et al.* 1994. Eur. J. Pharmacol. **265:** 53–59.

GR127935 Antagonizes the 5-HT_1-like Receptor-mediated External Carotid Vasoconstriction in Vagosympathectomized Dogs

C. M. VILLALÓN,[a] D. CENTURIÓN, AND J. A. TERRÓN

Department of Pharmacology
CINVESTAV-IPN
Apdo. Postal 22026
14000 México D.F., Mexico

5-Hydroxytryptamine (5-HT) increases the external carotid blood flow (ECBF) in dogs with intact sympathetic tone via prejunctional 5-HT_1-like receptors similar to the 5-HT_{1D} subtype which, upon stimulation, inhibit the carotid sympathetic outflow.[1]

In contrast, 5-HT *decreases* the ECBF in vagosympathectomized dogs via postjunctional 5-HT_1-like receptors, but other mechanisms can also be involved.[2] In this study, we have further investigated these receptors using GR127935, a selective 5-HT_{1D} receptor antagonist,[3] and ritanserin, a putative 5-$HT_{1D\alpha}$ receptor antagonist.[4] Thus, 16 vagosympathectomized dogs anesthetized with pentobarbital (baseline values of blood pressure and heart rate were 141 $\pm$ 6 mmHg and 174 $\pm$ 10 beats/min, respectively) received intracarotid (i.c.) infusions (for 1 min) of 5-HT (1, 3, and 10 μg/min) and sumatriptan (3, 10, and 30 μg/min). The animals were then divided into three groups and the responses to 5-HT and sumatriptan were induced after i.v. saline (0.015, 0.05, and 0.15 ml/kg; n = 6), GR127935 (1, 3, and 10 μg/kg; n = 6) or ritanserin (1 mg/kg; n = 4). The infusions of 5-HT and sumatriptan resulted in dose-dependent decreases in ECBF (TABLE 1) which remained unaltered after saline or ritanserin (not shown). In contrast, GR127935: i) abolished the vasoconstrictor responses to 5-HT and, notably, a dose-dependent vasodilator component was unmasked (TABLE 1, upper part); and ii) dose-dependently antagonized, but did not revert, the vasoconstrictor responses to the more selective agonist, sumatriptan[5] (TABLE 1, lower part); this antagonism was specific since GR127935 did not affect the vasoconstrictor responses to noradrenaline (not shown). Therefore, we suggest that the sumatriptan-sensitive 5-HT_1-like receptors mediating canine external carotid vasoconstriction resemble the 5-HT_{1D}-like receptor subtype. Although the resistance to antagonism by ritanserin might exclude the possible role of the 5-$HT_{1D\alpha}$ subtype, the involvement of the 5-$HT_{1D\beta}$ subtype will undoubtedly be reinforced by the advent of a highly selective 5-$HT_{1D\beta}$ receptor antagonist.

[a] Author to whom correspondence should be addressed.

TABLE 1. Effect of GR127935 (1, 3 and 10 μg/kg, i.v.) on the Decreases in External Carotid Blood Flow (ml/min) Elicited by i.c. 5-HT (1, 3 and 10 μg/min) or Sumatriptan (3, 10 and 30 μg/min)

		GR127935		
5-HT (μg/min)	Control	1 μg/kg	3 μg/kg	10 μg/kg
1	−16 ± 3	+8 ± 6*	+15 ± 6*	+13 ± 6*
3	−31 ± 4	+32 ± 15*	+39 ± 7*	+40 ± 8*
10	−53 ± 8	+68 ± 23*	+84 ± 11*	+82 ± 10*
Sumatriptan (μg/min)				
3	−23 ± 3	−16 ± 3*	−2 ± 2*	+5 ± 3*
10	−56 ± 6	−36 ± 2*	−10 ± 1*	−2 ± 1*
30	−84 ± 7	−66 ± 7*	−23 ± 3*	−7 ± 2*

* $p < 0.05$, control vs. GR127935.

REFERENCES

1. VILLALÓN, C. M. & J. A. TERRON. 1994. Br. J. Pharmacol. **113:** 13–20.
2. VILLALÓN, C. M. *et al.* 1995. Br. J. Pharmacol. **116:** 2778–2784.
3. SKINGLE, M. *et al.* 1993. Br. J. Pharmacol. **110:** 9P.
4. KAUMANN, A. J. *et al.* 1993. Cardiovasc. Res. **27:** 2094–2103.
5. HOYER, D. *et al.* 1994. Pharmacol. Rev. **46:** 157–203.

Pharmacological Analysis and Simulation of Ligand-Receptor Interactions in Microsoft Excel™

N. S. TILFORD, W. P. BOWEN, AND G. S. BAXTER

Neurology Research
SmithKline Beecham Pharmaceuticals
New Frontiers Science Park
Third Avenue
Harlow, Essex, CM19 5AW, United Kingdom

Microsoft Excel is a powerful spreadsheet which is widely used for manipulation and statistical analysis of numerical data. We have extended its use for the simulation of pharmacological models and routine data analysis by curve-fitting.

RoboFit is a macro-driven utility for Excel that incorporates curve-fitting to several equations for automated data analysis and presentation. RoboFit utilizes the Solver function in Microsoft Excel to iteratively fit models of ligand-receptor interactions to experimental data. The Solver function minimizes the difference between actual data and those predicted by a model (*e.g.*, the Hill equation) by least square analysis, by changing model parameters, and thus provides estimates of these parameters, *e.g.*, [A50], α and n. We also employ direct simultaneous fitting to the operational model of agonism to derive agonist equilibrium dissociation constants (K_A) and partial agonist affinity (K_P) from appropriate data.[1,2] RoboFit also incorporates simultaneous fitting for the estimation of competitive antagonist affinity by direct fitting to the Schild model.[3]

It is possible to simulate complex ligand-receptor interactions using Excel. For example, we have simulated the operational characteristics of an exogenous agonist, of varying efficacy, in the presence of an endogenous agonist. This involves the operational modeling of the interaction of two agonists with one receptor.[2] In the model below: E represents response, E_M, the theoretically achievable maximum response, K_A and K_B, the equilibrium dissociation constants, and τ_A and τ_B, the operational equivalents of efficacy for the exogenous (A) and endogenous (B) agonists, respectively.

$$E = \frac{E_M \tau_A[A]K_B + \tau_B[B]K_A}{[A]K_B + K_AK_B + K_A[B] + \tau_A[A]K_B + \tau_B[B]K_A}$$

By varying τ_A, in this model, it is possible to mimic all the operational characteristics of full agonists, partial agonists, neutral antagonists and inverse agonists (Fig. 1).

By expanding this model to the action of three ligand species at a single receptor, it is possible to mimic the behavior of both agonists and inverse agonists with a neutral antagonist.

Microsoft Excel can provide an efficient method for the simulation and analysis of ligand-receptor interactions for the analytical pharmacologist.

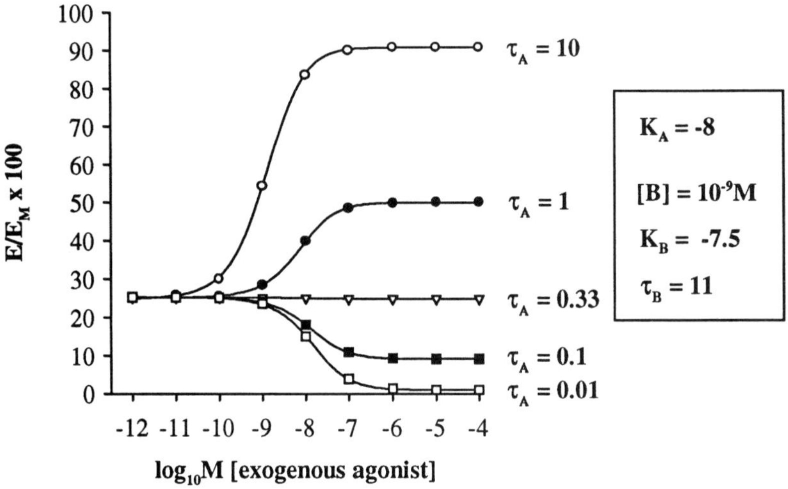

FIGURE 1. Model of the interaction of two agonists with one receptor.

REFERENCES

1. Black, J. W. & P. Leff. 1983. Proc. R. Soc. B. **220:** 141–162.
2. Leff, P., I. G. Dougall & D. Harper. 1993. Br. J. Pharmacol. **110:** 239–244.
3. Lazareno, S. & N. J. M. Birdsall. 1993. Br. J. Pharmacol. **109:** 1110–1119.

Atypical β-Adrenoceptors in the Rat Vasculature

S. SOOCH AND I. MARSHALL[a]

Department of Pharmacology
University College London
Gower Street
London WC1E 6BT, United Kingdom

β-adrenoceptors which differ from both β_1- and β_2-adrenoceptors have been referred to as atypical or β_3-adrenoceptors. Their common feature is resistance to blockade by most β-adrenoceptor antagonists. Furthermore, agonists selective for the β_3-adrenoceptor have been identified, *e.g.*., ZD 7114. Atypical or β_3-adrenoceptors have been shown to be present in adipocytes and a number of isolated tissues, *e.g.*, guinea pig ileum and rat distal colon (see Arch & Kaumann[1]). The present study provides evidence for the presence of atypical β-adrenoceptors in the rat vasculature.

Rat aortic, mesenteric and pulmonary arterial rings were suspended under 0.5 g tension in Krebs solution at 37°C and oxygenated with 95% O_2/5% CO_2 for isometric tension recording. Rings were preconstricted with phenylephrine and cumulative concentration-response curves to β-adrenoceptor agonists constructed. Antagonists were equilibrated for 30 min before a second dose-response curve was obtained. Results are expressed as mean ± S.E.M.

In the rat mesenteric artery, isoprenaline (10 nM–100 μM) produced a concentration-dependent relaxation (pD_2 6.5 ± 0.1) of phenylephrine-induced (1 μM) tone. Propranolol produced dose-ratios of 7.2 ± 1.1, 5.8 ± 2.0 and 5.5 ± 3.3 at 0.1, 1 and 3 μM, respectively with no depression of the maximum response. Isoprenaline-induced relaxations were not altered by the selective β_1-adrenoceptor antagonist, atenolol (0.1–3 μM), but the selective β_2-adrenoceptor antagonist, ICI 118551 produced dose-ratios of 10.7 ± 3.9 and 6.5 ± 1.7 at 0.1 and 3 μM, respectively. ZD 7114, a β_3-adrenoceptor agonist, relaxed rat mesenteric arterial rings (pD_2 5.3 ± 0.1). ICI 215001 (10 nM–100 μM[2]) and ZD 2079 10 nM–100 μM[3]) were less potent at relaxing arterial rings (relaxations of 78% and 52%, respectively). Responses to ZD 7114, ICI 215001 and ZD 2079 were not antagonized by propranolol (1 μM). Isoprenaline (10 nM–100 μM) also relaxed (pD_2 7.4 ± 0.1) phenylephrine-induced (0.3 μM) tone in rat thoracic aortic rings, with propranolol (1 μM) producing dose-ratios of 8.1 ± 1.8 and 7.3 ± 0.2 at 0.1 and 1 μM, respectively. In the rat pulmonary artery, isoprenaline (10 nM–1 μM) relaxed rings (pD_2 8.0 ± 0.1) preconstricted with phenylephrine (30 nM) with increasing concentrations of propranolol shifting control curves to the right. However, the antagonism was non-competitive (Schild plot slope significantly <1). ZD 7114 (10 nM–100 μM) produced concentration-dependent relaxations of both rat thoracic aortic (pD_2 5.4 ± 0.1) and pulmonary arterial rings (pD_2 6.2 ± 0.1). Responses to ZD 7114 were not antagonized by propranolol (1 μM) in either tissue.

[a] Author to whom correspondence should be addressed.

In conclusion, propranolol was not a competitive antagonist in any of the three blood vessels. Furthermore, selective β_3-adrenoceptor agonists produced relaxations, which were not antagonized by propranolol. Together these data suggest that atypical β-adrenoceptors exist in the rat vasculature and mediate vasorelaxation.

REFERENCES

1. ARCH, J. R. S. & A. J. KAUMANN. 1993. Med. Res. Rev. **13:** 663–729.
2. TESFAMARIAM, B. & G. T. ALLEN. 1994. Br. J. Pharmacol. **112:** 55–58.
3. GRANT, T. L., R. M. MAYERS, S. P. QUAYLE *et al.* 1994. Br. J. Pharmacol. **112:** 231P.

Interaction between H_3-Receptors and Other Prejunctional Receptor Systems in the Isolated Guinea Pig Duodenum

ENZO POLI, CRISTINA POZZOLI,
AND GIULIO BERTACCINI

Institute of Pharmacology
University of Parma, School of Medicine
43100 Parma, Italy

INTRODUCTION

Prejunctional (presynaptic) histamine H_3-receptors and α_2-adrenoceptors modulate the release of different neuromediators in the CNS and in peripheral tissues.[1] The activation of such receptor systems limits the availability of free Ca^{2+} at the axoplasmic level, possibly by reducing Ca^{2+}-conductance through N-type channels. This phenomenon results in an attenuation of the exocytotic neurotransmitter release and, as a consequence, of the post-synaptic responses associated to nerve stimulation.[2]

It has been recently demonstrated that histamine H_3-receptors and α_2-adrenoceptors, occurring at the adrenergic nerve endings of the mouse brain cortex, interact with each other, showing a negative cooperativity when simultaneously activated by endogenous or exogenous ligands.[3] Both H_3-receptors and α_2-adrenoceptor systems are also co-localized on the cholinergic terminals of the myenteric plexus, where they exert a prejunctional modulation of the excitatory pathways controlling the longitudinal muscle contractility.[2] Thus we wanted to test the hypothesis that these receptors may interact with each other also outside the CNS. The isolated, electrically driven, guinea pig duodenum was chosen as a functional model to study the modulatory activity of these receptors on the cholinergic transmission in peripheral nerves.

METHODS

Segments of the guinea pig duodenum were excised and placed into isolated organ chambers filled with Krebs' solution, and the isometric contraction of the longitudinal muscle was measured. A pair of coaxial platinum electrodes was placed into the chamber and electrical field stimulation (EFS) (1 Hz, 0.2 msec, supramax. voltage) was applied, obtaining muscle contractions sustained by endogenous Ach.[2] The action mediated by different presynaptic receptors was induced with specific agonists, administered cumulatively into the organ chamber, and measured as percent reduction of the amplitude of the cholinergic response to EFS. When interactions between receptors were tested, single concentration of agonists, corresponding to their ED_{50}, were simultaneously administered and the final effect compared to that expected for additivity of the individual effects. Student's t test for paired or unpaired observations was then applied for comparison.

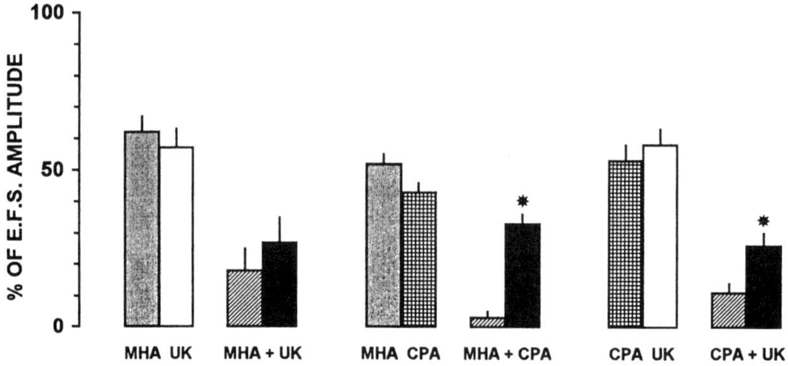

FIGURE 1. Inhibitory effect of cyclopentyladenosine (CPA, *grid columns*), UK 14,304 (UK, *white columns*) and R(α)methylhistamine (MHA, *grey columns*), given alone or in combination (*black columns*). Hatched columns represent the inhibition expected for additivity of the single effects. Ordinate: amplitude of the muscle contraction after drugs, in percent of the basal (pre-drug) level. Data are the mean ± S.E.M. of 5–8 observations. *: significantly different (p <0.05), compared with the expected value.

RESULTS

Specific agonists acting at H_3-receptors, α_2-adrenoceptors and adenosine A_1-receptors, R(α)methylhistamine (MHA), UK 14,304 and cyclopentyladenosine (CPA), respectively, elicited a concentration-dependent inhibition of muscle contractions evoked by EFS, but not by exogenously applied acetylcholine, suggesting a prejunctional site of action. The calculated ED_{50} were (nM): MHA: 17.4, UK 14,304: 28.0 and CPA 2.4. The inhibition of the cholinergic response was complete only with CPA and UK 14,304.

As shown in FIGURE 1, the effect elicited by the simultaneous activation of H_3-receptors and α_2-adrenoceptors did not differ from the sum of the effects elicited by the individual agonists, excluding the occurrence of cooperativity between these receptors. Conversely, the effect evoked by the co-administration of CPA and MHA or of CPA with UK 14,304 was significantly lower than the sum of the effects induced by the individual agonists, suggesting the occurrence of negative cooperativity between A_1- and α_2- or H_3-receptors. The negative cooperativity between H_3- and A_1-receptors, but not between A_1- and α_2-receptors, was lost in low- (1 mM) or high- (12 mM) K^+ media (not shown).

CONCLUSIONS

Data obtained in the present study reveal that histamine H_3-receptors and α_2-adrenoceptors do not show any cooperativity, in contrast with the observations performed in the CNS.[3] Whether this discrepancy is simply due to differences in the experimental conditions or reflects differences in α_2-adrenoceptor subtypes (α_{2A} vs. α_{2B}) occurring in peripheral and central neurons and/or their associated transduction mechanisms, cannot be established at present. Conversely, a negative cooperativity can be demonstrated between A_1- and α_2- or H_3-receptors. Since cross-reactions of the agonists at the recognition sites have been excluded (Poli *et al.*, unpublished), interactions between different second messengers or effectors, participating to the post-receptor signal transducing mechanisms associated with these prejunctional receptors, are likely to occur. Experiments performed in modifier K^+-media reveal that K^+-channels, which represent the primary effector associated with A_1-receptors (Poli *et al.*, unpublished), are involved in the interaction, together with other unknown factors.

REFERENCES

1. STARKE, K., M. GOTHERT & H. KILBINGER. 1989. Physiol. Rev. **69:** 864–989.
2. POLI, E., C. POZZOLI, G. CORUZZI & G. BERTACCINI. 1994. J. Pharmacol. Exp. Ther. **270:** 788–794.
3. SCHLICKER, E., A. BEHLING, G. LUMMEN, B. MALINOWSKA & M. GOTHERT. 1992. Naunyn-Schmiedeberg' Arch. Pharmacol. **345:** 639–646.

Cloned and Native Guinea Pig 5-ht$_7$ Receptors

Characterization Using an Integrated Approach

R. M. EGLEN, R. ALVAREZ, D. CARTER, E. LEUNG,
L. JAKEMAN, Z. TO, AND A-P. TSOU

Institute of Pharmacology, Neurobiology
Roche Biosciences
3401 Hillview Avenue
Palo Alto, California 94304

OPERATIONAL CLASSIFICATION

Functional Pharmacology—Guinea-pig Isolated Ileum

Substance P precontracted tissues were potently relaxed by 5-HT, 5-CT and several other indoles with the following rank order of potency: 5-CT (5-carboxamidotryptamine) > 5-methoxytryptamine > 5-HT > α-methyl-5-HT (4.7). 2-Methyl-5-HT, tryptamine, *N*,*N* dimethyltryptamine, *N*,*N* dimethyl-5-HT, dipropyl-5-CT, and sumatriptan were inactive at concentrations greater than 0.1 mM. 8-OH-DPAT (8-hydroxy-2-(di-n-propylamino)-tetralin) acted as a potent, but partial, agonist with respect to 5-CT (pEC_{50} = 6.3, intrinsic activity = 0.6). Responses to 5-CT were antagonized by several compounds, with the following rank order of affinity: lysergic acid diethylamide = mesulergine > methysergide = spiperone > clozapine $\gg$ pindolol > ketanserin = ondansetron = GR 113808.

Binding Pharmacology—Guinea-pig Cerebral Cortex and CHO-K1 Cells

Under conditions of 5-HT$_{1A}$, 5-HT$_{1D}$ blockade, [^{3}H] 5-CT bound specifically to a single population of binding sites (K_D = 0.18 nM) in CHO-K1 cells. The specific binding of [^{3}H] 5-CT was displaced by several compounds, with following rank order of affinity: 5-CT, 5-HT, 5-MeOT > DP-5-CT > lisuride, pergolide > LSD, methiothepin, metergoline, tryptamine > spiperone, ritanserin, ergotamine, clozapine > pindolol and (−)butaclamol. A similar rank order of affinity was seen at receptors specifically labeled in guinea pig cerebral cortical membranes, although the affinities were generally lower.

TRANSDUCTIONAL CRITERIA

Guinea Pig Hippocampus and CHO-K1 Cells

In both hippocampus and CHO-K1 cells, only augmentation of adenylyl cyclase was seen, with no evidence for inhibition of the enzyme or other transduction systems including phosphoinositide hydrolysis. The rank order of potency for several agonists of the endogenous hippocampal 5-HT receptor corresponded to that obtained for a cloned 5-ht$_7$ receptor stably transfected in CHO-K1 cells. Thus in both tissues, 5-CT > 5-MeOT, 5-HT > dipropyl 5-CT > 8-OH-DPAT.

STRUCTURAL CRITERIA

Predicted Amino Acid Sequence

The predicted sequence consisted of 446 amino acids, and exhibited all the hallmarks of a guanine nucleotide coupled receptor. Thus, the sequence was predicted to possess seven transmembrane spanning domains, potential N linked glycosylated sits, potential sites for phosphorylation by cAMP dependent protein kinase or by calmodulin-dependent kinase II. It was observed that the sequence was 40–48% homologous to the other 5-HT receptors, including the 5-HT$_{1A}$ receptor subtype.

Northern Blot Studies

A strongly hybridizing transcript was detected in the parietal cortex > hippocampus = frontal cortex ≫ cerebellum. In the peripheral tissues, only spleen and ileum weakly hybridized to the transcript.

In Situ *Hybridization Studies*

The highest levels of expression were found in the medial thalamic nuclei and hippocampal formation. Moderate expression was found throughout the layers of the cortex, medial geniculate nucleus, amygdala and hypothalamus. Low hybridization was associated with most midbrain and hindbrain regions.

SUMMARY

- **Structural criteria**, *i.e.*, primary sequence homology, indicates a unique 5-HT subtype.
- **Operational criteria** suggest that this is also true, although no selective agonist or antagonist is available to fully define the receptor, and thus its function *in vivo*.
- **Transductional** data provide perhaps the weakest criterion to define the receptor, since at least two other subtypes (5-HT$_4$ and 5-ht$_6$) signal via the same second messenger.

These criteria, taken together, suggest that the cloned sequence represents an endogenously expressed 5-HT receptor and should be referred to as "5-HT$_7$" receptors, rather than "5-ht$_7$".

Identification of a Single Residue Responsible for Agonist Selectivity in the Oxytocin-Vasopressin Receptors

B. CHINI,[a,d] B. MOUILLAC,[a] Y. ALA,[a] M. N. BALESTRE,[a]
S. TRUMPP-KALLMEYER,[b] J. HOFLACK,[b] J. ELANDS,[b]
M. HIBERT,[b] M. MANNING,[c] S. JARD,[a] AND C. BARBERIS[a]

[a]*INSERM U401, CCIPE*
rue de la Cardonille
34094 Montpellier, France

[b]*Marion Merrell Dow Research Institute*
16 rue d'Ankara
67009 Strasbourg, France

[c]*Department of Biochemistry*
Medical College of Ohio
P.O. Box 10008
Toledo, Ohio 43699-0008

Vasopressin (AVP) and oxytocin (OT) are two nonapeptides differing at position 3, in the cyclic part of the peptide, and at position 8, in the C-terminal tripeptide. In this study, we have evaluated the contribution of these positions to agonist specificity for the different receptor subtypes present in this family. The putative interactions between hormones and receptors were deduced by using a three dimensional model of the neurohypophysial hormones docked inside their receptors.[1] In this model, OT and AVP bind to the receptor in a region delimited by the upper part of the transmembrane core of the receptor. Several receptor residues, located in transmembrane regions 2, 3, 4 and 6, are involved in stabilizing the peptide inside the binding pocket. Furthermore, the terminal part of the peptide might interact with the extracellular regions of the receptor. Interestingly, the lateral chain of the non-conserved residue at position 8 of the peptide might interact with receptor residues located in the first extracellular loop of the receptor. By sequence comparison of the cloned receptors, we identified a non-conserved residue in the first extracellular loop of the receptor that might interact with the lateral chain of residue 8 in the hormone (FIG. 1). The naturally occurring Tyr residue present in the V1a receptor at this location was substituted by site-directed mutagenesis with the Asp residues present in the V2 receptor or with the Phe residue present in the OT receptor.[2] These two point mutations were able to convert the V1a receptor into a V2 or into an oxytocin receptor as far as agonist binding selectivity is concerned (TABLE 1). First, we mutated Tyr115 into a non-charged Leu residue, and we obtained a dramatic decrease in the affinity of AVP. This result indicates that a strong interaction should exist between the lateral chain of Arg in position 8 of

[d] Address correspondence to Bice Chini at CNR Cellular and Molecular Pharmacology Center, via Vanvitelli 32, 20129, Milano, Italy; Fax: 2-7490574; Tel.: 2-70146, ext. 271; E-mail: (Bice@Farma10.csfic.mi.cnr.it).

TABLE 1. Inhibition Constants (Ki in nM) for Wild-type and Mutant Receptors

Analogue	V1a Wild-type	V1a Y115L	V2 Wild-type	V1a Y115D	OTR Wild-type	V1a Y115F
AVP	4.2 ± 1.4	219 ± 61	1.06 ± 0.35*	2.2 ± 0.99	1.65 ± 0.49	1.04 ± 0.25
OT	963 ± 146	2067 ± 670	1544 ± 263	839 ± 28	0.79 ± 0.22	45 ± 13
dDAVP	2466 ± 402	4006 ± 554	31 ± 11	38 ± 7.8	203 ± 32	2212 ± 612
DAVP	871 ± 74	ND	25 ± 3.4	16 ± 4.5	ND	ND
[^{125}I]OH-LVA	0.063 ± 0.017•	0.048 ± 0.002	ND	0.100 ± 0.002•	ND	0.054 ± 0.005
[^{125}I]-OTA	ND	ND	ND	ND	0.095 ± 0.033♦	ND

For wild-type and mutant V1a receptors, Ki values were obtained by displacement of [^{125}I]OH-LVA, after determination of antagonist Kd values by Scatchard analysis (Kd values indicated by •) Similarly, Ki values for the V2 receptor were obtained by displacement of [^{3}H]AVP (Kd value indicated by *). Finally, the specific OT antagonist [^{125}I]-OTA (Kd value indicated by ♦) was used to determine the Ki values for the human OT receptor (OTR). In competition experiments, the labeled ligands were used at a concentration which approximately equals the Kd determined in the saturation experiments (100–200 pM for [^{125}I]OH-LVA and [^{125}I]-OTA 2–4 nM for [^{3}H]AVP). All values are the means ± S.E.M. of three different experiments performed in triplicate. ND: not determined.

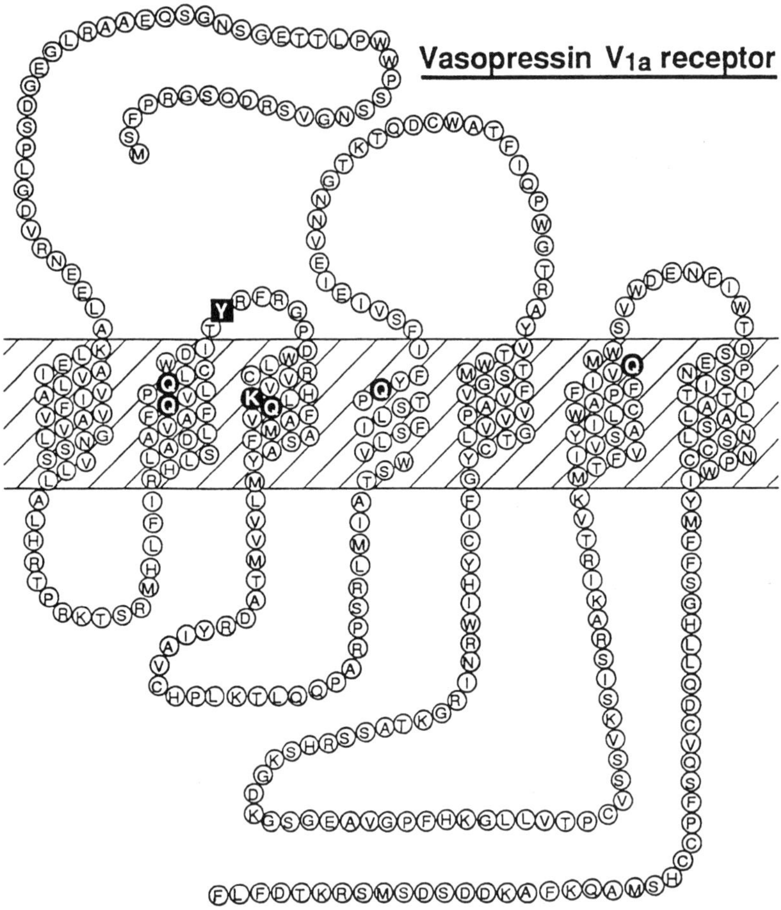

FIGURE 1. Two dimensional diagram of the neurohypophysial hormone receptors. Residues suggested to be involved in the binding of the natural peptide agonists[1] are shown in white on black. Tyrosine 115, the residue of the first extracellular loop responsible for agonist selectivity,[2] is shown in the black square.

AVP and Tyr115 of the V1a receptor, and that this interaction is suppressed in the V1a receptor mutant Tyr115Leu. Since an Asp residue is present in the human and rat V2 receptor at an equivalent position, we replaced Tyr115 of the rat V1a receptor by an Asp residue. In the V1a receptor mutant Tyr115Asp we observed a dramatic increase in the affinity for dDAVP and DAVP, two selective V2 agonists. These results indicate that the Arg in the D-form (as present in dDAVP and DAVP) interacts better with the first extracellular loop of the receptor when an Asp residue is present at position 115. Finally, we mutated Tyr115 into a Phe residue, the amino-acid which is naturally occurring in the oxytocin receptor. In the V1a receptor mutant Tyr115Phe, we observed an increase in the affinity for OT, the peptide that has a Leu residue at position 8. These data indicate that the lateral chain of the Leu might interact in a better way with the receptor when a more hydrophobic Phe residue is present in the first extracellular loop at position 115.

In conclusion, we found that in all the oxytocin/vasopressin receptor subtypes the interaction between position 8 and a non-conserved receptor residue located in the first extracellular loop is crucial for determining receptor specificity.

REFERENCES

1. MOUILLAC, B., B. CHINI, M. N. BALESTRE, J. ELANDS, S. TRUMPP-KALLMAYER, J. HOFLACK, M. HIBERT, S. JARD & C. BARBERIS. 1995. The binding site of neuropeptide vasopressin V1a receptor. J. Biol. Chem. **270:** 25771–25777.
2. CHINI, B., B. MOUILLAC, Y. ALA, M. N. BALESTRE, J. ELANDS, S. TRUMPP-KALLMAYER, J. HOFLACK, M. HIBERT, M. MANNING, S. JARD & C. BARBERIS. 1995. Tyr 115 is the key residue for determining agonist selectivity in the V1a vasopressin receptor. EMBO J. **14:** 2176–2182.

5-HT$_{1D}$ and 5-HT$_{2B}$ Receptors Mediate Contraction of Smooth Muscle in Human Small Intestine

R. A. BORMAN[a] AND D. E. BURLEIGH

Department of Pharmacology
Queen Mary & Westfield College
University of London
Mile End Road
London E1 4NS, United Kingdom

AIMS

Previous studies have indicated that 5-hydroxytryptamine (5-HT) may play a key role in the mediation of a number of gastrointestinal activities, including motility in the human colon.[1] In the small intestine, 5-HT has been shown to induce contraction of both circular and longitudinal muscle layers, however the receptors mediating these effects have not been identified.[2,3] In the present study, we have used selective antagonists to functionally classify the receptors responsible for the contractile actions of 5-HT in the human small intestine.

METHODS

Briefly, sections of human terminal ileum were obtained from macroscopically normal specimens of intestine removed at operation for carcinoma or Crohn's disease. After removal of mucosa and submucosa, circular or longitudinal muscle strips (1.5–2.5 × 15 mm) were prepared, and suspended under isotonic conditions and a preload of 1 g, in gassed (5% CO_2 in O_2) Krebs solution at 37°C. After 60 min equilibration, two cumulative concentration-response curves to 5-HT were obtained, the second response curve being obtained either in the absence (control) or presence of antagonist (incubated for 30 min). The response to 5-HT was expressed as a percentage of the maximum contraction induced by acetylcholine (30 μM), applied at the end of each experiment. Statistical analysis used the Mann-Whitney U-test, with $p < 0.05$ being taken to represent a significant difference.

RESULTS

Cumulative application of 5-HT produced a contraction of both circular and longitudinal muscle strips of human terminal ileum, with an EC_{50} value (95% C.L.) for 5-HT of 58.6 nM (21.5–86.5) and 34.9 nM (26.3–46.2) respectively. Application

[a] Address correspondence to R. A. Borman at the Clinical Pharmacology Group, Biomedical Sciences Building, Bassett Crescent East, Southhampton SO16 7PX UK.

of TTX (3.1 μM), ondansetron (10 μM) or DAU6285 (10 μM; a 5-HT_4 receptor antagonist) had no effect on the contractile response to 5-HT of either muscle layer. Methysergide was shown to be a non-surmountable antagonist of the contractile response to 5-HT of both muscle layers; at 10 μM it was shown to abolish the response to 5-HT in circular muscle layers and at 10 nM it abolished the response in longitudinal muscle layers. In addition, ketanserin was shown to antagonize the response to 5-HT in circular muscle layers, and antagonism which was non-surmountable at concentrations above 100 nM ketanserin. In contrast, application of ketanserin (1 μM) had no effect on the contractile response to 5-HT of human longitudinal muscle strips. The response to 5-HT of human circular muscle was antagonized, in an apparently competitive manner, by application of GR 127935T (10–300 nM, a selective 5-HT_{1D} receptor antagonist, n = 3–6), yielding a pA_2 estimate of 8.30 ± 0.12 with a Schild slope of 0.80 (95% C.L. 0.60–1.05). In longitudinal muscle, the contractile response to 5-HT was antagonized by application of either SB 200646A (0.3–3.0 μM, a selective 5-$HT_{2B/2C}$ receptor antagonist, n > 5) or yohimbine (30–300 nM, a 5-HT_1 and 5-HT_{2B} receptor antagonist, n > 5). The antagonism produced by SB 200646A gave a pA_2 estimate of 7.17 ± 0.16 with a Schild slope of 0.96 (95% C.L. 0.66–1.25), whereas the antagonism evoked by yohimbine gave a pA_2 estimate of 8.10 ± 0.06 with a Schild slope of 1.10 (95% C.L. 0.78–1.42).

SUMMARY

In conclusion, 5-HT has been shown to contract both circular and longitudinal muscle layers of human terminal ileum. By the use of selective antagonists, the receptors mediating these actions have been identified. In the circular muscle of human small intestine, 5-HT-induced contraction is mediated via a receptor of the 5-HT_{1D} sub-type, whereas a receptor of the 5-HT_{2B} sub-type mediates the contractile response to 5-HT of longitudinal muscle layers.

REFERENCES

1. BORMAN, R. A. & D. E. BURLEIGH. 1994. Br. J. Pharmacol. **112:** 558P.
2. BAXTER, G. S., J. PRITIKIN, H. YOUNG & D. E. CLARKE. 1991. Br. J. Pharmacol. **102:** 163P.
3. BORMAN, R. A. & D. E. BURLEIGH. 1994. Br. J. Pharmacol. **113:** 141P.

Heterogeneity of 5-HT Receptors Mediating Secretion in the Human Intestine[a]

R. A. BORMAN[b] AND D. E. BURLEIGH

Department of Pharmacology
Queen Mary & Westfield College
University of London
Mile End Road
London E1 4NS, United Kingdom

AIMS

In vitro, 5-hydroxytryptamine has been shown to influence intestinal fluid secretion across the mucosa of a number of species, including man.[1] The receptors mediating these effects differ, both between species and between different regions of the intestine. The present investigation attempts to characterize the receptors mediating 5-HT-induced fluid secretion across the mucosa of isolated human intestine, and to compare and contrast the receptors mediating this effect within different regions of the intestine.

METHODS

Sheets of mucosa complete with submucosa were prepared by sharp dissection from macroscopically normal specimens of human terminal ileum resected at operations for carcinoma or Crohn's disease. Mucosal sheets were mounted in Ussing chambers, bathed either side by gassed (5% CO_2 in O_2) Krebs solution and kept at 37°C, for measurement of short-circuit current (I_{sc}; a measure of electrogenic fluid secretion). Glucose in Krebs solution bathing the mucosal side of small intestinal preparations was replaced by an equimolar concentration of mannitol (11.5 mM). Following 60 min equilibration, 5-HT was applied to the serosal side of tissues, mucosal application was without effect. The response to 5-HT was tested against a variety of 5-HT receptor antagonists. The Mann-Whitney U-test was used for statistical comparisons, with $p < 0.05$ being taken to indicate a significant difference.

RESULTS

Serosal application of 5-HT (1 to 100 μM) was shown to induce a rise in I_{sc} in all regions of human intestine, the maximal rise being 65.4 ± 3.0 μAcm^{-2} (ileum,

[a] This work was supported by the Sir Jules Thorn Charitable Trust.
[b] Address correspondence to R. A. Borman at the Clinical Pharmacology Group, Biomedical Sciences Building, Bassett Crescent East, Southhampton SO16 7PX UK.

n = 16), 32.4 ± 2.5 μAcm^{-2} (ascending colon, n = 18) and 35.0 ± 3.0 μAcm^{-2} (sigmoid colon, n = 16). In the ileum, application of methysergide (10 μM), ketanserin (1 μM) or ondansetron (10 μM) had no effect on the I_{sc} response to 5-HT. Application of DAU 6285 (0.3–10 μM, a 5-HT_4 receptor antagonist, n > 5), however, produced a dextral shift of the concentration-response curve to 5-HT, with no alteration of the maximum response to 5-HT, yielding a pA_2 estimate of 6.17 ± 0.06 and a Schild slope of 1.00 (95% C.L. 0.67–1.33). In sigmoid colon, application of DAU 6285 (10 μM) or ondansetron (10 μM) had no effect on the secretory response to 5-HT. Application of ketanserin (1 μM, n = 4) was shown to abolish the response to 10 μM 5-HT (approximate EC_{75}), and even at a concentration of 30 nM caused a significant (88.2 ± 4.9%, $p < 0.05$) reduction of the response to 5-HT (n = 10). In ascending colon, application of methysergide (10 μM, n = 6) or ondansetron (10 μM, n = 6) had no effect on the secretory response to 5-HT. Application of DAU 6285 (10 μM), however, caused a rightward displacement of the concentration-response curve to 5-HT in 4 out of 6 specimens, with no alteration of the maximum response to 5-HT, yielding a single-concentration pA_2 estimate in those 4 tissues of 5.78 ± 0.20. In the remaining 2 specimens, the response to 5-HT was shown to be insensitive to prior application of DAU 6285 (10 μM), however application of ketanserin (1 μM) was shown to abolish the response to 5-HT. This concentration of ketanserin had no effect on the secretory response to 5-HT in the 4 specimens which were found to be sensitive to DAU 6285.

SUMMARY

In conclusion, application of 5-HT has been shown to induce fluid secretion in both the small and large intestine of man. By the use of selective antagonists, the receptors mediating these effects have been identified and characterized. 5-HT induces secretion across human ileal mucosa via a receptor of the 5-HT_4 sub-type, whereas a receptor of the 5-HT_{2A} sub-type appears to mediate the effect in human sigmoid colon. The response in human ascending colonic mucosa cannot be definitively classified at this time, but may reflect the presence of a heterogenous receptor population.

REFERENCE

1. Brown, D. R. & R. J. Miller. 1991. *In* Handbook of Physiology. S. G. Schultz, Ed. Section 6, Volume 4: 527–589. Oxford University Press. New York.

13-Tert-Butylergoline Derivatives

Assessment of Functional Serotonergic $5\text{-}HT_{1A}$ Component

M. G. FORNARETTO,[a] C. CACCIA, G. MARCHI,
E. BRAMBILLA, S. MANTEGANI, AND C. POST

Pharmacia, R&D-CNS
20014 Nerviano (Milano), Italy

INTRODUCTION

There is considerable evidence suggesting that dysfunctions of serotonin (5-HT) neurons are involved in the pathophysiology of anxiety and depressive disorders.[1] The identification of 5-HT receptor subtypes has led to the development of several classes of compounds with selective actions on these receptor subtypes. The anxiolytic and/or antidepressive activity of buspirone was linked to its specific activation of $5\text{-}HT_{1A}$ receptors. As a consequence, synthetic efforts have been devoted to developing potent compounds with high selectivity and affinity for this receptor.

Many ergoline derivatives lack selectivity and show strong affinity not only for serotonin, but also for some other catecholaminergic receptors. Our work was focused on exploring different ergoline derivatives, to identify compounds displaying selective and high affinity for the $5\text{-}HT_{1A}$ receptor.

Receptor profile was assessed by *in vitro* binding techniques. Studies on second messenger system coupled with $5\text{-}HT_{1A}$ receptors (inhibition of forskolin-stimulated adenylyl cyclase) were carried out in $h5\text{-}HT_{1A}$ transfected HeLa cells, to functionally define the agonist component of the compounds synthesized.[2]

CHEMICAL APPROACH

In order to obtain selectivity, consideration of the general SAR requirements for catecholamine receptors has led to the hypothesis that the dopamine receptor binding might be prevented by the presence of a bulky group in the phenyl ring of the ergoline skeleton. The strategy was based on the introduction of a tert-butyl group in position 13 of the ergoline nucleus.[3] The approach was assessed on festuclavine (Cp. 1), chosen for its unselective binding profile. Encouraging results regarding selectivity led us to try to increase affinity. The role of the substituent in position 8 β was explored by changing its nature especially within the class of amides.

RESULTS AND CONCLUSIONS

The aromatic ring substitution (R_2) imparted a remarkable selectivity especially with regard to the dopaminergic component as shown by the binding profile of

[a] Author to whom correspondence should be addressed.

TABLE 1. Receptor Binding Affinity of 13-Tert-Butylergoline Compounds (K_i, μM)

Cp	R_1	R_2	R_3	α_1	α_2	D_1	D_2	**5-HT$_{1A}$**	5-HT$_2$
1	H	H	H	0.70	0.14	3.68	0.14	**0.26**	0.97
2	H	$(CH_3)_3C$	H	2.48	1.35	>10	5.11	**0.18**	2.82
3	NH_2	H	H	0.37	0.19	>10	0.09	**0.20**	0.10
4	NH_2	$(CH_3)_3C$	H	1.99	0.48	>10	>10	**0.038**	>10
5	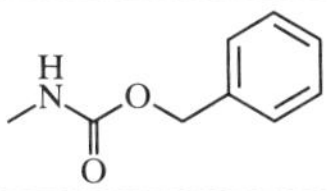	H	H	0.22	0.063	1.17	0.040	**0.022**	0.009
6	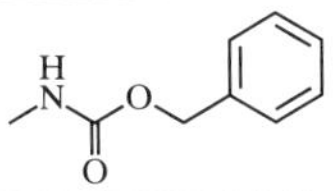	$(CH_3)_3C$	H	2.00	1.73	>10	0.77	**0.014**	0.32
7	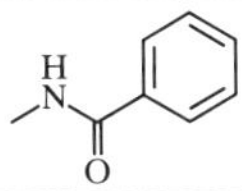	H	H	0.35	0.36	2.95	0.045	**0.0098**	0.29
8	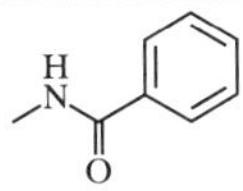	$(CH_3)_3C$	H	1.72	>10	2.41	3.12	**0.022**	4.51
9	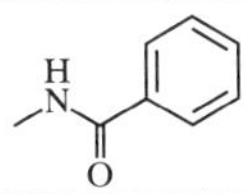	$(CH_3)_3C$	CH_3S	2.11	0.62	>10	1.42	**0.80**	0.87
10	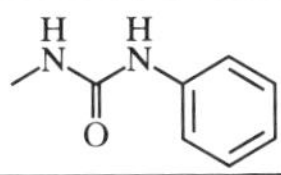	$(CH_3)_3C$	H	>10	2.08	>10	1.75	**0.077**	1.22
11	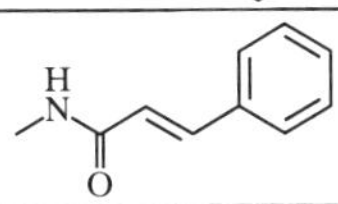	$(CH_3)_3C$	H	0.38	0.65	>10	1.89	**0.043**	2.30
12	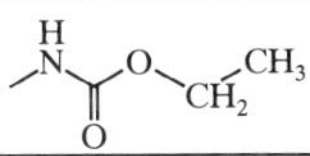	$(CH_3)_3C$	H	1.50	0.14	>10	3.07	**0.0093**	0.74
13	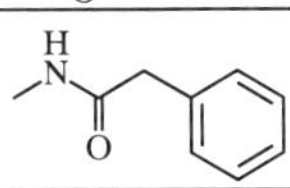	$(CH_3)_3C$	H	1.56	0.59	>10	1.71	**0.0096**	1.17
14	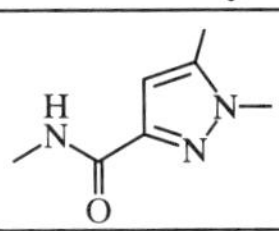	$(CH_3)_3C$	H	1.09	0.34	>10	1.23	**0.010**	3.88
15	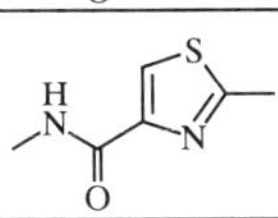	$(CH_3)_3C$	H	0.82	0.12	>10	1.03	**0.0041**	0.67
16	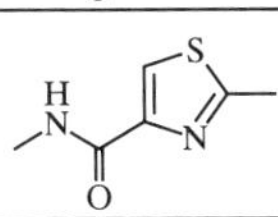	$(CH_3)_3C$	H	1.01	0.56	>10	0.83	**0.0075**	0.14
Buspirone				>10	>10	>10	0.18	**0.028**	>10

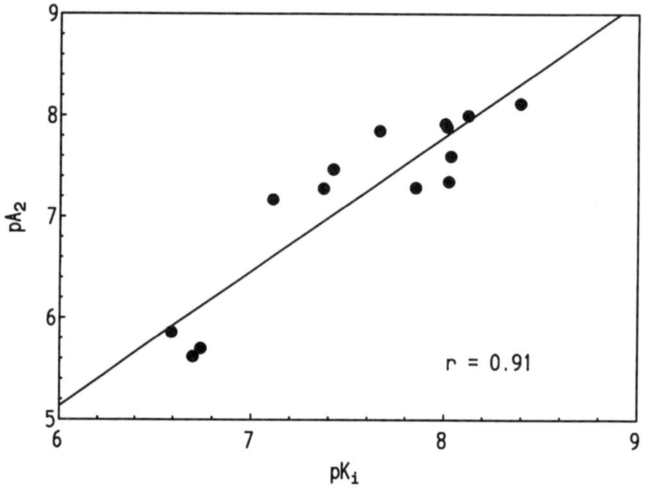

FIGURE 1. Correlation between binding affinity and potency on adenylyl cyclase.

compounds in TABLE 1. In fact the Cps. 1, 3, 5 and 7, where R_2 was a hydrogen, were considerably less selective than their analogues, substituted in R_2 with a tert-butyl (Cps. 2, 4, 6, 8).

Because of the highest selectivity and affinity displayed by Cp. 8, the class of amides was further investigated. Within this class all compounds synthesized displayed a medium to high affinity (submicromolar to nanomolar range). It was worthwhile to note that the replacement of the phenyl group with the heterocyclic ring thiazoyl (Cp. 15) led to the highest affinity.

The substitution in R_3 was detrimental in terms of affinity and selectivity (Cp. 9). When the affinity to 5-HT_{1A} receptors was present, all compounds acted as agonists (inhibition of forskolin stimulated adenylyl cyclase) and the affinity for receptor binding sites correlated with the potency as agonists (FIG. 1).

A new class of 13-tert-butyl ergoline derivatives was identified to display high and selective 5-HT_{1A} affinity over the α_1, α_2, D_1, D_2, 5-HT_2 receptor sites, with some of them displaying a better binding profile than the standard compound buspirone. This result demonstrated that ligand interaction at the 5-HT_{1A} receptor, unlike that of dopamine receptor, is compatible with the presence of a bulky group in position 13 of the ergoline skeleton. These compounds, owing to their agonist activity, could be valuable for the potential therapeutical application in the treatment of depression, anxiety and stroke/ischemia.

REFERENCES

1. CHARNEY, D. S., S. W. WOODS, J. H. KRYSTAL & G. R. HENINGER. 1990. Ann. N. Y. Acad. Sci. **600:** 558–573.
2. FARGIN, A., J. R. RAYMOND, J. W. REGAN, S. COTECCHIA, R. J. LEFKOWITZ & M. G. CARON. 1989. J. Biol. Chem. **264**(25): 14848–14852.
3. BRAMBILLA, E., C. CACCIA, M. G. FORNARETTO, S. MANTEGANI & M. VARASI. 1995. II Joint Meeting Italian-Spanish Pharm. Chem.

High Level Expression of the Human 5-$HT_{1D\alpha}$ Serotonin Receptor Using the Semliki Forest Virus Expression System

J. STABLES, S. REES, K. LUNDSTROM, D. LIVINGSTONE, AND M. G. LEE

Biology Division
Glaxo-Wellcome
Gunnels Wood Road
Stevenage SG1 2NY, United Kingdom

The preparation of cell membranes containing high levels of recombinant 7-transmembrane receptor is necessary for ligand binding studies, particularly for high-throughput screening. Current methods involve expression in mammalian cell lines following either stable or transient transfection of expression vector. Stable cell lines expressing 7-transmembrane receptor exhibit levels of expression of up to 5 pmol/mg protein, although less than 1 pmol/mg is more common; transient transfection of mammalian cells, usually CosM6, can yield up to 10 pmol recombinant receptor/mg protein.[1,2] However, both these methods are labor intensive, and a higher level of expression is desirable.

Recently, a simple, novel method of expression of recombinant protein in mammalian cells has been described using the Semliki Forest Virus (SFV) expression system.[3] This process involves *in vivo* packaging of recombinant RNA into infectious SFV particles followed by infection of host cells with the resulting virus stock. This is achieved by *in vitro* transcription of a recombinant vector (containing the non-structural SFV genes, and the 5-$HT_{1D\alpha}$ receptor gene), and a helper vector (containing the structural genes of SFV, which encode the capsid protein and viral coat glycoprotein). The resulting RNAs are coelectroporated into baby hamster kidney cells. The medium, containing recombinant virus, can be collected after 24 hours, and either stored, or used to infect host cells.

We have used the Semliki Forest Virus expression system to prepare cell membranes containing the human 5-$HT_{1D\alpha}$ serotonin receptor in the Chinese hamster ovary cell line (CHO). Binding of [^{3}H]-5-CT to homogenates prepared from CHO cells infected with 5-$HT_{1D\alpha}$ virus stocks yielded very high levels of receptor expression (21 pmol receptor/mg protein). Non-specific binding accounted for 0.5–5% of total binding. Following transient transfection into CosM6 cells (DEAE/dextran method), we saw 1.1 pmol receptor/mg protein. Thus the SFV expression system gives a 20-fold increase in receptor expression.

We have also expressed the human 5-HT_{1A}, 5-HT_{1D}, and 5-HT_{1E} receptors with similar results. The pharmacological characteristics of receptor prepared from the two expression systems appear to be identical. The SFV expression system offers a rapid and simple procedure for the expression of receptor for ligand binding analysis. Furthermore, co-infection with receptor, G-protein, and second messenger-containing virus stocks may allow the rapid analysis of second messenger responses.

REFERENCES

1. WEINSHANK, R. L., J. M. ZGOMBICK, M. J. MACCHI, T. A. BRANCHEK & P. R. HARTIG. 1992. Proc. Natl. Acad. Sci. USA **89:** 3630–3634.
2. BACH, A. W. J., L. UNGER, R. SPRENGEL, G. MENGOD, J. PALACIOS, P. SEEBURG & M. M. VOIGT. 1993. J. Receptor Res. **13**(1–4), 479–502.
3. LILJESTRÖM, P. & H. GAROFF. 1991. Bio/technology **9:** 1356–1361.

Different Levels of Receptor Expression as a New Procedure to Estimate Agonist Affinity Constant

Application to the Metabotropic Receptors

C. CORTI, P. CAVANNI, E. CAVEGGION, F. FERRAGUTI, M. CORSI, AND D. G. TRIST

Pharmacology Research Department
GlaxoWellcome S.p.A.
Via A. Fleming 4
37135 Verona, Italy

The estimation of the affinity constant (K_A) for agonists is of primary importance in the definition of drug parameters to be used in receptor classification. While for partial agonists the application of comparative methods against agonists of higher efficacy can be always applied, the determination of K_A for a full agonist requires the reduction of the number of receptors by using an irreversible antagonist. Hence, the availability of such a compound limit the K_A calculation. Metabotropic glutamate receptors (mGluR) represent a large family of G protein-coupled receptors which have been further subdivided into three main groups according to their sequence similarities and coupling to intracellular transduction mechanisms. Definition of pharmacological parameters for these receptors has been complicated by the lack of potent and selective radioligands and of irreversible antagonists. The use of mGluR transfected in cell lines at different expression levels can be used to estimate these parameters. We have selected for this study mGluR1 and mGluR5, which activate phospholipase C, both at high and low levels of receptor expression.

MATERIALS AND METHODS

Transfection, selection and characterization of mGluR1α-chinese hamster ovary (CHO) cell lines were as described by Ferraguti *et al.*[1] Rat mGluR5b cDNA was inserted into the eukaryotic expression vector pBK-CMV (Stratagene) which drives the expression of the inserted gene by the CMV promoter. This plasmid was transfected by electroporation into CHO-K1 cells. Transfectants were selected as described by Ferraguti *et al.*[1] Dot and Northern blot analysis were performed as described by Sambrook *et al.*[2] on total RNA and poly(A)+ mRNA, extracted from transfected cells, respectively. The mGluR1α SacI/XbaI and mGluR5b HindIII/StuI fragments were labeled with ^{33}P dCTP through a Rediprime DNA labeling kit (Amersham) and used as probes. Paired concentration response curves to glutamate obtained on the same day on the cell lines at high (H) and low (L) levels of receptor expression, were fitted to the inactivation methods described by Black and Leff[3] and James *et al.*[4] Values of pK_A ($-\log K_A$) are given as mean ± S.E.M.

RESULTS AND DISCUSSION

Selected clonal cell lines showed different ability to stimulate the phospholipase C pathway when treated with agonists for mGluR. Among these, the cell line showing the highest response was selected as that with H receptor expression levels. On the other hand, a cell line with a lower response was identified as expressing L levels of receptors. Glutamate, t-ACPD and quisqualate were able to produce a dose-related increase of cytidin diphospho diacylglycerol (CDP-DAG) accumulation in both mGluR1α-H&L and mGluR5b-H&L transfected cells. Dot blot and Northern blot analysis confirmed high and low amount of mRNA levels for mGluR1α and mGluR5b cell lines consistent with their relative functional activity. The different levels of receptor expression, both in mGluR1α and mGluR5b, has allowed to transform glutamate, t-ACPD and quisqualate into partial agonists, thus

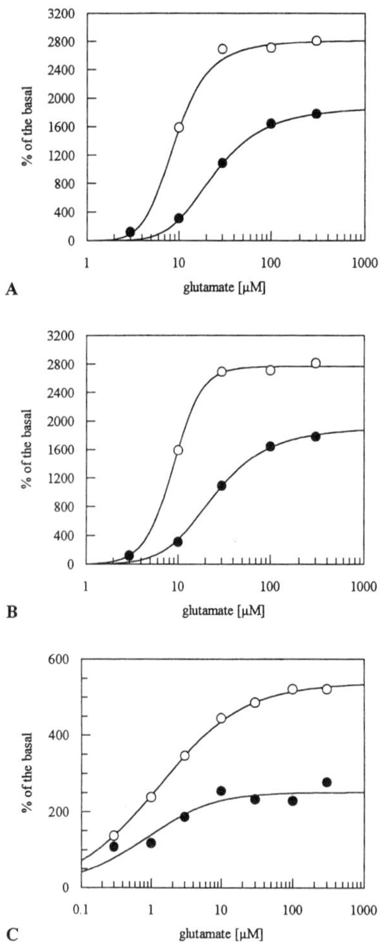

FIGURE 1. Example of glutamate dose response curves on mGluR1α- and mGluR5b-CHO cells with H (○) and L (●) receptor expression levels. **A**: mGluR1α: data fitted with the Black & Leff model. The estimated parameters were: $\tau 1 = 1.7 \pm 0.4$, $\tau 2 = 1.1 \pm 0.06$, $n = 5.3 \pm 2.4$, $pK_A = 5.17 \pm 0.26$, $Em = 2989 \pm 100.4$; **B**: mGluR1α: data fitted with the James model. The estimated parameters were: $Em = 2766 \pm 34$, $n = 2.83 \pm 0.3$, $A_{50} = 8.98 \pm 0.3$ μM, $pK_A = 5.36 \pm 0.26$, $q = 0.73 \pm 0.11$; **C**: mGluR5b: data fitted with the James model. The estimated parameters were: $Em = 538 \pm 18$, $n = 0.72 \pm 0.09$, $A_{50} = 1.32 \pm 0.22$ μM, $pK_A = 5.83 \pm 0.28$, $q = 0.43 \pm 0.14$.

TABLE 1. pK_A ± S.E.M. Calculated for Glutamate on mGluR1α and mGluR5b Receptor Subtypes

Method	mGluR1α	mGluR5b
Black & Leff	5.33 ± 0.05 (4)	n.d. (3)
James	5.67 ± 0.15 (4)	6.07 ± 0.12 (3)

The number of replications is reported in parentheses.
n.d. = not determinable.

overcoming the lack of irreversible antagonists. Glutamate, the most efficacious agonist, induced a maximal response (expressed as folds over basal) of: 20 for H-; 11 for L-mGluR1α; 5.5 for H- and 2.7 for L-mGluR5b. In cell lines with L-receptor expression maximal responses obtained for all agonists appeared considerably reduced. Glutamate dose response curves were fitted to the inactivation methods to obtain the pK_A values. Data from mGluR1α-H and -L could be fitted to both the Black and Leff and James models (FIGS. 1A and 1B). The resulting pK_As estimated are reported in TABLE 1. Dose response curves to glutamate on mGluR5b-H and mGluR5b-L could be fitted only by using the James model (FIG. 1C) from which a pK_A value could be calculated (TABLE 1). On the other hand, by using the Black and Leff method, an over estimation of Em (> 10 times the maximal effect reached by glutamate) and standard errors of the estimated parameters of about 50–80% were observed suggesting a poor quality of fit. This may suggest that the number of receptors has been reduced to a value smaller than the transducer and therefore the ternary complex mechanism may have affected the K_A estimation for the agonist. In conclusion, the use of cell lines with different levels of expression for a particular receptor, allows the estimation of pharmacological parameters required for receptor classification.

REFERENCES

1. FERRAGUTI *et al.* 1994. Mol. Cell. Neurosci. **5:** 269–276.
2. SAMBROOK *et al.* 1989. Molecular Cloning. A Laboratory Manual. C. S. H. Laboratory Press.
3. BLACK, J. W. & P. LEFF. 1983. P. Proc. R. Soc. B. **220:** 141–162.
4. JAMES, M. K. *et al.* 1989. J. Pharmacol. Exp. Ther. **249:** 61–69.

Information in Agonist Curve Shape for Receptor Classification

D. J. PRENTICE AND S. M. O. HOURANI

School of Biological Sciences
University of Surrey
Guildford, Surrey GU2 5XH, United Kingdom

In many systems agonist concentration effect (E/[A]) curves are sigmoidal so that fitting to the Hill equation (given below) is appropriate, providing estimates of midpoint location ($[A_{50}]$), upper asymptote (α) and midpoint slope parameter (n_H).

$$E = \frac{\alpha[A]^{n_H}}{[A]_{50}^{n_H} + [A]^{n_H}}$$

The utility of $[A_{50}]$ and α estimates is well established. However, often little attention is paid to n_H estimates, which can provide clues as to the nature of the drug-receptor interaction under study. This point is exemplified in the phenylephrine-contracted isolated rat aorta where relaxatory E/[A] curves to four adenosine receptor agonists, 5′-(N-ethyl)carboxamido adenosine (NECA), adenosine, N^6-R-phenylisopropyl-adenosine (R-PIA) and N^6-cyclopentyladenosine (CPA) yielded E/[A] curves with different n_H values (see FIG. 1).

These data could be explained by the presence of two sites, one coupled to a steep transducer function (activated by R-PIA and CPA) and one coupled to a rectangular hyperbolic transducer function (activated by adenosine and NECA). Studies with the non-selective adenosine receptor antagonist 8-sulphophenyltheophylline (8-SPT) and the A_{2a}-selective antagonist PD 115,199 indicated that E/[A] curves to NECA were blocked by these antagonists suggesting activation of A_{2a} receptors (see TABLE 1). In contrast, curves to CPA and R-PIA were resistant to blockade suggesting activation of a different site by these ligands. However, whilst the $[A_{50}]$ values for adenosine E/[A] curves were not significantly altered in the presence of the antagonists, the n_H values were increased (see TABLE 1).

These data suggest activation of both sites by adenosine so that in the presence of A_{2a} receptor blockade the residual E/[A] curve is characteristically steep. Again the utility of curve shape information is evident in the interpretation of these interactions. Such information is easily accessible and should not be ignored.

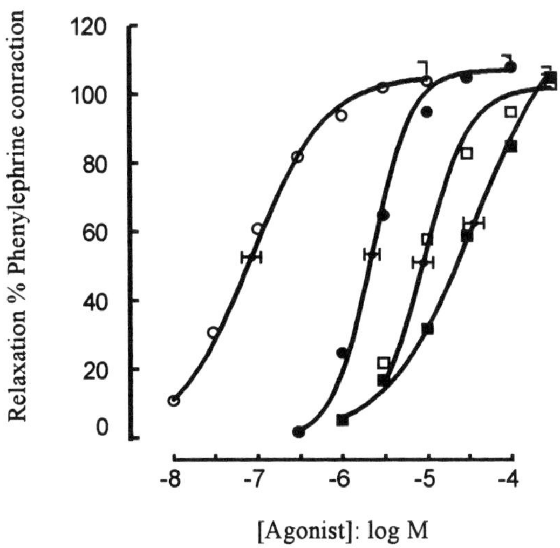

FIGURE 1. Agonist E/[A] curves generated using average curve fitting parameters and superimposed on the average data points. 5′-(N-ethyl)-carboxamido adenosine, (○); adenosine (■), N^6-R-phenylisopropyladenosine (●) and N^6-cyclopentyladenosine (□) gave E/[A] curves with n_H values as follows: 0.99 ± 0.08, 0.83 ± 0.06, 1.88 ± 0.18, and 1.57 ± 0.18, respectively, n = 5–8 ± S.E.M.

TABLE 1. Average Curve Fitting Parameters for E/[A] Curves to NECA and Adenosine Performed in the Absence and Presence of 8-SPT (50 μM) or PD 115,199 (0.1 μM)

Parameter	Agonist	Control	+PD115,199	Control	+8-SPT
$p[A]_{50}$	NECA	7.07 ± 0.11	6.41 ± 0.13*	7.05 ± 0.17	6.03 ± 0.16*
α		105.5 ± 5.4	98.0 ± 0.9	86.5 ± 6.4	85.0 ± 7.2
n_H		0.99 ± 0.08	1.18 ± 0.11	1.18 ± 0.17	1.45 ± 0.14
$p[A]_{50}$	Adenosine	4.44 ± 0.12	4.16 ± 0.14	4.42 ± 0.21	4.17 ± 0.18
α		125.2 ± 3.0	106.7 ± 7.6	110.2 ± 4.1	129.8 ± 16.4
n_H		0.83 ± 0.06	1.76 ± 0.16*	0.86 ± 0.10	1.44 ± 0.13*

* Indicates significantly different from control ($p < 0.05$).

Sodium Dependency of Constitutive Activity at 5-$HT_{1D\alpha/1D\beta}$ Receptors

D. W. GRAY, H. GILES, V. BARRETT, AND G. R. MARTIN[a]

Receptor Pharmacology
Glaxo-Wellcome Research & Development
Gunnels Wood Road
Stevenage SG1 2NY, United Kingdom

INTRODUCTION

It is generally accepted that the first step in many signal transduction pathways is the binding of ligands to 7-transmembrane receptors resulting in a change to, or stabilization of, an "active" agonist/receptor complex. This complex may facilitate the binding of GTP to G proteins causing separation of the α subunit from the $\beta\gamma$ complex, both of which can activate various transduction mechanisms. The GDP/GTP exchange process can be used to study receptor/G protein interactions using binding of [^{35}S]GTPγS as an index of receptor activation. In this study the properties of two pharmacologically similar receptors, the human cloned 5-$HT_{1D\alpha}$ and 5-$HT_{1D\beta}$ receptors, were studied.

METHODS

Membranes were prepared from CHO cells expressing human 5-$HT_{1D\alpha}$ or 5-$HT_{1D\beta}$ receptors. High affinity agonist binding sites for [^{3}H]5-HT were demonstrated with a B_{max} of approximately 5 pmol/mg protein for both receptor subtypes. Membranes were resuspended in either low [Na^+] buffer containing in mM: Tris 10, $MgCl_2$ 10, EDTA 0.1, and GDP 0.003 at pH 7.4 or high [Na^+] buffer (to encourage the separation of receptor from G protein; see Tian and Deth[1]) which, in addition to the above, contained NaCl 150 mM. [^{35}S]GTPγS binding was performed at 27°C using 200 pM [^{35}S]GTPγS. Reactions were started by addition of 30 μg membrane protein and terminated after 30 min by filtration. Where agonists and antagonists were used in competition, membranes were preincubated with antagonist for 30 min. 5-HT curves in the presence of antagonist were performed in high [Na^+] buffer. Curves to antagonist alone were performed in low [Na^+] buffer.

RESULTS

Effect of [Na^+] on Basal [^{35}S]GTPγS Binding

In 5-$HT_{1D\alpha}$ membranes, basal [^{35}S]GTPγS binding was unaltered by the presence of Na^+ (FIG. 1A). In contrast, basal [^{35}S]GTPγS binding to 5-$HT_{1D\beta}$ membranes was approximately 3-fold higher in the absence of Na^+ (FIG. 1B).

[a] Address for G. R. Martin: Institute of Pharmacology, Roche Bioscience, 3401 Hillview Avenue, Palo Alto, CA 94304-9819 USA.

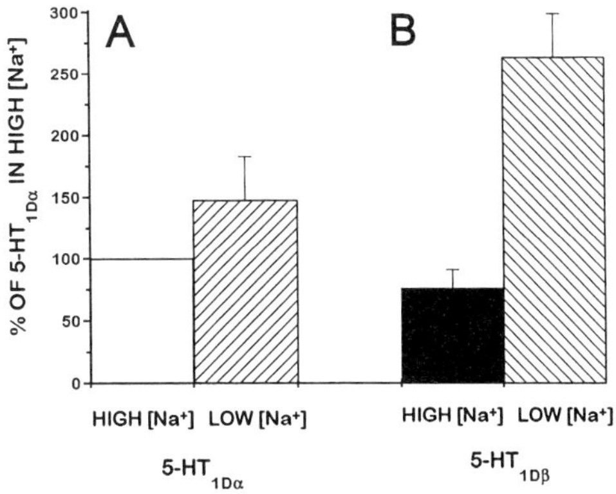

FIGURE 1. Effect of [Na^+] on basal [^{35}S]GTPγS binding in CHO membranes containing **(A)** human 5-$HT_{1D\alpha}$ or **(B)** human 5-$HT_{1D\beta}$ receptors. Results are the mean ± S.E.M. of 5 separate experiments performed on different membrane preparations. Data are normalized for each membrane preparation setting CHO5-$HT_{1D\alpha}$ membranes in the presence of 150 mM Na^+ as 100%.

Antagonist Effects in Low [Na^+] Buffer

In both 5-$HT_{1D\alpha}$ and 5-$HT_{1D\beta}$ membranes methiothepin (0.1 nM–100 μM) caused concentration dependent and saturable decreases in basal [^{35}S]GTPγS binding (p$[A]_{50}$ 5-$HT_{1D\alpha}$ 7.9 ± 0.3, 5-$HT_{1D\beta}$ 7.9 ± 0.2; n = 3–5) (FIGS. 2A and 2C). Ketanserin (10 nM–100 μM) had no significant effect on basal [^{35}S]GTPγS binding to 5-$HT_{1D\beta}$ receptors. However, in 5-$HT_{1D\alpha}$ receptors it elicited concentration-dependent decreases in binding (p$[A]_{50}$ 6.4 ± 0.5, n = 3) producing a similar maximal effect to that of methiothepin.

Agonist and Antagonist Effects in High [Na^+] Buffer

In 5-$HT_{1D\alpha}$ membranes 5-HT (0.1 nM–100 μM) was without significant effect on 2 membrane preparations (FIG. 2B). In the other 3 preparations increases in binding were observed (p$[A]_{50}$ 8.1 ± 0.4, n = 3). Both methiothepin (0.1 μM) and ketanserin (100 μM) caused profound decreases in basal [^{35}S]GTPγS binding (FIG. 2A) and, in the presence of these antagonists robust 5-HT concentration-effect curves could be constructed (FIG. 2B). The resulting pA_2 estimates were: methiothepin 9.10 ± 0.12 and ketanserin 6.3 ± 0.4; n = 3. In contrast, in 5-$HT_{1D\beta}$ membranes, methiothepin or ketanserin induced only small decreases in basal [^{35}S]GTPγS binding (FIG. 2D). Furthermore, robust concentration-effect curves to 5-HT could be constructed in the absence of either antagonist (FIG. 2D, p$[A]_{50}$ 8.2 ± 0.4, n = 5). The presence of methiothepin (0.1 μM) or ketanserin (100 μM) caused a rightward shift of the 5-HT curve (FIG. 2D) the pA_2 estimates being: methiothepin 8.6 ± 0.3 and ketanserin 5.2 ± 0.3; n = 5.

CONCLUSIONS

The presence of sodium had little effect on the basal level of [^{35}S]GTPγS binding to 5-$HT_{1D\alpha}$ membranes, but binding to 5-$HT_{1D\beta}$ membranes was increased 3-fold in the absence of sodium. Presumably this reflects receptor-dependent differences in the ability of sodium ions to uncouple the receptor from its G-protein. An alternative

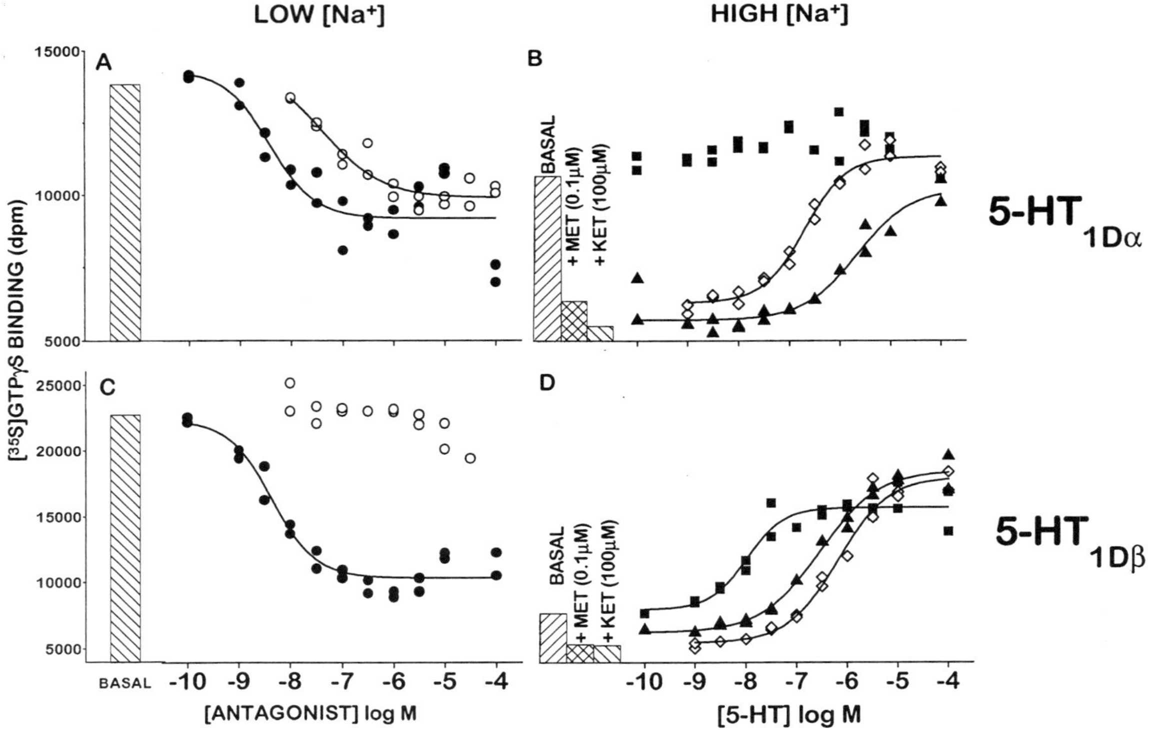

FIGURE 2. Effect of methiothepin and ketanserin on basal and agonist driven [^{35}S]GTPγS binding in CHO membranes containing human 5-HT$_{1D\alpha}$ (**A** & **B**) or human 5-HT$_{1D\beta}$ receptors (**C** & **D**). The effect of methiothepin (●) or ketanserin (○) in low [Na$^+$] is shown in **A** & **C**. In high [Na$^+$] (**B** & **D**) (■) is the 5-HT control with (✧) and (▲) representing the 5-HT curve in the presence of methiothepin (0.1 μM) or ketanserin (100 μM), respectively. Results are from an individual experiment performed in duplicate and are representative of 5 similar experiments performed on different membrane preparations.

explanation, that the membranes synthesize significant concentrations of 5-HT in a sodium-sensitive manner seems unlikely. The data suggests that both 5-HT$_{1D\alpha}$ and 5-HT$_{1D\beta}$ receptors displayed constitutive activity and that methiothepin behaved as an inverse agonist at both receptors. Ketanserin was an inverse agonist at 5-HT$_{1D\alpha}$ receptors but had little effect on 5-HT$_{1D\beta}$ receptors, consistent with the lower pA_2 estimate at this receptor.

The negligible response to 5-HT in 5-HT$_{1D\alpha}$ membranes was probably due to the high level of constitutive activity, resulting in near maximal [^{35}S]GTPγS binding even in the absence of agonist. Agonism was only revealed when constitutive activity was decreased by an inverse agonist. Clearly the absence of agonist activity does not necessarily indicate the absence of receptor or receptor/effector coupling but may indicate constitutive activity. Consequently, inverse agonists may be useful in discriminating between constitutive activity and absence of receptors in tissues or native cells.

REFERENCE

1. TIAN, W.-N. & R. C. DETH. 1993. Life Sci. **52:** 1899–1907.

NMDA Binding Sites in Rat Hippocampus Under Oxidative Stress

Effect of Antioxidants and NMDA Antagonists

Z. KVALTÍNOVÁ AND S. ŠTOLC[a]

Institute of Experimental Pharmacology SASc
Dúbravska cesta 9
SK-842 16 Bratislava, Slovak Republic

Oxidative stress and/or overactivation of glutamate receptors may lead to neuronal degeneration in a manner consistent with the course of several degenerative disorders in the brain. Understanding of the relation between glutamate and oxidative stress is important for knowing the pharmacological strategy in the treatment of neurodegeneration. With this aim in mind, we measured [^{3}H]glutamate binding to *N*-methyl-D-aspartate binding sites (NMDA b.s.) in the rat hippocampus and studied the action of two NMDA antagonists, 2-amino-5-phosphonovalerate (APV) and Mg^{2+}, and of two antioxidants, stobadine (STB) and trolox, as to their effect on the decrease in the density of NMDA b.s. in the rat hippocampus due to oxidative stress induced by hypoxia followed by reoxygenation.

Wistar male rats (150–200 g) were anesthetized with ether and decapitated. The hippocampus was removed and cut into 400 μm slices which were placed into artificial cerebrospinal fluid (ACSF) in covered chambers and incubated at 37°C under normoxia (ACSF with 10 mM glucose was equilibrated with 95% O_2 and 5% CO_2) or hypoxia (ACSF with 4 mM glucose was equilibrated with 95% N_2 and 5% CO_2). During incubation with $MgSO_4$ (10 mM), $CaCl_2$ was omitted. The membrane fraction prepared from the slices was used in a binding assay with [^{3}H]glutamate or [^{3}H]STB, nonspecific binding was measured in the presence of 100 μM NMDA or 50 μM STB, respectively. We found that the antioxidant STB (10 μM), applied 10 min before and during 10-min hypoxia followed by 10-min reoxygenation prevented an oxidative stress–induced decrease in the density of NMDA b.s., while trolox (200 μM) under the same conditions was without effect. A partial protective effect of trolox against the decrease of NMDA b.s. was however found after 30 min preincubation and a complete protective effect was reached after 60 min preincubation with trolox. The protective effect of STB was manifested also when the drug was applied only during reoxygenation. Further, the noncompetitive NMDA antagonist Mg^{2+} (10 mM) was more potent than the competitive NMDA antagonist APV (10 μM) in inhibiting the density decrease of NMDA binding sites induced by oxidative stress. High levels of excitatory amino acids released in oxidative stress are known to exert neurotoxic effects. We studied STB as to its effect on the inhibition of glutamate receptors also in normoxic slices. We found that STB (10^{-9} – 10^{-3} M) did not compete [^{3}H]glutamate binding, similarly to the noncompetitive

[a] Correspondence should be addressed to S. Štolc.

NMDA antagonist (+)MK-801 (10^{-9} – 10^{-3} M). Other drugs tested inhibited [^{3}H]glutamate binding in the order: glutamic acid > NMDA > APV. [^{3}H]STB was bound to the hippocampal membrane fraction, but this binding was not competed by glutamic acid, APV or Mg^{2+}. Thus the protective effect of the antioxidant STB was not operative by direct interaction with the glutamate receptors on binding sites for the given compounds.

Index of Contributors